Low Power Designs in Nanodevices and Circuits for Emerging Applications

This reference textbook discusses low-power designs for emerging applications. This book focuses on the research challenges associated with theory, design, and applications toward emerging microelectronics and VLSI device design and developments about low power consumption. The advancements in large-scale integration technologies are principally responsible for the growth of the electronics industry.

This book is focused on senior undergraduates, graduate students, and professionals in the field of electrical and electronics engineering, nanotechnology. This book:

- discusses various low-power techniques and applications for designing efficient circuits;
- covers advanced nanodevices such as FinFETs, TFETs, and CNTFETs;
- covers various emerging areas like quantum-dot cellular automata circuits, FPGAs, and sensors; and
- discusses applications like memory design for low-power applications using nanodevices.

The number of options for ICs in control applications, telecommunications, high-performance computing, and consumer electronics continues to grow with the emergence of VLSI designs. Nanodevices have revolutionized the electronics market and human life; it has impacted individual life to make it more convenient. They are ruling every sector such as electronics, energy, biomedicine, food, environment, and communication. This book discusses various emerging low-power applications using CMOS and other emerging nanodevices.

Low Power Designs in Nanodevices and Circuits for Emerging Applications

Edited by
Shilpi Birla
Shashi Kant Dargar
Neha Singh
P. Sivakumar

CRC Press
Taylor & Francis Group
Boca Raton London New York

CRC Press is an imprint of the
Taylor & Francis Group, an **informa** business

Front Cover Image: Titima Ongkantong/Shutterstock

First edition published 2024
by CRC Press
2385 NW Executive Center Dr, Suite 320, Boca Raton, FL 33431

and by CRC Press
4 Park Square, Milton Park, Abingdon, Oxon, OX14 4RN

CRC Press is an imprint of Taylor & Francis Group, LLC

© 2024 selection and editorial matter, Shilpi Birla, Shashi Kant Dargar, Neha Singh and P. Sivakumar individual chapters.

Library of Congress Cataloging-in-Publication Data
Names: Birla, Shilpi, editor. | Dargar, Shashi Kant, editor. | Singh, Neha (Electronic engineering professor), editor. | Sivakumar, P. (Pothiraj), editor.
Title: Low power designs in nanodevices and circuits for emerging applications / edited by Shilpi Birla, Shashi Kant Dargar, Neha Singh and P. Sivakumar.
Description: Boca Raton : CRC Press, 2024. | Includes bibliographical references and index.
Identifiers: LCCN 2023024334 (print) | LCCN 2023024335 (ebook) | ISBN 9781032412771 (hardback) | ISBN 9781032604602 (paperback) | ISBN 9781003459231 (ebook)
Subjects: LCSH: Nanoelectronics. | Low voltage integrated circuits.
Classification: LCC TK7874.84 .L697 2024 (print) | LCC TK7874.84 (ebook) | DDC 621.3815–dc23/eng/20230729
LC record available at https://lccn.loc.gov/2023024334
LC ebook record available at https://lccn.loc.gov/2023024335

ISBN: 978-1-032-41277-1 (hbk)
ISBN: 978-1-032-60460-2 (pbk)
ISBN: 978-1-003-45923-1 (ebk)

DOI: 10.1201/9781003459231

Typeset in Sabon
by codeMantra

Contents

14 Low power designs for enhanced CMOS performance   213

GOWSIKA DHARMARAJ, ASHWIN KUMAR S, CHANDRA PRAKASH S, AND REBA P

15 Recent advances in carbon nanotubes-based sensors   254

AMANDEEP KAUR, JITENDER KUMAR, AND AVTAR SINGH

Preface

The electronics industry has grown at a breakneck pace during the past few decades, primarily owing to significant advancements in integration technologies and large-scale systems design—in other words, the introduction of VLSI. Nanodevices have revolutionized the electronics market and human life; they have impacted individual lives to make them more convenient. They are ruling every sector, whether electronics, energy, biomedicine, food, environment, and communication. Low-power VLSI device designs have gotten much attention in recent years, thanks to the growing desire for high-density portable systems with low power consumption. Compact applications demand low power dissipation, and high efficiency in terms of speed is increasing, which in turn brings out new technologies in this domain. Low power consumption must be satisfied in most of these scenarios, and the equally challenging high chip density and throughput goals. As a result, low-power digital integrated circuit design has evolved into a very dynamic and quickly evolving CMOS design sector. Chip density and operation speed increase, resulting in increasingly complex devices with high frequency. The power dissipation of the device, and hence the temperature, increases linearly as the clock frequency of the chip increases. Furthermore, smaller integrated circuits (ICs) with better functions and faster reactions boost the system's compactness. As a result, in the last 30 years, the metal-oxide-semiconductor field-effect-transistor (MOSFET), a critical component of integrated circuits, has been dramatically reduced to lower technological nodes.

This book covers the recent nanodevices which include thin film transistors (TFT), tunnel field effects transistors, and carbon nanotubes, as well as new emerging devices such as nanowires, HEM transistors, and their applications. This book will give insights into various low-power applications for emerging applications such as IoT, biomedical-related area, graphene-based devices, and recent trends in flexible electronics. Overall, this book covers the nanodevices and their modeling along with various applications related to these devices. This book also focuses on various low-power techniques for designing low-power applications.

BOOK ORGANIZATION

This book is organized into 18 chapters as follows:

Chapter 1 provides a quick overview of clock-gating techniques for low-power applications. The clock-gated shift register has been proposed with clock-gating techniques to reduce power consumption.

Chapter 2 presents emerging applications of low power device designs (LPDDs) in flexible and stretchable devices. This chapter discusses various techniques for designing low-power applications.

Chapter 3 analyzes a highly stable and reliable ultralow-power 11T near-threshold SRAM in 10-nm FinFET technology. This chapter presents a novel device-level and circuit-level techniques-based write-read-enhanced 11T (WRE11T) near-threshold SRAM cell and compares it with FinFET devices in 10-nm technology and at a 0.45 V.

Chapter 4 focuses on understanding the investigation and optimization of high-stability 6T CNTFET SRAM cell with low power. The noise and power performances of the 6T CN-SRAM cell are improved by optimizing the CN-FET parameters.

Chapter 5 discusses the impact of study of transistor sizing techniques for low power design in FinFET technology. An analysis of static D-type flip-flops was conducted in this study. Simulations are performed in 16 nm FinFET technology with a shorted-gate (SG) mode configuration.

Chapter 6 presents the emerging nanodevices for low-power applications. It also discusses issues with nanoscale and also proposes solutions for power issues.

Chapter 7 presents the emerging application of graphene. It focuses on potential uses for graphene and related nanostructures. Methods for the undeviating and controlled deposition of large-area abridged graphene oxide thin films are focused.

Chapter 8 contains the analysis of low-power SRAM cells. This chapter focused on leakage power reduction techniques and stability improvement. The self-controllable voltage level (SVL) technique shows the best results with the improvement in the hold, read, and write static noise margins, respectively, for FinFET-based cells.

Chapter 9 discusses various sources of power dissipation in SRAM and different leakage power minimization techniques in SRAM. A proper comparison has been done for various leakage minimization techniques in SRAM cell.

Chapter 10 introduces a next-generation non-volatile field-programmable gate array (nvFPGA) where non-volatility (NV) is introduced in the lookup tables (LUTs), D flip-flops (DFFs), and switching blocks (SwBs). NV is implemented in these circuits through resistive random-access memory (ReRAM).

In Chapter 11, all the novel clocking schemes in the literature are analyzed, and layout techniques are proposed to develop efficient QCA circuits. The proposed layout techniques are validated by designing multiplexer using universal scalable and efficient clocking schemes and are equated with the designs in the literature.

Chapter 12 introduces a low-power CNTFET SRAM cell to reduce power loss due to leakage. The proposed step-down transistor method improved leakage power and delay while the voltage divider method improved leakage power.

Chapter 13 presents a novel MOSFET structure that minimizes all kinds of leakage currents, consumes less area, simplifies some process steps, and minimizes the cost of fabrication while demonstrating almost similar performance.

Chapter 14 focuses on different low-power CMOS circuit techniques that can be generally applied to almost any kind of digital circuit. First, it gives the basic understanding of the sources of power dissipation. Second, the fundamentals of different static and dynamic logic techniques are discussed.

Chapter 15 studies carbon nanotube-based sensors that have been widely investigated because of their nanosize morphology and high surface-to-volume ratio. It offers a large surface area for species to be detected. The structure, properties, and limitations of nanotubes are explained in this chapter.

Chapter 16 sheds light on proposal to use the operations such as right shift, left shift, and carry save adder to develop FIR filter. The multiplication and division are performed using the right shift and left shift operation. The FIR filter using the modified constant shift method (MCSM) multiplier block instead of a conventional multiplier is proposed.

Chapter 17, focused on one-sided Schmitt-Trigger (ST)-based 10T (OSST10T) SRAM cell, is presented in this work to overcome those issues. The read stability is improved through a robust back-to-back structure made of normal and ST inverters. To enhance writability, a data-dependent feedback-cutting technique is employed.

Chapter 18 provides a comprehensive study of the advanced low-power devices, and it also covers the recent developments and their applications in the field of FinFETs, nanowire field-effect transistors (NW-FETs), and CNTs. The detailed different aspects of the technology, physics, and modeling of these nanoscale devices will build fundamentals for future research.

About the Editors

Dr. Shilpi Birla is currently working as an Associate Professor at Manipal University, Jaipur. She has completed B.Tech in Electronics & Communication Engineering, M.Tech in VLSI Design, and Ph.D. in Low power VLSI Design from Uttarakhand Technical University, Dehradun, India. She has 14 years of experience, which includes teaching and industrial experience. She is a senior member of IEEE. She has authored and co-authored 80 research papers in journals of repute and international conferences. She is a reviewer and TPC member of several International Conferences. She is currently teaching Microelectronics and VLSI Design courses at UG & PG level. She is also a reviewer of several journals like *Microelectronics Journal*, *International Journal of Electronics*, and many more IEEE-sponsored conferences. Two Doctoral degrees have been awarded under her guidance. She is guiding four Ph.D. students in low power VLSI & nanodevices. She has set up VLSI lab in different institutes. She has conducted short-term courses in TCAD, HSPICE, and fabrication and workshops at national level in VLSI domain. Her research area is VLSI design, digital system design, analog VLSI and image processing.

Dr. Shashi Kant Dargar is currently working as an Associate Professor in the Electronics & Communication Engineering Department at Kalasalingam Academy of Research and Education, Tamil Nadu, India. He received his Ph.D. (2017) in Microelectronics, Master's (2013) in Digital Communication, and Bachelor's (2005) in Electronics and Communication Engineering. He completed his 3-year post-doctoral research in Electronic Engineering at the University of KwaZulu-Natal, Durban, South Africa. He has contributed to research and academics for 16 years, including microelectronics, thin-film device design, VLSI, wireless communication, and microwave & communication engineering. He taught analog integrated circuits, VLSI signal processing, VLSI technology, analog devices, circuits, signals and systems, digital signal processing, communications, and low power VLSI design. At the postgraduate level, he taught thin film electronics, flexible electronics, sensor devices, microwave and communication, semiconductor device modeling, and quantum electronics. He has authored 41 scientific research articles in reputed journals and conferences. He is a senior member of IEEE and Life Member of IEEE HKN- Mu Eta Chapter.

Dr. Neha Singh is a well-rounded and dynamic teaching professional with 19 years of in-depth experience in educating undergraduate and postgraduate engineering students in Electronics & Communication Engineering. She is currently working as Assistant Professor, in the Department of Electronics & Communication Engineering at Manipal University Jaipur, Rajasthan, India. She completed her Ph.D. in 2020. Her areas of research interest include image processing, machine learning, VLSI design, and nanodevices. She has several papers and book chapters published in Journals and conferences of repute. She has co-authored engineering textbooks and edited a book on *Nanotechnology: Device Design and Applications*. She has served as a reviewer in various international and peer-reviewed journals and conferences. She has also worked as Convener,

Session Chair, and organizer of various international conferences, summer internships in Diode Fabrication and Faculty Development Programs. She has guided several M.Tech Dissertations and B Tech projects and guiding Ph.D. scholars as well.

Dr. P Sivakumar is presently working as a Professor & Dean School of Electronics, Electrical and Biomedical Technology at the Kalasalingam Academy of Research and Education in India. He received his undergraduate degree from Madurai Kamaraj University, India, in 2001, and his Ph.D. degree in VLSI design from Anna University, Chennai, India, in 2013. He has published 80 papers in peer and refereed journals and chapters in books. He participated in his research in various forums of VLSI Design. He is currently teaching RTL logic and system design course. He also presented various academic and research-based papers related to VLSI, image processing and optimization at several national and international conferences. He has teaching experience of almost 20 years in Electronics and Communication. He is in the reviewer panel of many journals for reviewing scientific manuscripts. And he has supervised more than 30 PG and Ph.D. scholars. He has also been involved as an investigator and co-investigator in several projects & consultancy supported by government agencies in India.

List of Contributors

Erfan Abbasian
Department of Electrical and Computer Engineering
Babol Noshirvani University of Technology
Babol, Iran

Haider A.F. Almurib
Department of Electrical and Electronic Engineering
University of Nottingham Malaysia
Semenyih, Malaysia

A. Jose Anand
Department of Electronics and Communication Engineering
KCG College of Technology
Chennai, India

Shilpi Birla
Department of Electronics & Communication Engineering
Manipal University Jaipur
Jaipur, India

Gowsika Dharmaraj
Electronics and Communication Engineering department
PSG Institute of Technology and Applied Research
Coimbatore, India

M. Elangovan
Electronics and Communication Engineering
Government College of Engineering Srirangam
Tiruchirappalli, India

Morteza Gholipour
Department of Electrical and Computer Engineering
Babol Noshirvani University of Technology
Babol, Iran

Lakshminarayanan Gopalakrishnan
Electronics and Communication Engineering
National Institute of Technology-Tiruchirappalli
Tiruchirappalli, India

Anjali Gupta
School of Basic and Applied Science
Galgotias University
Greater Noida, India

J. L Mazher Iqbal
Department of Electronics and Communication Engineering
Vel Tech Rangarajan Dr.Sagunthala R&D Institute of Technology and
    Science
Chennai, India

V.S. Janani
Department of Electronics and Communication Engineering
Easwari Engineering College
Chennai, India

S. Jayanthi
Department of Electronics and Communication Engineering
Sri Manakula Vinayagar Engineering College
Pondicherry, India

T. Jayasankar
Department of ECE
University College of Engineering, Anna University
Tiruchirappalli, India

Raja Sekar K
Research & Development
Centre for Development of Advanced Computing
Bengaluru, India

G. Karthy
Department, School of Engineering and Technology
Dhanalakshmi Srinivasan University
Tiruchirappalli, India

Amandeep Kaur
Department of Instrumentation
Bhaskaracharya College of Applied Sciences, University of Delhi
Delhi, India

Seok-Bum Ko
Electrical and Computer Engineering
University of Saskatchewan
Saskatoon, Canada

Appikatla Phani Kumar
School of Electronics Engineering
VIT-AP University
Amaravati, India

Jitender Kumar
Department of Electronics
Bhaskaracharya College of Applied Sciences, University of Delhi
Delhi, India

Sujit Kumar
Department of Electrical and Electronics Engineering,
Dayananda Sagar College of Engineering
Bengaluru, India

T. Nandha Kumar
Department of Electrical and Electronic Engineering
University of Nottingham Malaysia
Semenyih, Malaysia

Chee Hock Leong
Department of Electrical and Electronic Engineering
University of Nottingham Malaysia
Semenyih, Malaysia

Rohit Lorenzo
School of Electronics Engineering
VIT-AP University
Amaravati, India

Ehsan Mahmoodi
Department of Electrical and Computer Engineering
Babol Noshirvani University of Technology
Babol, Iran

T. Manikandan
Department of Electronics and Communication Engineering
Rajalakshmi Engineering College
Chennai, India

M. Meena
Department of Electronics and Communication Engineering
Vels institute of science, Technology & Advanced Studies
Chennai, India

Debasis Mukherjee
Department of ECE
Brainware University
Kolkata, India

Anurag Nain
School of Basic and Applied Science
Galgotias University
Greater Noida, India

G. Narayan
School of Electronics Engineering
Vellore Institute of Technology
Vellore, India

Mahdieh Nayeri
Young Researchers and Elite Club, Yazd Branch
Islamic Azad University
Yazd, Iran

Maryam Nayeri
Department of Electrical Engineering, Yazd Branch
Islamic Azad University
Yazd, Iran

Reba P
Electronics and Communication Engineering Department
PSG Institute of Technology and Applied Research
Coimbatore, India

A. Selwin Mich Priyadharson
Department of Electronics and Communication Engineering,
Vel Tech Rangarajan Dr.Sagunthala R&D Institute of Technology and
    Science
Chennai, India

Marshal R
Cyber Security
Indian Computer Emergency Response Team
New Delhi, India

N. Raghu
Department of Electrical and Electronics Engineering
Jain (Deemed-To-Be-University)
Bengaluru, India

G. Boopathi Raja
Department of ECE
Velalar College of Engineering and Technology
Erode, India

P. Raja
Department of Electronics and Communication Engineering
Sri Manakula Vinayagar Engineering College
Pondicherry, India

Hannah Jessie Rani
Department of Electrical and Electronics Engineering
Jain (Deemed-To-Be-University)
Bengaluru, India

Ashwin Kumar S
Electronics and Communication Engineering Department
PSG Institute of Technology and Applied Research
Coimbatore, India

Chandra Prakash S
Electronics and Communication Engineering department
PSG Institute of Technology and Applied Research
Coimbatore, India

Deepika Sharma
Manipal University Jaipur
Jaipur, India

Savitesh Madhulika Sharma
Chinmaya Vishwa Vidyapeeth
Puttur, India

Vinny Sharma
School of Basic and Applied Science
Galgotias University
Greater Noida, India

Avtar Singh
Department of Electronics and Communication Engineering
Adama Science and Technology University
Adama, Ethiopia

P. Sivakumar
ECE Department
Kalasalingam Academy of Research and Education
Krishnankoil, India

Divya Bajpai Tripathy
School of Basic and Applied Science
Galgotias University
Greater Noida, India

V N Trupti
Department of Electrical and Electronics Engineering
Jain (Deemed-To-Be-University)
Bengaluru, India

Anantharaj Thalaimalai Vanaraj
Senior Technologist, R& D, Western Digital
Milpitas, California

H.K Yashaswini
Department of Electrical and Electronics Engineering
Jain (Deemed-To-Be-University)
Bengaluru, India

Chapter 1

Low-power VLSI design using clock-gated technique

J.L Mazher Iqbal and A. Selwin Mich Priyadharson
Vel Tech Rangarajan Dr.Sagunthala R&D
Institute of Technology and Science

T. Manikandan
Rajalakshmi Engineering College

V.S. Janani
Easwari Engineering College

A. Jose Anand
KCG College of Technology

1.1 INTRODUCTION

Reconfigurable computing platform enhances the computation using pro-grammable logic. Field programmable gate arrays (FPGAs) are able to be programmed an indefinite amount of times with many different hardware algorithms. The innovative development of FPGAs and the quickness of straight hardware implementation on FPGAs attracted the attention of very large-scale integrated circuit digital signal processing (VLSI DSP) meth-odology designers. The reconfigurable computing using FPGAs outclasses the DSP-built processor in terms of various parameters such as area, speed and power consumption, and massive instruction-level parallelism. FPGA is used in low-capacity applications. The programmable hardware in FPGAs is modified to a specific algorithm, which significantly increases the process-ing speed. FPGAs are configured using look-up table (LUT) and multiplexer operations that appear exactly in the algorithm. The configuration of FPGAs can be modified by loading different bit file data into the internal memory, just as different programs can be uploaded onto a microcontroller or DSP processors. The application mapping for the FPGAs using high computation algorithms could achieve the speed as ASICs with the flexibility of software.

The need for power delicate designs has risen dramatically in recent years. This remarkable demand has been spurred by the firm prolifera-tion of battery-functioned movable applications containing notepads and

DOI: 10.1201/9781003459231-1

CPUs, mobile phones, and additional movable communication maneuvers. Semiconductor maneuvers are belligerently mounted in every technical cohort to realize enhanced performance and enhanced integration density. The power consumption in a die is increasing with each technology generation due to greater frequencies of operation and increased transistor density in a die. The supply voltage is adjusted to retain the power consumption under control. The high-performance requirement, however, limits supply voltage scaling. As a result, just scaling the supply voltage is not enough to keep the density of power within acceptable limits, which is essential for power delicate applications. To accomplish low-power designs, circuit and system-level approaches, as well as supply voltage scaling, are necessary. Leakage currents account for a huge section of the entire power consumption in high-performance digital system in the nanometer regime. Due to rise in demand of high performance devices, the systems necessity follow to a strict power plan, outflow power limits the amount of power available, lowering performance. It also adds to the power imbalance. It also increases power usage while in standby mode, lowering battery life. As a result, strategies to reduce leakage power while keeping good performance are required. Furthermore, as multiple leakage components become more relevant as technology scales, each leakage reduction strategy must be reevaluated in scaled expertise where sub-threshold conveyance is not the only leaky contrivance.

To decrease complete leak in high-performance nano-scale circuits, low-power circuit approaches are necessary. To reduce dynamic power, a variety of circuit approaches, such as transistor size, clock gating, and dynamic supply voltage, are available. The dissimilar circuit approaches, such as dual V_{th}, advancing/converse bias, dynamically altering the V_{th} throughout run time, and natural stacking, are available for low-leakage design. The complexity and speed of circuits increase as VLSI technology progresses, resulting in significant power consumption. Small size and great performance are two opposing restrictions in VLSI design. The activities of the integrated circuit (IC) exclusive have been elaborated in the transaction of these restrictions. There are numerous design issues, and power efficiency has become increasingly crucial as a result. The furthermost moveable systems in use today, which are power-driven by batteries, are capable of doing jobs that need a large number of calculations. The most significant feature of Moore's Law is that it has turned into a widespread forecast of semiconductor industry growth. Figure 1.1 shows the clock gating using AND gate and Figure 1.2 shows the gated D FF and gated clock design sequential circuit.

1.1.1 Power consumption in VLSI digital circuits

The entire power consumption in a VLSI circuit usually comprises two modules, specifically, the static and dynamic power consumption.

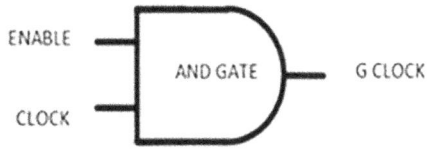

ENABLE

AND GATE

G CLOCK

CLOCK

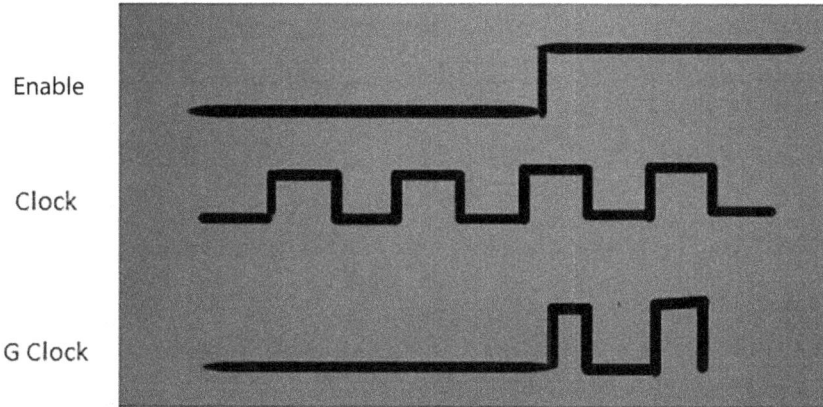

Enable

Clock

G Clock

Figure 1.1 Clock gating using AND gate.

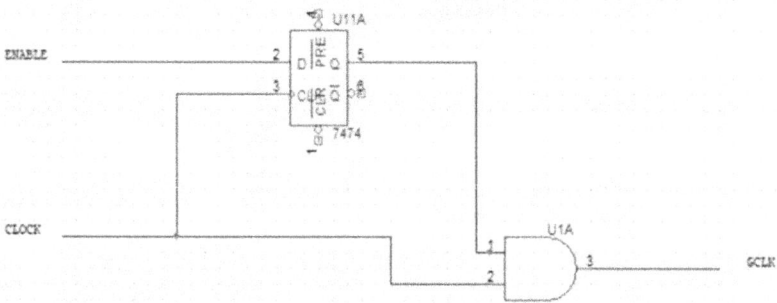

ENABLE

CLOCK

GCLK

Figure 1.2 (a) Gated D FF (b) Gated clock design sequential circuit.

1.1.1.1 Dynamic power

There are three mechanisms for dynamic power dissipation: load capacitance charging, discharging power, and switching power. The other is short circuit power, which `is triggered by the input waveforms nonzero peak and fall times. A single gate's switching power can be represented as

$$P = C_l * V_{dd} * 2f$$

where C_l is the capacitance, V_{dd} is the supply voltage, and f is the operational frequency

1.1.2 Leakage power

In the nanoscale regime, there are three major sources of leak in a MOSFET:

1. Sub-threshold leak that leaks from the drain to the source (Isub).
2. Straight tunneling gate leakage is caused by electrons (or holes) tunneling from the majority of silicon over the gate oxide potential barrier and into the gate.
3. Reverse-biased p-n junction leak at the substrate or source or drain.

1.1.3 Dynamic power lessening methodology

Despite the fact that leak power grows dramatically through every group of expertise, dynamic power still dominates the overall power dissipation of wide-ranging microprocessors and microcontrollers. Transistor scaling optimization, gated clock, numerous supply voltages, and active source voltage management are all effective circuit strategies for lowering dynamic power usage. The dynamic power dissipation of nano-scale circuits may be greatly lowered by incorporating the preceding ideas into their design. Other strategies for reducing dynamic power dissipation include memory access minimization and low-complexity algorithms. It minimizes dynamic power consumption in logic and memory.

1.1.3.1 Transistor optimization in terms of size and interconnect

There are two basic types of sizing procedures.

- Optimization based on a path.
- Optimization on a global scale

The off-critical path gates are reduced in size to save power. Entirely gates in a circuit are comprehensively optimized for a quantified interval in overall optimization.

1.1.3.2 Clock gating

The clock gating in digital circuits is a good approach to decrease vibrant power dissipation. Only a piece of a typical synchronous circuit, such as a general-purpose microprocessor, is active at any given moment. As a result, excessive power consumption can be avoided by turning off the circuit's idle part. Covering the clock that becomes the sluggish section of the circuit is one method to accomplish this. Clock gating works by using the enable criteria associated with registers to gate the clocks. To practice and benefit from clock gating, a design must include these enable signal requirements. The clock-gating method can save substantial die area and power because it eliminates the need for a huge number of multiplexers. Clock-gating circuit in IC is normally used to implement clock-gating logic. The clock-gating circuits, on the other hand, will affect the clock tree structure because it will be located within the circuit structure of clock tree.

The circuits with clock gating are implemented in various ways:

1. Allow circumstances that repeatedly transformed through the synthesis tools that are coded into the Register Transfer Level (RTL) code (fine grain clock gating) in the clock-gating logic.
2. RTL designers manually include reference library explicit combined clock-gating circuits into the design (usually as module-level clock gating).
3. Automatic clock-gating tools insert semi-automatically into the RTL code. These tools put the ICG circuit into it. These usually include sequential clock-gating optimizations as well. Lowering the dynamic power minimizes needless changes in the inputs to the indolent circuit module. The registers, which are normally made up of sequential devices like D flip-flops, provide the input to the combinational logic. Altering the clocking configuration results in a gated clock design. When the combinational block is not in use, a control switch signal is used to choose and halt the local clock (LCLK). When the control switch signal is high, the local clock is stopped. When the global clock (GCLK) is high, the latch is required to avoid any errors in control switch signal from disseminating to the AND logic gate. During the clock's positive edge, only the control switch signal is valid. When the global clock is low, the latch is transparent, but it has no effect on the AND gate. If control switch signal is high during the global clock's low-to-high transition, the AND gate will block the global clock, keeping the local clock low. The effective synthesis and modification of clock-preventing circuit for digital circuits are critical for power savings using the clock-gating methodology.

Effective clock gating necessitates a system for determining which circuits are gated, when they are gated, and how long they are gated. Clock-gating techniques result in recurrent switching of the clock-gated circuit among allowed and restricted states with overhead. Because of this overhead, the power dissipation may be huge without clock gating.

1.1.3.3 Low-voltage process

The low-voltage process is established for switching power reduction in VLSI circuits. The objectives of the proposed work are:

- To decrease the dynamic power dissipation by removing clock whenever it is not required.
- To avoid the excess switching power dissipation for the duration of the ideal period.
- To avoid power waste, the clocks are restricted to ideal period.
- As soon as the clock is on throughout the ideal state, it affects the clock loading.
- A lock-gating XOR is presented to each accessible flip-flop, the input and output are analyzed by the gating circuit in the clock-gating method.

In dynamic source voltage scaling technique, the maximum source voltage provides the maximum performance at the quick intended occurrence of process. When the performance ultimatum is little, source voltage and clock frequency are dropped, just bringing the vital performance with significant power lessening. Figure 1.1 shows a clock gating using AND gate. Figure 1.2 illustrates a gated D Flip-flop and a gated clock design in sequential circuit.

1.1.3.4 Power gating

In an application, most of the slice of the ICs is not used for most of the time. If we cut off power in the IC, this can lead to power saving. This saves the static power and dynamic power in the VLSI circuit. The following points are considered for such techniques: Complexity for Application Development: The portion needs to be reconfigured after each state. In large cases, this is negligible and is used in most of the cases.

1.1.4 Procedure to decrease leakage power

The methods to decrease leakage power by applying the sagging deprived of impacting performance and classified established on the existing timing sagging in the design period decrease the leakage power. The methods that exploit the sagging during the run time were alienated into two clusters reliant on the decrease in stand-in leakage or vigorous leakage. Stand-in leakage attenuation methods put the complete scheme in a small leakage approach when computation is not necessary. Dynamic leak reduction methods slow down the system by vigorously decreasing leak when maximum performance is not required.

1.1.4.1 Design phase

Design phase method proceeds with sagging in non-critical channels to avoid leak. These approaches are stationary; once set, they may be altered with energy, although the circuit is operational.

CMOS logic with two thresholds: In the logic with two thresholds, high V_{th} transistors in non-critical channels may be used to decrease sub-threshold leak current, though small V_{th} transistors in the crucial circuit do not harm performance(s). There is no need for additional circuitry, and high performance and low leakage may be attained instantaneously. One of the twofold threshold CMOS methodologies is to alter the doping profile.

- Greater oxide
- Longer channel length

1.1.4.2 Run interval method

Stand-in leak-decreasing approaches abode parts of the circuitry in stand-in approach while they aren't in use (low leakage mode).

- One sort of standby leakage reduction technique is natural transistor stacks.
- A transistor with forced stacking
- Biasing of the body in both forward and backward directions. When maximum performance isn't necessary, the dynamic leakage reduction method slows down the faster circuit and decreases leakage and dynamic power usage. The lowest V_{th} is offered if maximum performance is desired. If the finest potential outcomes are desired when one's performance deteriorates. When performance requirements are modest, the run-time leak power dissipation is reduced by dropping the clock frequency and improving the V_{th}. When no workload is present, the V_{th} can be set to its highest setting to dramatically decrease stand-in power leakage.

1.1.4.3 Cache memories

- Source biasing scheme is a circuit approach for reducing leakage in cache memory.
- Body biasing system (forward/reverse)
- VDD system that is dynamic
- A skewed leakage system
- The use of a negative word line scheme

1.2 LITERATURE SURVEY

Gluzer and Wimer discussed the low-power multi-bit flip-flop design using a data-driven clock-gated method [1]. By the clock-gating method, the clock to an indolent portion is deactivated, thus evading power dissipation because of needless charging and discharging of the unemployed circuit. Lin et al. described the design of clocking master-slave flip-flop using CMOS logic and optimization of transistor count to achieve high performance in terms of power and speed [2]. Aditya et al. presented the realization of clock-gating-based sequential circuit such as '4' bit Shift Register [3]. It selects a subgroup of flip-flops to be gated to decrease dynamic power consumption using the clock-gating technique. The limitation is that it upsurges total logic and henceforth increases area. Kamaraju et al. described a technique called power gated '2' threshold source voltage in order to decrease power consumption in digital circuits [4]. Lin et al. presented the CMOS-based clock-gating synchronous counter drives with no redundant transitions [5]. Xan and Kimura [6] discussed the switching action at the nodes of a '16' bit multiplier and accumulator module for the filters model.

Rizvandi et al. presented the multiplier-and-accumulator unit, the carry save adder designed using adder-altered Booth's algorithm [7]. Their design uses customary CMOS library with different nm technology. Chaudhary et al. described the clock-gating techniques for sequential design with high-frequency clock [8]. Mazher Iqbal and Varadarajan described an algorithm considering the condensed adder graph method for a set of stationary coefficients [9]. The condensed adder graph method is helpful in numerous filter applications such as FIR or infinite impulse response (IIR) filters, where the multipliers can be clustered. The algorithm selects common adder, which are exceedingly shared with the remaining coefficients. Mazher Iqbal and Varadarajan discussed the comparison of various realizations of dynamically reconfigurable FIR filter architecture using computation sharing multiplier [10]. The shift and add unit produces {1, 3, 5, 7, 9, 11, 13, 15} and {0, 1, 2, 3, 4, 5, 6, 7} Binary Common Sub-expression (BCS). Individually these '8' BCSs are then fed to the data selector unit. Mazher Iqbal and Varadarajan presented the parallel and cascade architectures of FIR architectures with shift and add unit which accomplishes the multiplication [11]. The shift and add unit produces {1, 3, 5, 7, 9, 11, 13, 15} BCS. Mazher Iqbal and Varadarajan presented FIR filler with memory-based multiplication [12]. The LUT stores pre-computed multiplication and summation performed in FIR filter and the stored value is retrieved through memory indexing. Mazher Iqbal and Narayan discussed the Least Mean Square (LMS) adaptive filter architecture using reversible processing element architecture with less power dissipation because of reversible methodology adapted in the processing elements [13]. Mazher Iqbal and Manikandan discussed the FPGA-based reconfigurable common sub-expression elimination (CSE) architectures using LUT and serial processing for DSP applications

[14]. Mazher Iqbal and Munagapati Siva Kishore described the employment of sensors using Programmable System on Chip (PSoC) microcontroller and printed circuit board (PCB) strip [15]. Karthik and Mazher Iqbal presented the speech recognition in the presence of numerous noises. The paper proposes the convolutional Encoder and Decoder to eliminate the noise [16]. The article [9–16] utilized the DSP architectures without clock gating. The clock-gating circuit inserted in the architectures will reduce the power consumption of the given architectures.

1.3 CLOCK-GATING METHODOLOGY

The clocking section is an important section in microprocessor's power as the clock is required for almost the entire circuit module in the microprocessor, and the clock changes during each cycle. As a result, overall clock power accounts for a significant portion of total processor power requirement. Clock gating is a method of lowering clock power consumption. Because separate circuit utilization differs inside and between applications, circuits are not completely active throughout the entire period, resulting in a potential for power reduction [17]. Clock gating effectively shuts the clock in a circuit anytime the circuit is not in use by AND gate to the clock with a gate-control signal, preventing power consumption due to needless charging and discharging of the circuits that aren't in use. Clock gating is a technique for reducing clock power consumption in sequential pipeline circuit and active CMOS logic, which are employed for speediness and less area in fixed logic. Clock-gating methods have a lot of overhead. Because of this overhead, power dissipation may be larger than without clock gating.

1.4 PROPOSED CLOCK-GATED SHIFT REGISTER

A microprocessor's clock setup offers clock to sequential elements like shift registers with logic gates cast off in high-performance implementation. The proposed work decreases the power consumed in the VLSI circuits such as shift register, counters and S27 using clock-gated technique. The proposed work is related with the power dissipated of predictable VLSI circuits and the clock-gated circuits [18]. The demand for a longer life battery has made it trendy to enhance power dissipation. The switching action can be restricted in a VLSI scheme through a method called Clock Gating by removing the indolent sequences of D flip-flop. We propose a clock-gating technique in sequential circuits that successfully decreases the power consumption in digital circuits. A Verilog/VHDL-based method is used to supplement a clock-gated digital circuit. The clock gating for NOR Gate-based clock gating is affected by hazards. The

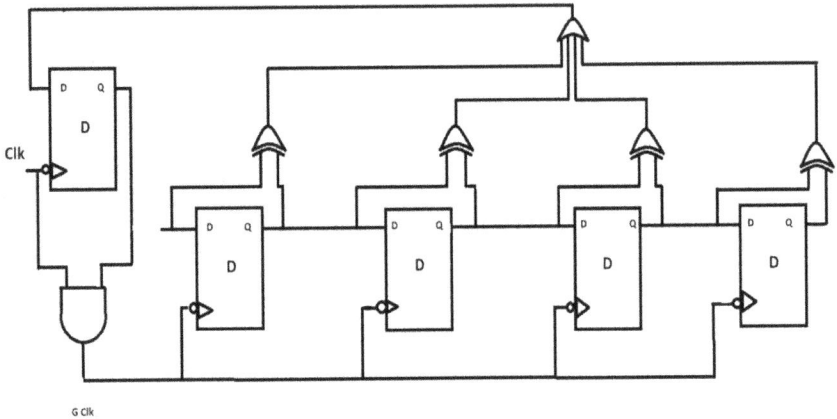

Figure 1.3 Gated clock shift register.

latch-based Nor Gate Clock Gating takes one additional clock cycle delay to change its state. Figure 1.3 shows the shift register with clock gating. The benefits of clock gating are:

- The upgradation in technology, enhanced performance, and more functionality require minimized power dissipation in the circuit design.
- The sequential and combinational circuits are used in VLSI design. The clock is the key source of dynamic power consumption in sequential circuits.
- In VLSI circuits, the clock power can be key for higher performance; hence, the deterministic clock-gating technique can effectively reduce power consumption.

Clock gate is shut off when the flip-flop is not required to switch state, and the clock is not permitted, conserving the power of the clock. As a result, we can make net electricity savings.

1.4.1 Low-power strategy

The most significant constraint in terms of mobility is power consumption. As a result, the system's entire power usage must be evaluated. It is preferred to maximize run time while minimizing total power consumption. As a result, the low-power design is the most critical factor to consider when creating a system on chip for portable devices. For a reduced price, mobile customers often want more features and a longer battery life. Power efficiency is the most important need for 4G operators. Customers are continuously on the lookout for mobile gadgets that are smaller, trimmer, and more graceful. High degrees of silicon incorporation are required in current procedures, but distinguished procedures have a greater power indulgence by definition. As a result, with low-power devices, design is critical.

1.4.2 Influence of power dissipation

When power dissipation exists, the chip temperature rises in a predictable manner. When devices are switched on and off when the temperature rises. When the device is off, power dissipation increases the amount of intrinsic carriers, ni. It is obvious from the preceding equation that when the temperature rises, inherent carriers rise with it. If the heat dissipation is not appropriately evacuated, the equipment may eventually fail. These characteristics decrease when the temperature rises, resulting in a shift in drain current (ID). As a result, the device's performance may fall short of the required standards. Also, in battery-powered applications, power dissipation is more significant because the more power dissipated, the shorter the battery life.

1.4.3 Reduction of temperature

The heat created by power dissipation is dissipated using heat sinks. The heat sink has a lower thermal resistance than the package. As a result, the heat is drawn to the heat sink. To effectively eradicate heat, the amount of heat transmission to the surroundings must be larger than the rate of heat generation.

1.4.4 Low-power design methodology

Circuit speed has traditionally been utilized as a performance metric by VLSI designers. In fact, in certain portable applications, power considerations have been the most important design criteria. The major goal of these applications was to get the longest battery life possible while using the least amount of electricity. Low-power design is also essential in systems with high incorporation density in order to decrease power consumption and increase operation speed. It is critical to understand how power is distributed. As a result, the blocks or portions that consume a significant amount of power might be visibly enhanced to save energy.

1.4.5 Power lessening via process technology

Most operational ways to decrease power dissipation are to decrease the device's supply voltage. The disadvantage of this strategy is that as VDD approaches the threshold voltage, latency may increase dramatically. To address this issue, devices must be correctly scaled. The following are some of the benefits of scaling: Enhance the device's capabilities. Reduce junction

```
┌─────────────────────────┐
│         System          │
└─────────────────────────┘
            │
┌─────────────────────────┐
│        Algorithm        │
└─────────────────────────┘
            │
┌─────────────────────────┐
│      Architecture       │
└─────────────────────────┘
            │
┌─────────────────────────┐
│      Logic / Circuit    │
└─────────────────────────┘
            │
┌─────────────────────────┐
│     Device / Process    │
└─────────────────────────┘
```

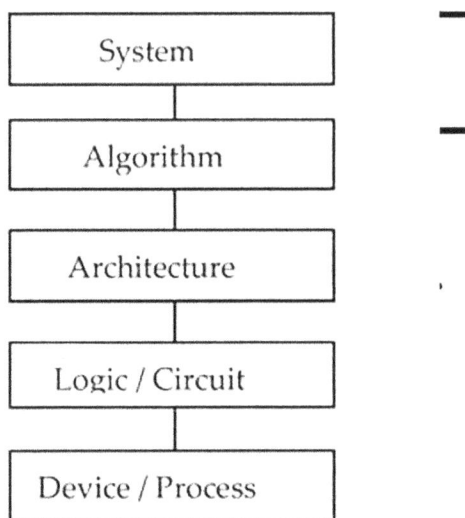

Figure 1.4 Power reduction design aspects.

and geometric capacitances. Improvements in interconnect technology. Integration density is high (Figure 1.4).

1.4.6 Power reduction through circuit/logic design

The power reduction in VLSI circuits is achieved by

- Optimized algorithm reduces switching activity.
- More static circuits are used than dynamic circuits.
- Time and bus loading should be optimized.
- Smart circuit approaches reduce the number of devices in a circuit.
- Custom design has the potential to increase the power.
- VDD is reduced in non-critical pathways, and correct transistor size is achieved.
- Multi-VT circuits are used.
- Sequential circuits re-encoding

1.5 RESULTS AND DISCUSSION

The results show that the digital circuits with the clock-gating method considerably decrease the dynamic power consumption. The clock-gating method considerably decreases the dynamic power of the sequential circuit but might upsurge the amount of logics and therefore the area enlarged. The result of the proposed circuit with a gated clock shift register is shown

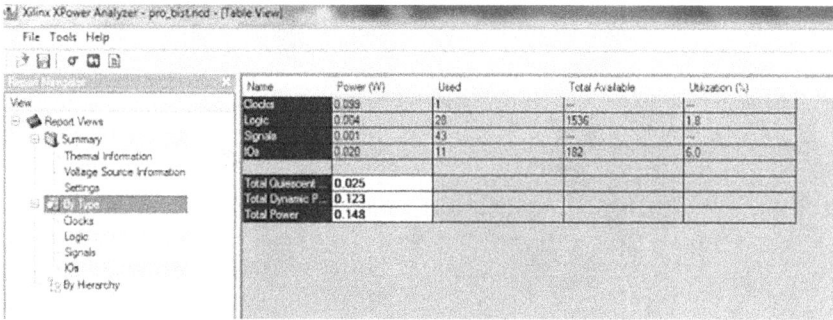

Figure 1.5 SISO power analysis.

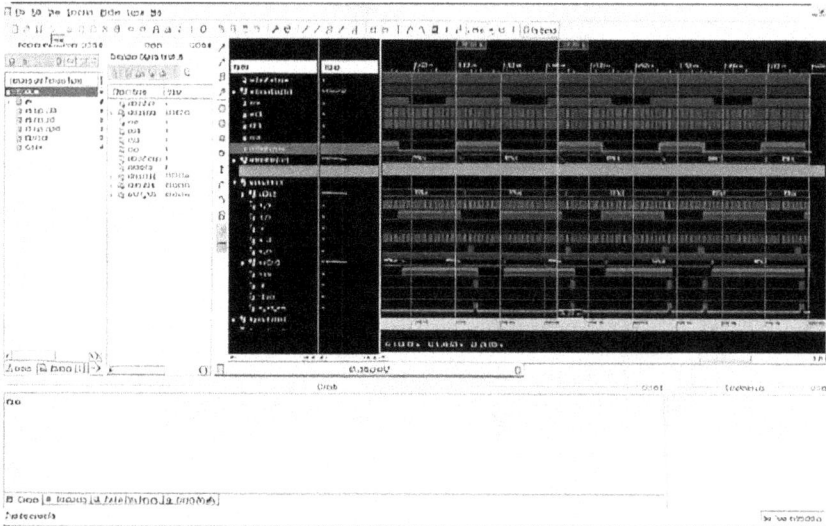

Figure 1.6 Waveform of SISO.

in Figures 1.5–1.8. Table 1.1 shows the comparison of power analysis of SISO for different voltages with and without clock gating.

1.6 CONCLUSION

The necessity for a low-power VLSI design system is determined by numerous market sections. But the development of low-power VLSI design enhances additional dimension to the current demanding design and ought to be improved in power as well as performance and area. So, the

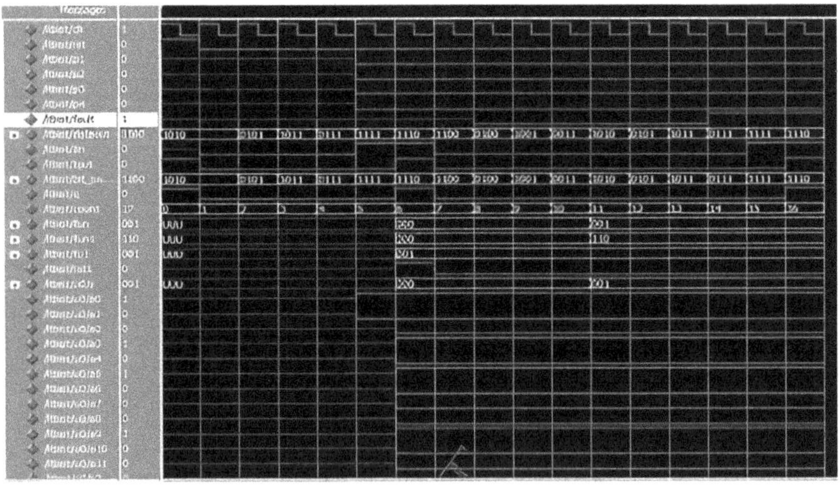

Figure 1.7 SISO without clock gating.

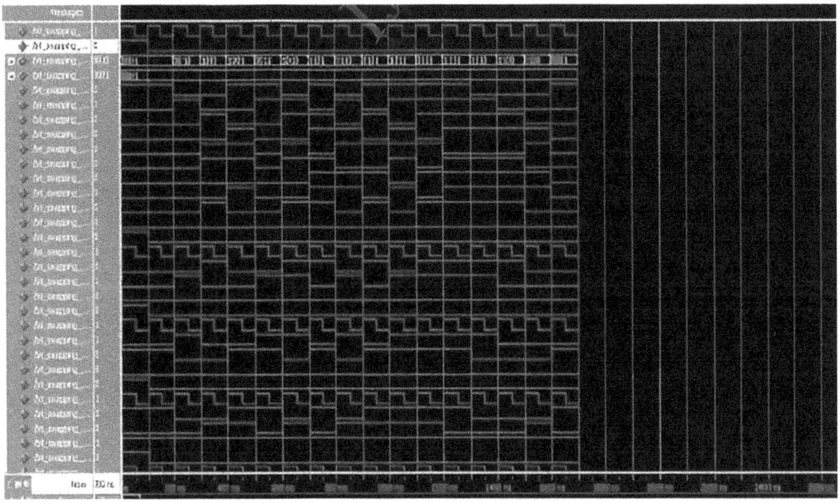

Figure 1.8 SISO with clock gating.

Table 1.1 Stimulation results of SISO circuit

Voltage (V)	Power without clock gating (µw)	Power with clock gating (µw)	Power reduction (%)
0.8	1.3	1.12	91
1	129.35	74.7	42
1.3	662	257	58
1.5	1882.4	355	81

clock-gating method is unique as a proficient methodology to decrease the power in the sequential circuits. In this chapter, a modest technique to decrease active power consumption is presented. The proposed system is established on clock-gating methodology. And gate-based clock gating is cast off in the proposed system. Result analysis shows that the clock-gating method considerably decreases the active power consumption.

REFERENCES

1. D. Gluzer and S. Wimer, "Probability-Driven Multibit Flip-Flop Integration with Clock Gating," *IEEE Transactions on Very Large Scale Integration (VLSI) Systems*, 25(3), 1173–1177, 2017.
2. I J. Lin, M. Sheu, Y. Hwang, C. Wong and M. Tsai, "Low-Power 19-Transistor True Single-Phase Clocking Flip-Flop Design Based on Logic Structure Reduction Schemes," *IEEE Transactions on Very Large Scale Integration (VLSI) Systems*, 25(11), 3033–3044, 2019.
3. K. V. S. S. Aditya, B. B. Kotaru and B. B. Naik, "Design of Low Power Shift Register using Activity-Driven Optimized Clock Gating and Run-Time Power Gating", *International Conference on Green Computing Communication and Electrical Engineering (ICGCCEE)*, Coimbatore, pp. 1–7, 2014.
4. M. Kamaraju, V. Satyavolu and K. L. Kishore, "Design and Realization of CMOS Circuits using Dual Integrated Technique to Reduce Power Dissipation," *International Conference on Signal processing and Communication Engineering Systems*, Guntur, pp. 112–117, 2015.
5. Chen-Hsien Lin, Shih-Hsu Huang, Jia-Hong Jian and Xin-Jia Chen, "New Activity-Driven Clock Tree Design Methodology for Low Power Clock Gating," *6th International Symposium on Next Generation Electronics (ISNE)*, Keelung, pp. 1–3, 2017.
6. Xan and S. Kimura, "Comparison of Optimized Multi-Stage Clock Gating with Structural Gating Approach," *TENCON 2011, IEEE Region 10 Conference*, Bali, pp. 651–656, 2011.
7. N. B. Rizvandi, S.A.M. Barandagh and A. Khademzadeh, "Power Dissipation and Gate Number Reduction of a Utilized, Register Replaced by Equivalent Counters," *24th International Conference on Microelectronics (IEEE Cat. No.04TH8716)*, vol. 2, Nis, Serbia, pp. 789–791, 2004.
8. H. Chaudhary, N. Goyal and N. Sah, "Dynamic Power Reduction using Clock Gating: A Review," *International Journal of Electronics and Communication Technology*, 6(1), 22–26, 2015.
9. J L Mazher Iqbal and S. Varadarajan, "A New Algorithm for FIR Digital filter Synthesis for a Set of Fxed Coefficients," *European Journal of Scientific Research, EJSR*, 59(1), 104–114, 2011.
10. J L Mazher Iqbal and S. Varadarajan, *Performance Comparison of Reconfigurable Low Complexity FIR Filter Architectures*, vol. 250, pp. "637–642. Berlin Heidelberg: Springer-Verlag, 2011. https://doi.org/10.1007/978-3-642-25734-6_151.

11. J L Mazher Iqbal and S. Varadarajan, "High Performance Reconfigurable FIR Filter Architecture using Optimized Multipliers, Circuits System and Signal Processing," *Springer Journal*, 32(34), 663682, 2013, https://doi.org/10.1007/s00034-012-9473-3.

12. J L Mazher Iqbal and S. Varadarajan, "Memory Based and Memory Less Computation for Low Complexity Reconfigurable Digital FIR Filter," *WSEAS Transactions on Systems*, 12(3), 142–153, 2014.

13. J L Mazher Iqbal and G. Narayan, "Design and Implementation of Efficient Adaptive Filter using High Performance Reversible Adder," *Journal of Advanced Research in Dynamical & Control Systems*, 10, 1494–1499, 2018.

14. J.L. Mazher Iqbal and T. Manikandan, "FPGA Based Reconfigurable Architectures for DSP Computations," *Advances in Intelligent Systems and Computing*, Springer, Singapore, vol. 1163, pp 587–594, 2020, https://doi.org/10.1007/978-981-15-5029-4.

15. J.L. Mazher Iqbal, Munagapati Siva Kishore, Arulkumaran Ganeshan and G. Narayan, "Design and Implementation of SOC-Based Noncontact-Type Level Sensing for Conductive and Nonconductive Liquids, Hindawi," *Advances in Materials Science and Engineering*, 2021, A1rticle ID 7630008, 2021, https://doi.org/10.1155/2021/7630008.

16. A. Karthik and J. L. Mazher Iqbal, "Efficient Speech Enhancement Using Recurrent Convolution Encoder and Decoder," *Wireless Personal Communications*, 2021, https://doi.org/10.1007/s11277-021-08313-6.

17. N. V. Haritha and Jose Anand, "Solar Power Forecasting Using Long Short-Term Memory Algorithm in Tamil Nadu State", In: *Anirbid Sircar*, Goutami Tripathi, Namrata Bist, Kashish Ara Shakil, Mithileysh Sathiyanarayanan (Eds.). *Emerging Technologies for Sustainable and Smart Energy*, 1st Edition, CRC Press, pp. 73–96, 2022.

18. T. Thomas Leonid, M. Mary Grace Neela, and Jose Anand, "Signed Pipelined Multiplier using High Speed Compressors" *International Journal of Research in Computer Applications and Robotics*, 1(6), 29–38, 2013.

Chapter 2

Emerging applications and challenges in low-power device designs in flexible and stretchable low-power devices and strategies

Anurag Nain, Divya Bajpai Tripathy,
Anjali Gupta, and Vinny Sharma
Galgotias University

2.1 INTRODUCTION

All integrated circuits today must have low-power designs. Battery backup time became crucial when businesses began putting more and more functions and applications on battery-operated devices (mobile, handheld, and laptops). Power usage gradually started to matter more and more to the customers. All chip firms that provide products for battery-operated gadgets are concentrating on lowering power consumption as a result. Both static and dynamic power usage are being reduced. Companies started to lower the chip's nominal voltages, but the technology's limitations also applied to this. The reduction of electronic circuits power consumption is one of the main difficulties facing circuit designers today. The creation of micron and nano-size transistors has been prompted by the rising demand for more processing power in order to permit a higher component density in a device. The circuits' dependability is, however, being constrained by issues like Short Channel Effects (SCE) and process fluctuations as a result of their shrinking size [1]. Overheating caused by power dissipation reduces the durability of the circuits. The increasing current density in silicon also becomes a problem for manufacturers [2]. The nano-range transistors are increasing the severity of this issue. In addition to the power density rising as more transistors are placed in smaller areas, the power leakage of transistors is also approaching the active power usage, which means that a significant amount of the power consumed is dissipated only in generation of heat. To avoid overheating the circuits, it can be solved by investing in more research in cooling and packaging; however, this is an expensive and time-consuming approach.

As the Internet of Things (IoT) booms these days, a new emphasis is being placed on reducing the size of smartphones, smart watches, tablets,

DOI: 10.1201/9781003459231-2

17

and other electronic devices while improving the computing capacity. Furthermore, with the modern involvement in the Internet of Bio-Nano Things (IoBNT) [3], the emphasis has moved from just increasing computation power to developing incredibly tiny and low-power designs that allow sensors and actuators to function without the requirement for wired data connections or outside power sources. Bulk batteries are rarely an option when it comes to implanted sensors because of the size restriction's critical need for biocompatibility. Instead, they will be run on ambient energy collected from their surroundings. The medical industry, where numerous research studies are being done on implants and other wearable gadgets to track bodily reactions, finds this approach to be particularly intriguing. In addition, there is a sizable market for the production of various biosensors used in sporting equipment to track athletes' physical condition.

When scaled to the next generation, advanced CMOS technology increases three key metrics: (1) transistor and interconnect performance, (2) transistor density, and (3) energy used during switching transitions. Three main objectives of technology scaling with a 30% decrease in minimum feature size every generation are: (1) a 30% reduction in gate latency; (2) a doubling of transistor density; and (3) a 30%–65% reduction in energy per transition, depending on the supply voltage reduction. The clock frequency, die size, functional integration, and power consumption of high-performance microprocessors are predicted to continue following historical patterns thanks to these technological advancements as well as improvements in circuits and microarchitecture [4].

2.2 THE NEED FOR LOW-POWER DESIGN

Businesses are always pushing the envelope with regard to the latest features and usefulness that are fit in handheld battery-operated products. For such products, long-lasting battery life by reducing power consumption is a significant differentiation and crucial to the applications used by their end users. To supply the end user with a smooth experience and longer battery life, it is equally crucial to reduce the time taken by a device to transit from the OFF/SLEEP state to the ON/ACTIVE state.

Power consumption is crucial for "plug-in" devices since it can raise system costs by necessitating sophisticated cooling systems and heat sinks, for example, or by demanding more energy [4].

There are some low-power design techniques that are currently under study, some of them are:

- **Clock Gating:** Circuits can be turned off thanks to a power-saving function in semiconductor microelectronics. To decrease dynamic power usage, many electronic devices employ clock gating to switch off buses, controllers, bridges, and portions of CPUs.

- **Power Gating:** Power Gating is a method used in the design of integrated circuits to cut down on power use by cutting off the current to circuit blocks that are not in use. When the system is not in use, power gating is employed to preserve the leaking power. To do this, a switch is added to the VDD or VSS supply.
- **Dynamic Voltage and Frequency Scaling (DVFS):** Dynamic voltage scaling is a power-management strategy used in computer architecture, where a component's voltage is changed based on the situation. Undervolting is the result of dynamic voltage scaling while overvolting is the result of dynamic voltage scaling. Undervolting is done to save energy, especially in laptops and other portable electronics to improve dependability. In order to accommodate higher frequencies for performance, overvolting is done.
- **Retention Power Gating (RPG):** Gating Power Retention When the cell's primary power source is turned off, a special flip is utilized to keep the cell in its current condition. In some circumstances, it is necessary to maintain the status of key control flops throughout power off. SRPG flops are used to hasten power-up recovery. If certain control signaling criteria are satisfied, they maintain their condition even when the power is turned off.
- **Save and Restore power gating:** The benefit of the save and restore power gating approach over power gating is that one may save the state machine's value in the gated power domain. This method extends the continuous power domain with a RAM. The gated power domain must use an extra state machine to save the state of the state machine in RAM before entering the power-off state. Before regular activities can resume when the gated domain is woken up from the power-off state, it must first read the data from the RAM and return to its initial state.

2.2.1 Clock gating

As stated in the definition of clock gating, "Clock gating is a technique/ methodology to switch off the clock to select elements of the digital architecture when not needed."

Clock gating is one of the simplest and most common techniques for reducing the dynamic power of the design, which is crucial because most SoCs are severely restricted by power budgets. Simple in its conception. To preserve functioning, turn off the design when not in use. The design's switching activity and hence dynamic power are effectively decreased by turning off the clock to the system or design. The tiniest component in a digital architecture, a single FF, can have clock gating applied to it, all the way up to whole SoCs or subsystems [5] (Figure 2.1).

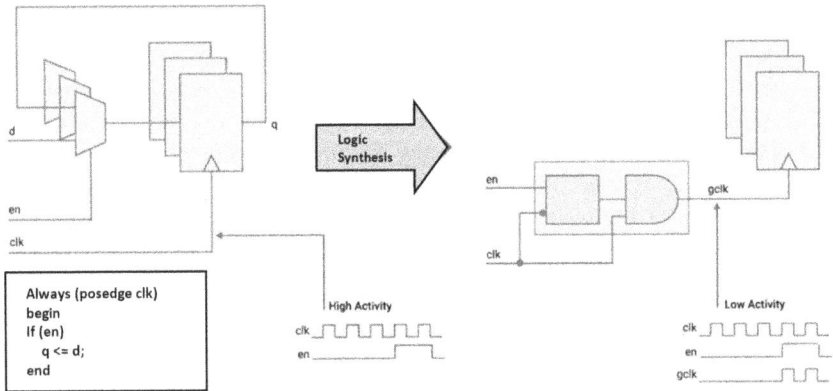

Figure 2.1 Clock gating.

2.2.1.1 Techniques of clock gating

In general, there are two methods for putting clock gating into practice.

1. **Intent-Based Clock Gating:** RTL adds usefulness to the design by including this form of clock gating.
2. **Tool Produced Clock Gating:** During synthesis, tools detect all the flip-flops that share the same control logic and enable all of those FFs as necessary.

2.2.1.2 Clock gating overview

The easiest way to implement clock gating is to use an AND gate, as seen in the image below:

When to supply the clock to the downstream logic is determined by the clock enable signal, which is produced by combination logic (FF in the above Figure 2.2). When enable is set to 1, FF will receive a clock; when enable is set to 0, the clock will be switched off, rendering FF inactive.

However, this most basic sort of clock gating approach has the drawback of causing undesirable irregularities in the clock that the FF receives.

2.2.1.3 Clock gating tips

1. The first design phase is the time to specify and decide on the clock/power gating method. This will enable designers to choose the right options for their budget, location, and performance requirements.

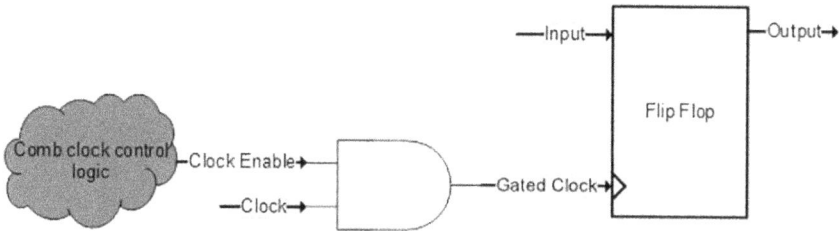

Figure 2.2 AND gate-based clock gating.

2. Consider both static and dynamic power, especially when working with smaller technological nodes.
3. Avoid overusing clock/power gating since this consumes more space and makes verification more difficult.

2.2.2 Power gating

By shutting off components of the design that are not in use or that are in an inactive mode, power gating is a strategy for reducing the power consumption of ASIC and SoC. In addition, it is a particularly effective method for minimizing leakage power in ASIC designs [6].

The fundamental idea is to have just two power modes:

1. **Low-Power Mode:** Power-saving mode is a common feature of contemporary laptops. This mode, which the computer will use to enter a very low-powered condition, is typically referred to as "Hibernate" [7].
2. **Dynamic Mode:** Dynamic mode refers to a system's operation in response to either (a) previous or simultaneous user input or (b) prior or simultaneous network instructions [8].

The design should alternate between these two modes as needed throughout the operation in order to maximize power savings and minimize performance effect.

One method for turning off the power supply to inactive blocks is software, specifically driver software, another method is using hardware timers.

By including a specific power gating controller into the plan.

The need for power gating includes several points:

1. No matter if it is a battery-operated device or a device that has to dissipate heat efficiently, the bulk of modern technologies have modest power dissipation requirements.
2. At lower technology nodes, sub-threshold leakage power is becoming relatively equivalent to dynamic power dissipation, which used to be rather large in older technology nodes compared to leakage

power. Because of this, concentrating on lowering leakage power in the design is just as crucial as reducing dynamic power.

3. Dynamic power may be conserved by lowering the supply voltage for the entire design or select subsystems, and leakage power can be significantly decreased by cutting off the power to the design's inactive components.

4. Power gating is becoming increasingly common in mobile devices because it is particularly effective at limiting the leakage power in the design.

5. There is a very high likelihood that some of these functional blocks may be dormant for extended periods of time during operation given the amount of functionality that is being integrated into current System on Chips (SoCs). This offers several chances to power gate the design.

Power gating is becoming increasingly common in mobile devices because it is particularly effective at limiting the leakage power in the design.

There is a very high likelihood that some of these functional blocks may be dormant for extended periods of time during operation given the amount of functionality that is being integrated into current SoCs. This offers several chances to power gate the design (Figure 2.3).

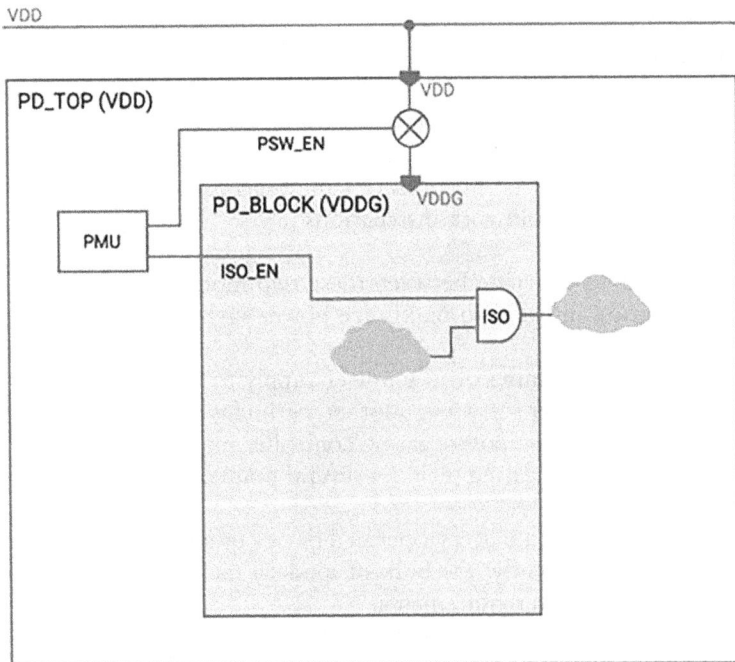

Figure 2.3 Power gating.

2.2.2.1 Leakage power and power gating in CMOS

Typically, p-channel transistors (pull-up network) and n-channel transistors are used in conjunction to create the majority of CMOS logic circuits (pull-down network). Due to the leakage current flowing from VDD to the ground, the CMOS circuit continues to dissipate power even when there is no switching activity [9] (Figure 2.4).

Disconnecting the route to the power and ground terminals is one straightforward approach in such a circumstance. To accomplish this, a power gating circuit that can cut off the route while the circuit is dormant can be included.

Here, a SLEEP signal that is employed for active or inactive mode of operation controls the sleep transistors, also known as switch cells.

SLEEP=OFF when (0),

The NMOS and PMOS sleep transistors are both turned on.

The circuit runs in normal mode since the pull-up and pull-down networks are linked to virtual power and ground, respectively.

When SLEEP=ON (1)

Sleep transistors for PMOS and NMOS switch off.

As a result of cutting off the direct channel from VDD to GND, leakage power is decreased (Figure 2.5).

By using switch cells, the design's power supply was divided into two networks: one that is permanently linked to the global power supply and the other that powers the gated logic and may be shut off when not in use.

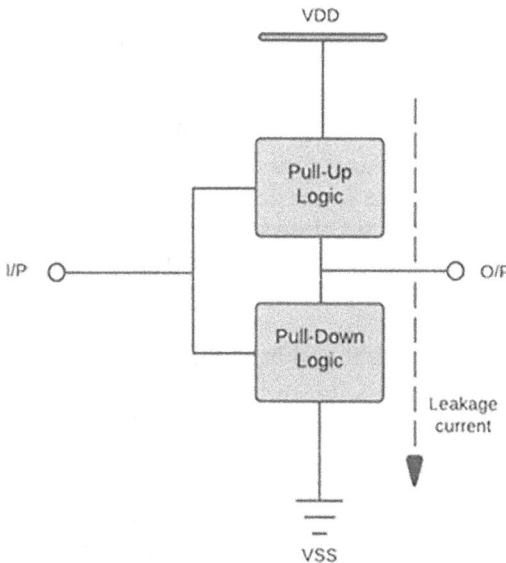

Figure 2.4 Leakage in CMOS circuit.

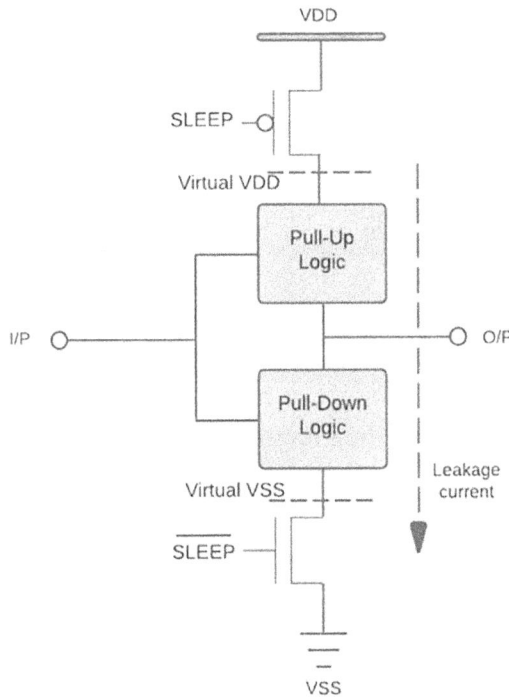

Figure 2.5 Power gating in CMOS.

2.2.2.2 Switch cells

Switch cells, also known as sleep transistors, fall into one of two categories:

- **Header Switch Cell:** To gate the VDD rails, the header switch is constructed using PMOS transistors. PMOS transistors are often less leaky than NMOS transistors of the same size. However, since the PMOS transistor has a lower driving current than an NMOS transistor of the same size, the drawback in this situation is that a header switch often occupies more space than a footer switch (Figure 2.6).
- **Footer Switch Cell:** To gate the VSS rails, NMOS transistors are used in the footer switch. In comparison to header switch cells, footer switch cells may be deployed in smaller spaces since they have a high drive output. The NMOS transistor is leakier than the PMOS transistor, which increases the design's susceptibility to ground noise on the virtual ground (Figure 2.7).

The size of these switch cell designs must be extremely precise. They should have little delay when the circuit is operating and little leakage when it is in sleep mode.

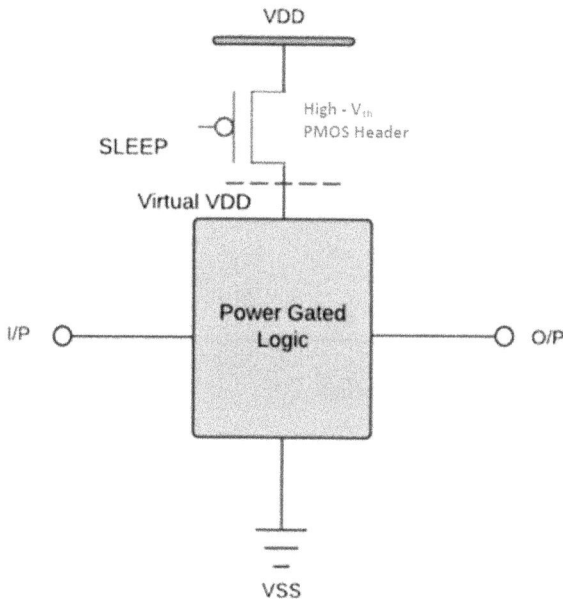

Figure 2.6 Header switch cell.

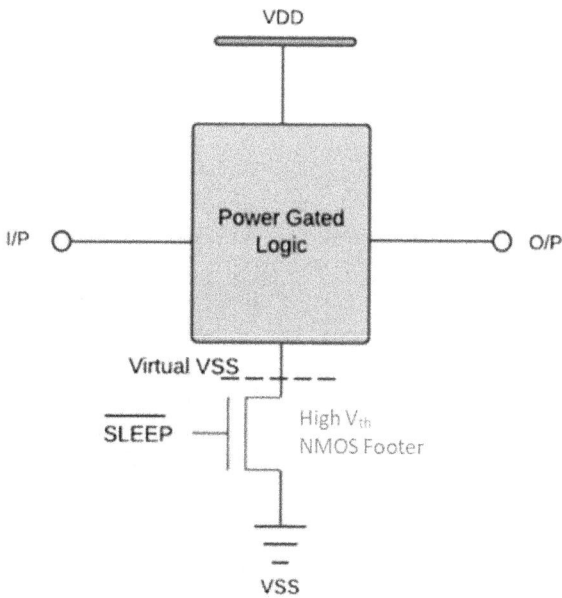

Figure 2.7 Footer switch cell.

In actual designs, a single switch cell is by no means adequate to power the whole circuitry. To achieve minimal voltage ramp-up time and prevent IR-drop-related problems, a network of switch cells is utilized.

Any header cell or footer cell in a design can be utilized for power gating. As an alternative, a header and footer cell combination can be used for the same purpose.

The power gating control block in the design generates the control to the switch cell network.

2.2.2.3 Challenges in power gating

Power gating has its own set of difficulties while being a very effective method for reducing leakage power [10].

1. A trade-off analysis must be performed to compare the energy required to enter and depart low power modes with the amount of leakage power reduction in these modes.
2. Complex wake-up is required by power gating in order to enter and exit the power shut-off state securely, which can add several clock cycles or more when powering up or down an area.
3. The area penalty will increase with the inclusion of state retention cells, isolation cells, and power control logic.
4. Larger power grids may be needed to accommodate current surges. This will increase the overhead for routing resources.

2.2.3 Dynamic voltage and frequency scaling

DVFS modifies the power and speed settings on a computer's different CPUs, controller chips, and peripheral devices to maximize power savings when those resources are not required. With the exception of programs and malware, an idle smartphone, for instance, ought to switch to a low-power mode. Since multimedia uses more power, heavier processing activities like video and games cause the device to run at a higher power level and generate more heat. Many passively cooled devices would need active cooling if it weren't for DVFS. However, active cooling is impracticable for smaller devices due to the noise, mass, and power requirements. With more mobility, DVFS aids in maintaining operable parameters [11].

Not just mobile technologies can use DVFS. The power savings of DVFS also apply to servers, desktops, and virtual environments. When resource demands are low, DVFS in VMware vSphere enables host CPUs to dynamically shift power modes, lowering a host's energy use.

Devices can carry out necessary activities using the least amount of power possible thanks to DVFS. To optimize power savings, battery life, and device lifetime while yet retaining immediate compute performance available, the technique is employed in nearly all current computer hardware.

Recently, as processor speeds have surpassed gigahertz, power dissipation has increased significantly to the order of 10 W, making it a crucial factor to take into account when designing microprocessors, particularly for battery-operated portable systems. It has also emerged as a key technology in the design of VLSI systems.

Around 50% of the total energy used by computer systems is spent by the processors.

The majority of modern digital circuits, particularly processors, are built using CMOS circuits; hence it is crucial to analyze energy release in CMOS circuits in order to understand the relation between clock frequency, supply voltage, and power.

The sum of dynamic power, static power, and short circuit power is the power dissipation for CMOS circuits.

$P_{dynamic}$:- due to charging and discharging capacitors {1}.
P_{static}:- due to reverse biased diodes {2}.
$P_{shortcircuit}$:- due to switching direct path between Vdd-GND {3}.
Mathematically,

$$P_{cmos} = P_{dynamic} + P_{static} + P_{shortcircuit}$$

The dynamic power of the CMOS power dissipation can be expressed as:

$$P_{dynmaic} \alpha C_L V_{dd}^2 f_{clk} \{2\}$$

Here, C_L is the collective capacitance, V_{dd} is supply voltage, and f_{clk} is clock frequency (Figure 2.8).

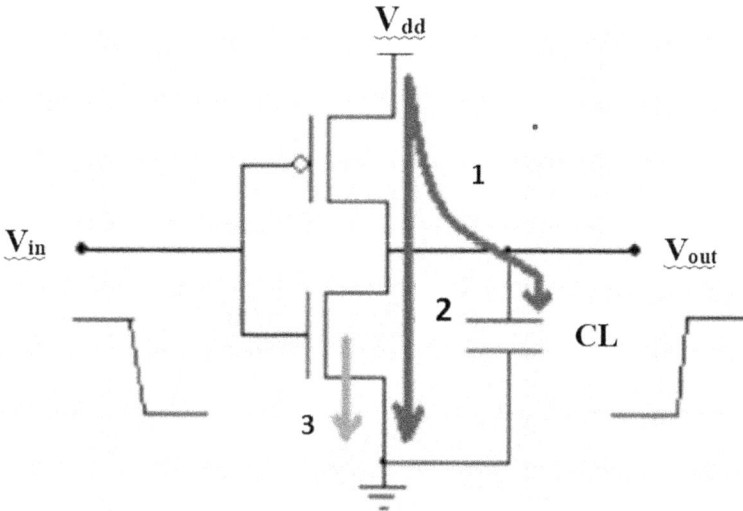

Figure 2.8 Power dissipation for a CMOS inverter.

2.2.3.1 Drawbacks of high-power dissipation

A CPU's high-power dissipation has at least the following drawbacks:

1. High-power systems frequently run hot, which leads to the failure of the processor and other system components. Every 10°C increase in temperature doubles the failure rate of a CPU.
2. It raises the cost of manufacture by complicating integrated circuit cooling methods for heat removal. Once the processor power dissipation approaches 35–40 W, Intel predicts that the cost per processing chip will increase to more than $1/W.
3. It raises the expense of running the business, such as the power used to cool the computer and system rooms. The Internet was responsible for 8% of US power in 1998; by 2020, that percentage is expected to reach 30%.
4. The life of the battery or UPS is shortened. Every 4 years, processing power doubles, which reduces the average battery or UPS life.

2.2.4 Retention power gating

Compared to power gating, this method has the benefit that the state machine's value may be preserved in the gated power domain. However, this method involves more overhead in terms of space and implementation because it is more complicated. Special flip-flops are utilized in this situation in the power-gated realm. These flip flips will also have continuous power in addition to the power of the gated domain. In addition, they feature inbuilt circuitry to save the flip-flop's state in case the power is turned off. This is only utilized in extremely unique circumstances where a quick wake-up of the gated domain is required and it is preferred not to reconfigure the gated domain due to area and routing overheads. This approach provides the quickest wake-up time compared to previous approaches while still preserving state machine data [12].

In some circumstances, it is necessary to maintain the status of key control flops throughout power-off. State retention power gating (SRPG) can be used to speed the power-up recovery. If certain control signaling criteria are satisfied, they hold their condition even when turned off. These particular state retention cells are still available in cell libraries today. After checking that these library-specific constraints are met and the flop keeping its state is a crucial component of verification [13] (Figure 2.9).

To perform power gating, certain state retention cells are required in order to store the previous state of the blocks prior to power-off. The typical flip-flop has been modified in SRPG such that the master latch employs a separate power source while the slave latch makes use of the same power source (Vdd) (Vcc). Flip-flops retain the system's state after power-down, and all combinational logic is turned off during sleep mode [14].

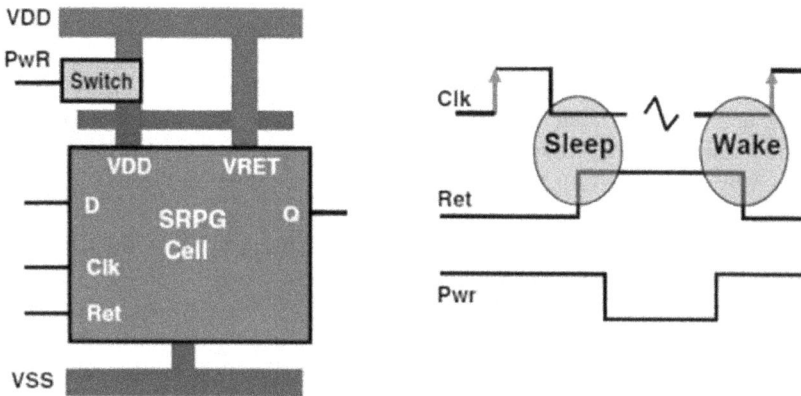

Figure 2.9 Retention.

An always-on power source and a switchable power supply are both necessary for state retention registers. Power routing area standards become more difficult and penalized as a result. To make room for this additional power wiring, the physical designer or implementation tool must allot more space.

Shutdown leakage savings are one of SRPG's benefits, and they may be unaffected by changes in the procedure. The state is maintained in the slave latch, allowing for quicker system start-ups.

The drawbacks include larger area and die sizes, timing penalties like increased routing resources, longer signal and clocking delays, higher power consumption in the active mode, specialized library models for SRPG cells, and effects on functional verification.

2.2.5 Save and restore power *approach*

In comparison to power gating, the save and restore approach offers the benefit that the state machine's value may be preserved. The higher space overhead is a drawback of these low-power design strategies. The continuous power domain is essentially expanded in this method by adding a RAM [15]. The gated power domain must use an extra state machine to save the state of the state machine in RAM prior to entering the power-off state. Before regular activities may resume when a power-off state is entered, the gated domain must read the data from the RAM and return to its initial state. This low-power design method has a space overhead, is complicated, takes time to enter the power-off state, and also takes time to depart the power-off state. Such a method is extremely helpful, nevertheless, when it is known that the gated domain must be turned off for a comparatively longer period of time and that maintaining the value of state machine is also crucial.

2.3 APPLICATIONS OF LOW-POWER DEVICE DESIGN

Applications for low-power device technologies are crucial during the current energy crisis. Its applications are able to lessen the usage of power while still meeting societal demands. And this can help us use energy efficiently. And when it comes to applications, there are a ton of different ones, especially given the vast frequency range of low-power devices [16, 17].

For India, the low-power device frequency spans from 2.4 to 2.4835 GHz; however, this may differ for other nations. In contrast, a few illustrations of low-power device uses are shown below [18]:

1. **Automotive Industry:**
   Low-power technology has the advantage of being less damaging than competing technologies. It will thus be ideal for usage in automobiles. There are several low-power device technology uses in the automobile industry.

2. **Security Purposes:**
   The security system is one of the Low-Power Technology Applications. Here, the alarm system may be supported by a low-power device. The alarm system will be more reliable and unbreakable when used in tandem with a low-power device. It's interesting to note that this technology supports movement detectors as well. When used in conjunction with the alarm, this technology's combination will result in a better security system. Additionally, Low Power may be utilized in CCTV.

3. **Networking Technology:**
   We can observe how this technology contributes to the support of local area networks in the networking system. Also included in the cordless connections are applications for low-power device technology. These are the connectors that are occasionally used in cordless audio systems, wireless microphones, and many other devices. Users will be able to link two or more wireless devices with the help of low-power devices utilizing this type of connection. That means they are no longer bothered by tangled cords.

4. **Medical Applications:**
   The field of medicine is one more area where low-power devices are being used. This technology is occasionally used by doctors to assist pneumatic tube systems. As is well known, RFID, a low-power device, is used in pneumatic tube systems. It will be much simpler to transfer medical supplies and drugs with the help of this gadget. Additional low-power device technology applications include medical implants, the organization of medical records, and many more.

Given the above-mentioned applications, a low-power gadget would undoubtedly be extremely beneficial in daily life. The nice part is that applications for low-power device technologies are always growing.

2.4 CONCLUSION AND FUTURE ASPECTS

The difficulties in designing integrated circuits for future wearable and implantable devices without batteries were discussed in this chapter. Additionally, several potential methods to address these issues at the device, circuit, and architectural levels were discussed. Overall, without the introduction of ground-breaking technologies like FinFET etc. as a replica for CMOS technology as well as circuit and architectural strategies to drastically reduce power consumption, battery-free devices would not be practical due to a lack of sufficient energy from harvesters [19, 20]. The solution to generating more power for such gadgets, on the other hand, will be to combine numerous harvesting techniques in a compressed manner. In contrast to static power dissipation, dynamic power dissipation accounts for the bulk of power loss in CMOS devices. In CMOS devices, static power dissipation is on the order of nanowatts. The transition activities of the circuits are the main cause of dynamic power dissipation. A higher operating frequency causes the circuits to undergo more transitional activities, which increases power loss. The circuit's switching activity may be decreased by using appropriate encoding methods. As a result, the dynamic power dissipation in VLSI circuits may be efficiently minimized [21].

REFERENCES

[1] Pal, A. *Low-Power VLSI Circuits and Systems*. Springer: New Delhi, 2014.
[2] Low Power Design Technique 1: Clock Gating Truechip. n.d. Available at: https://www.truechip.net/articles-details/low-power-design-techniques-basics-concepts-in-chip-design/26234 (Accessed: January 7, 2023).
[3] Akyildiz, I.F.; Pierobon, M.; Balasubramaniam, S.; Koucheryavy, Y. "The Internet of Bio-Nano Things," *IEEE Communications Magazine*, 53, 32–40, 2015.
[4] What Is Low Power Design? - Techniques, Methodology & Tools. Synopsys. n.d. https://www.synopsys.com/glossary/what-is-low-power-design.html (Accessed: January 7, 2023).
[5] The Ultimate Guide to Clock Gating. AnySilicon. 2022. https://anysilicon.com/the-ultimate-guide-to-clock-gating/ (Accessed: January 6, 2023).
[6] Bord, T. D. *Energy-Efficient Processor System Design*. Ph. D. Dissertation. Berkeley: University of California, 2001.
[7] Randy. Laptop Says Entering Power Save Mode. What to Do? WhatsaByte. 2022. https://whatsabyte.com/laptop-says-entering-power-save-mode (Accessed: January 9, 2023).

[8] Active Mode Definition. Law Insider. n.d. https://www.lawinsider.com/dictionary/active-mode (Accessed: January 5, 2023).

[9] Chandrakasan, A. P. "Low-Power CMOS Digital Design," *IEEE Journal of Solid-State Circuits*, 27(4), 473–484, 1992.

[10] Hong, I.; Kirovski, D.; Qu, G.; Potkonjak, M.; Srivastava, M. B. "Power Optimization of VariableVoltage Core-Based Systems," *IEEE Transactions on Computer-Aided Design of Integrated Circuits and Systems*, 18(12), 1702–1714, 1999.

[11] Ergin, O. *Circuit Techniques for Power-Aware Microprocessors*. Master Thesis. New York: The State University of New York, 2003.

[12] Pering, T.; Burd, T.; Brodersen, R. *Dynamic Voltage Scaling and the Design of a Low-Power Microprocessor System*, University of California Berkeley, Electronics Research Laboratory. https://infopad.eecs.berkeley.edu/~pering/lpsw.

[13] Tiwari, V.; Singh, D.; Rajgopal, S.; Mehta, G.; Patel, R.; Baez, F. "Reducing Power in High-Performance Microprocessors," In *The 35th ACM/IEEE-CAS/EDAC Design Automation Conference*, San Francisco, California, USA, June 15–19, 1998.

[14] Hsu, C. H. *Compiler-Directed Dynamic Voltage and Frequency Scaling for CPU Power and Energy Reduction*. Ph.D. Dissertation. The State University of New Jersey, 2003.

[15] Power Gating Retention. Semiconductor Engineering. 2019. Available at: https://semiengineering.com/knowledge_centers/low-power/techniques/power-gating/power-gating-retention/.

[16] Rajput, S.S.; Jamuar, S.S. "Low Voltage Analog Circuit Design Techniques," *IEEE Circuits and Systems Magazine*, 2, 24–42, 2002.

[17] Rajput, S.S.; Jamuar, S.S. "Design Techniques for Low Voltage Analog Circuit Structures," In *Proceedings of the 2001 IEEE National Symposium on Microelectronics*, Genting Highlands, Malaysia, 12–13 November, 2001.

[18] Low Power Device Technology Applications in Life & Industry. Narmadi.com, 2022. https://narmadi.com/low-power-device-technology-applications/ (Accessed: January 7, 2023).

[19] Ueno, K.; Hirose, T.; Asai, T.; Amemiya, Y. "A 300 nW, 15 ppm/°C, 20 ppm/V CMOS Voltage Reference Circuit Consisting of Subthreshold MOSFETs," *IEEE Journal of Solid-State Circuits*, 44, 2047–2054, 2009.

[20] Wuytack, S.; Catthoor, F.; Franssen, F.; Nachtergaele, L.; De Man, H. "Global Communication and Memory Optimizing Transformations for Low Power Systems," In *Proceedings of the IEEE International Workshop on Low Power Design*, Napa Valley, CA, 24–27 April, pp. 203–208, 1994.

[21] Lundager, K.; Zeinali, B.; Tohidi, M.; Madsen, J. K.; Moradi, F. Low power design for future wearable and implantable devices. 2016. https://www.mdpi.com/161894 (Accessed: January 2, 2023).

Chapter 3

A highly stable, reliable ultralow-power design for 11T near-threshold FINFET SRAM

Erfan Abbasian
Babol Noshirvani University of Technology

Maryam Nayeri
Islamic Azad University

3.1 INTRODUCTION

The principal demand of system-on-chip (SoC) applications and battery-powered devices is to design ultra-low power systems to improve the battery's health to operate as long as possible [1]. Static random access memory (SRAM) covers a huge part of SoC's overall space area. Therefore, SRAM is the dominant factor for overall power consumption. To reduce the total power consumption in the SoC, ultra-low power SRAM is required to be designed. The most-efficient and popular technique to diminish the overall power is to drop the operating supply voltage (V_{DD}) [2]. However, with V_{DD} reduction and technology scaling, the variations in process-voltage-temperature (PVT) highly degrade the transistor's attributes, as well as the performance of an SRAM [3]. This issue is resolved by replacing the traditional Silicon-based complementary metal–oxide–semiconductor (Si-CMOS) technology with the fin-shaped field-effect transistor (FinFET). This is because FinFET devices offer extraordinary properties including excellent gate control on the channel and subthreshold slope [4–6]. Anyway, the traditional 6T bitcell designed with FinFET devices still offers unacceptable amounts of stability for read/write operation in low V_{DD} due to the existence of intrinsic fights between different transistors with each other [7]. Due to the quantization of width in the FinFET devices, transistor sizing will be not an optimum option for traditional 6T SRAM cell performance improvements [8]. Therefore, the conventional 6T design should be restructured and some design-level techniques should be taken into account. In this regard, a write-read-enhanced 11T (WRE11T) SRAM cell is introduced in this chapter. The suggested WRE11T improves read stability and writability through decoupling the reading path from the latch core and cutting one of the feedback paths of back-to-back inverters, respectively. It also reduces leakage power dissipation and dynamic power consumption by means of

DOI: 10.1201/9781003459231-3

the single-bitline structure. The remaining parts of the chapter are structured as below. In Section 3.2, the structure and working of the suggested WRE11T SRAM cell are described. In Section 3.3, the results obtained from the simulations are presented and discussed. Finally, a conclusion for this chapter is derived in Section 3.4.

3.2 THE SUGGESTED WREI1T DESIGN: STRUCTURE AND WORKING

3.2.1 Bitcell structure

Figure 3.1 indicates the transistor-level structure of the suggested WRE11T SRAM cell. The structure of the suggested design consists of robust back-to-back inverters, M1 to M3 form the inverter on the left side and M4 to M7 form the inverter on the right side, M9, gated by the wordline responsible for reading (RWL), is enabled for executing a read operation only, M11, gated by the wordline responsible for writing (WWL), is enabled for executing a write operation only, M8, gated by WWL, is responsible for removing the feedback path during the write operation, and bitline discharging

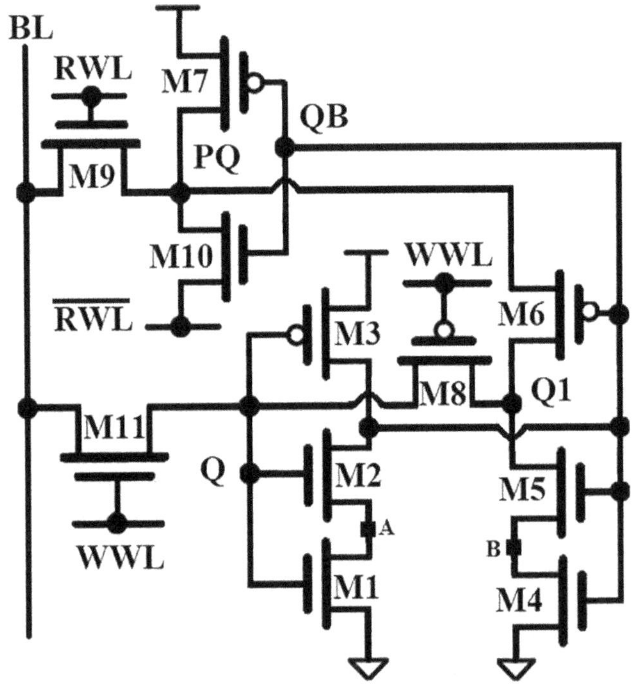

Figure 3.1 The suggested WREI1T bitcell's structure at transistor level.

Table 3.1 Status level of various signals employed in the suggested WRE11T bitcell

Control signals	Hold	Read	Write '1'/'0' @ Q
BL	V_{DD}	Floating	V_{DD}/GND
RWL	GND	V_{DD}	GND
WWL	GND	GND	V_{DD}

transistor M10. The source of the M10 is connected to the complement of the *RWL*. The read/write operation is executed by one bitline (*BL*). Various signals employed in the suggested design, along with their status level, are given in Table 3.1.

As can be seen in Figure 3.1, the transistor M9 shares the bitline with the transistor M11. This issue increases the bitline's overall capacitance. The capacitance enlarges to a smaller amount of 10% for every 2^{10} bitcells because of the main reasons mentioned in [4]. Minimum-size transistors can be used in an SRAM design to reduce the layout area, provided that SRAM offers acceptable read stability and writability. The suggested WRE11T design provides good stability in all the operations using minimum-size transistors.

3.2.2 Hold mode

Two signals *RWL* and *WWL* are pulled down to the ground to eliminate all the paths from bitlines to the latch core and establish the feedback path. Therefore, the data accumulated in the latch core is maintained by the back-to-back inverters.

3.2.3 Single-ended read operation with bitline precharging free

The suggested WRE11T design employs the built-in inverter formed by M7 and M10 to write the '0' and '1' on the *BL*. Therefore, the suggested design is free from the bitline precharge operation before the beginning of the read cycle. By pulling up the *RWL* to V_{DD} the read operation is commenced, then, *BL* is linked to the drain terminal of transistors M7 and M10 (i.e. node *PQ*). As a result, a large capacitance can be charged to V_{DD} or discharged to the ground very well. Since M9 passes a strong '0' and a degraded '1', the *BL* will be completely discharged to the ground and charged up to "$V_{DD} - V_{th, M9}$". This value can be differentiated by the amplifier. In the suggested WRE11T design with the single-bitline structure, the read performance can be enhanced by using the positive feedback sensing keeper scheme proposed in [9].

In the read operation, when *Q* (*QB*) accumulates '0' ('1'), *BL* is completely separated from *Q* through M9 and M6. This prevents to flow of the reading current through the storing nodes. Consequently, the read-disturb problem is resolved in the suggested design and the read stability will be equal to the hold stability.

3.2.4 Write operation

WWL is pulled up to enable the M11 to establish the writing path. The transistor M8 is disabled to eliminate the connection between the two internal nodes Q1 and Q. Also, RWL is pulled down to the ground. BL is either kept at '0' or '1' based on the new data, which will update the cell's content. Suppose that Q/QB stores '0'/'1' at the beginning. For writing a '1' to node Q, BL should be kept at V_{DD}. The applied '1' logic on the BL is transferred by the M11 to the Q node and charges this node up to "$V_{DD} - V_{th, M11}$". Finally, M1 and M2 are enabled and discharge the node QB to the ground. For writing a '0' to node Q, which accumulates a '1' at the beginning, a '0' logic value should be applied on BL. Then, node Q is fully discharged to the ground through M11-BL. This turns on the M3 to connect the QB node to the power V_{DD} rail.

3.3 SIMULATION SETUP, RESULTS, AND ANALYSES

The suggested WRE11T design's performance is measured and compared with those of the conventional 6T [10], read-decoupled feedback-cutting 11T (RDFC11T) [11], and stability- and power-improved 11T (SPI11T) [7] SRAM cells, as shown in Figure 3.2. The aforementioned designs are implemented with FinFET technology in the 10-nm channel length [12] and re-simulated in the HSPICE software environment. The FinFET technology parameters and characteristics are reported in Table 3.2. As given in Table 3.2, the magnitude of the threshold voltage for both the n-type and p-type transistors is 0.32 V. Thus, the power supply voltage V_{DD} is set to 0.45 V to perform the simulation in the near-threshold voltage. Moreover, the temperature is set to 27°C. The capacitance of bitlines is supposed to be 10 fF and 1 fin is considered for all the transistors [11].

All the investigated designs are studied and evaluated in an 8×8 SRAM array for measuring operations delay and power consumptions, and one SRAM bitcell for measuring hold/read/write static noise margin (HSNM/RSNM/WSNM) [11].

The transistor's attribute variations are significant in the sub-100 nm technology node. Therefore, to study the impact of these variations, widespread simulations of the Monte-Carlo (MC) are carried out following the setups specified in [8,13].

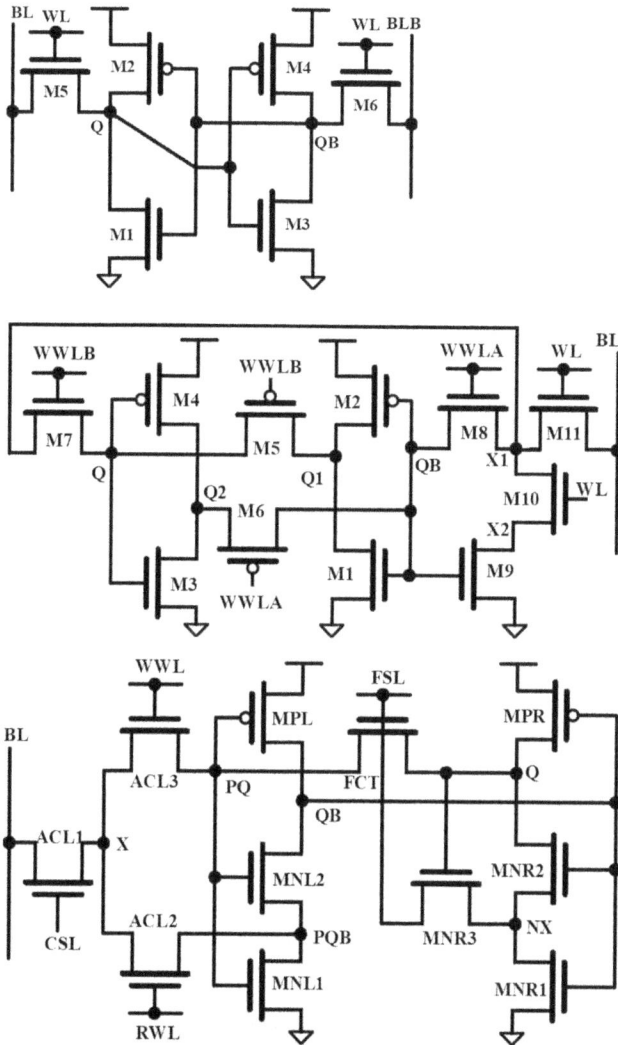

Figure 3.2 The investigated SRAM designs' structure at transistor level. (a) 6T, (b) RDFC1IT, and (c) SPIIIT.

3.3.1 Stabilities

The static noise margin (SNM) characterizes the SRAM design's stability. The read/write SNM of an SRAM design is computed by drawing the voltage transfer characteristics (VTCs) of both the latched inverters, known as the butterfly curve, while the SRAM cell is put in the read/write operation [14].

Table 3.2 FinFET technology parameters characteristics [12]

Parameters	N-type	P-type
Length of channel, nm	14	14
Height of fin, nm	21	21
Thickness of fin, nm	9	9
Doping level of body, cm^{-3}	2.5×10^{16}	2.5×10^{16}
Doping level of source/ Drain, cm^{-3}	3×10^{20}	3×10^{20}
Threshold voltage, V	0.32	−0.32
Work function of gate, eV	4.604	4.565

The butterfly curve for the read mode of operation of an SRAM design can be extracted in the ways mentioned below: (1) the SRAM cell is put in the reading mode of operation by setting its control signals, (2) a DC voltage source is applied on node Q, and then, is changed from '0' to 'V_{DD}' while monitoring the QB node's voltage variation, (3) a DC voltage source is applied on the node QB, and then, is varied from '0' to 'V_{DD}' while monitoring the voltage variation at node Q, and (4) voltages measured in the steps (2) and (3) are combined for generating the butterfly curve. The largest square's side length, drawn in the tighter lobe of the reading butterfly curve provides a quantitative value of the SNM in the read mode of operation [8].

To plot the butterfly curve of an SRAM cell in the write '1' mode of operation, the following steps should be done: (1) putting the SRAM cell on the write '1' mode of operation, (2) a DC voltage source is connected to the storage node Q (this node will be flipped from '0' to '1'), and then, is varied from '0' to 'V_{DD}' while monitoring voltage variation at storage node QB, and (3) this measured voltage (writing VTC) is utilized in joining with the reading VTC. The smallest square's side length plotted between and lower half of the reading and writing VTCs of the SRAM provides a quantitative value of the SNM in the write mode of operation [3].

Figure 3.3 indicates the reading butterfly curve for each SRAM cell at 0.45 V. The traditional 6T design has the read-disturb problem and, consequently, offers the lowest amount of RSNM. The SPI11T SRAM cell mitigates the read-disturb problem using a robust back-to-back structure of normal and Schmitt-trigger inverters. Moreover, considering a unique path for discharging the bitline capacitance, which does not contain true internal storing nodes, further mitigates the read-disturb problem, thereby, indicating a 2.66× higher RSNM than that of the 6T. The RDFC11T and suggested WRE11T designs indicate high RSNM value because Q and QB are entirely separated from bitline. However, the proposed WRE11T SRAM cell offers the highest RSNM (at least 1.07×) among the studied SRAM cell because the used stacked n-type transistors M1 and M4 increase the pull-down path's resistance, so the inverter's switching voltage shifts toward

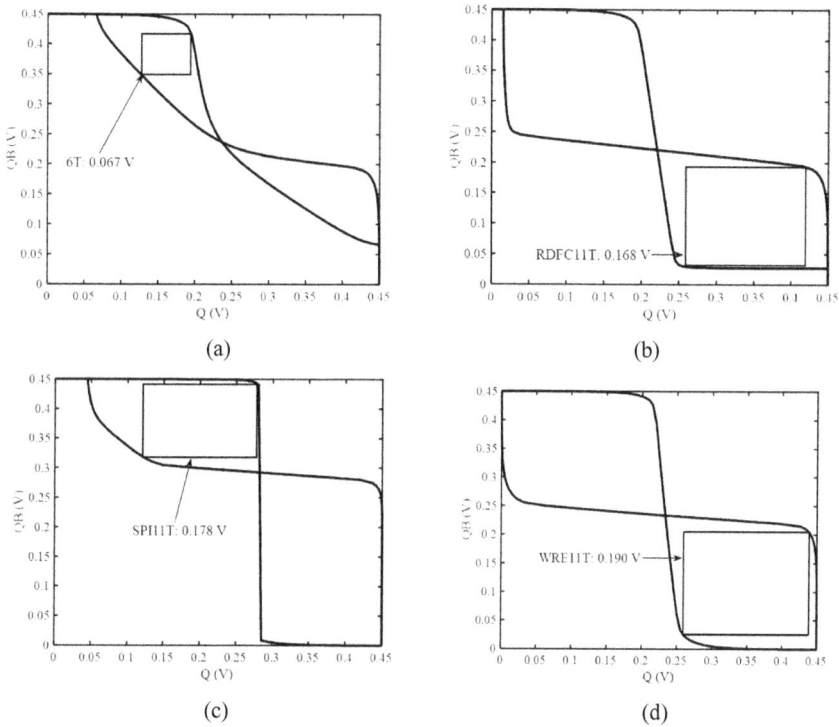

Figure 3.3 RSNM of (a) 6T, (b) RDFC11T, (c) SPI11T, and (d) WRE11T at 0.45 V.

right. Consequently, an expanded lobe is formed in the reading butterfly curves [15], offering RSNM enhancement.

The VTCs associated with reading and writing, along with WSNM values, at 0.45 V are indicated in Figure 3.4. The 6T cell is the worst design from the WSNM point of view because of its intrinsic conflicts between different transistors with each other and does not employ any write-assist scheme for solving this problem. Other SRAM designs indicate an enhanced WSNM because they utilize the feedback-cutting technique. However, the highest and second-highest amount of WSNM are associated with SPI11T and suggested WRE11T designs, correspondingly, because of their sharp reading VTCs. In the SPI11T design, unlike the proposed WRE11T SRAM cell, the write VTC intersects the horizontal axis at 0.038 V because of the degraded logic '1' passed by two serially NFETs presented in the path of writing. The two serially arranged NFETs presented in the writing path of the RDFC11T design do not degrade the writing VTC because they are passing a strong '0' logic value, applied on the bitline, in the write '0'/'1' mode of operation. The suggested design enhances WSNM by 1.49×/1.06× in comparison with 6T/RDFC11T and indicates 1.03× lower WSNM in contrast to SPI11T.

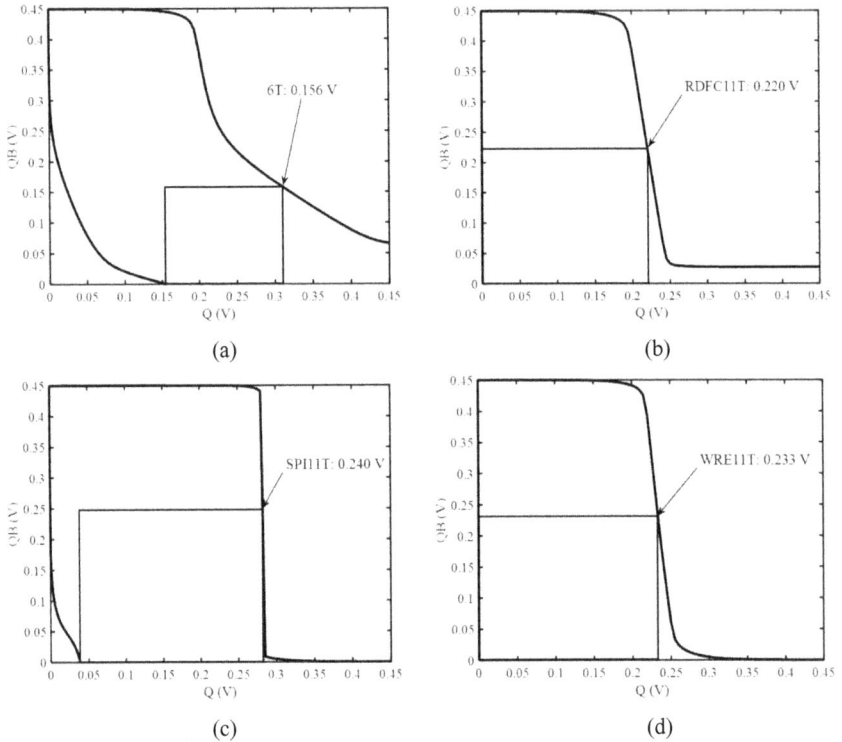

Figure 3.4 WSNM of (a) 6T, (b) RDFCI1T, (c) SPI11T, and (d) WREI1T at 0.45 V.

Figure 3.5 exhibits the statistical plots for RSNM and WSNM of the investigated designs at 0.45 V. The mean, standard deviation, and variability, symbolized by μ, σ, and σ/μ, respectively, for RSNM and WSNM of each SRAM cell, are calculated and annotated in Figure 3.5. As observed in Figure 3.5a, the traditional 6T design indicates the highest amount of variability in the RSNM metric because of the read-disturb problem. The robust back-to-back structure, coupled with a separate reading path, mitigates the read-disturb problem in the SPI11T design, consequently, this cell offers the second-lowest RSNM variability. The suggested WRE11T design utilizes a read buffer to fully decouple the reading path, consequently, resolving the read-disturb problem entirely. This results in process variations mitigation. The n-type stacked structure utilized in the proposed design further mitigates these variations. As a result, the suggested WRE11T SRAM cell indicates the lowest amount of RSNM variability (at least 1.58×). All the investigated SRAM cells, except the traditional 6T SRAM cell, offer the lowest amount of variability in the WSNM metric owing to utilizing the feedback-cutting scheme, as indicated in Figure 3.5b. However, the suggested WRE11T design reduces WSNM variability by at least 1.12×.

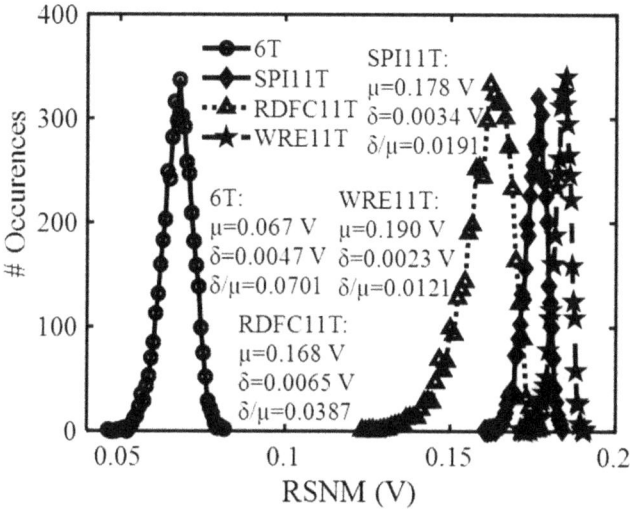

(a)

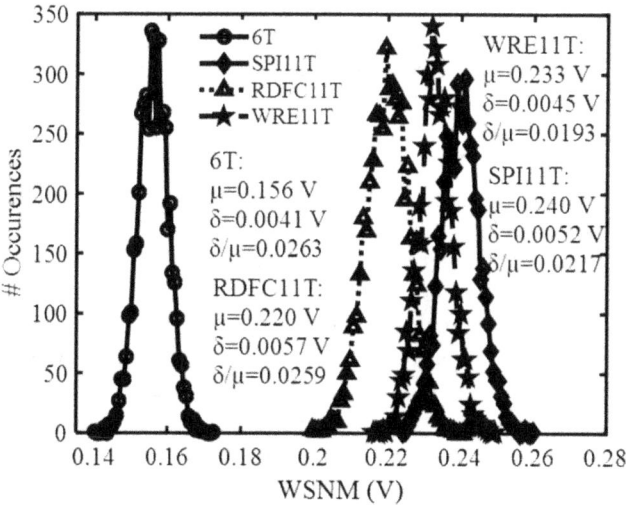

(b)

Figure 3.5 Statistical plots of (a) RSNM and (b) WSNM at 0.45 V.

3.3.2 Speed performance

Read delay and write delay are measures of the speed of an SRAM in the reading and writing modes of operation, respectively. To measure the read delay of the studied SRAMs, two different methods have been used depending on their reading structures (differential or single-ended). The traditional

6T design has a dual-ended bitline reading structure, consequently, the read delay is considered as a time duration between when the wordline is pulled up to V_{DD} and when pair of bitlines (BL and BLB) have a voltage gap of 50 mV [16]. During the reading of the single-ended reading scheme SRAMs, the delay is considered as the length of time between when the wordline responsible for reading is pulled up to V_{DD} and when the read-bitline is discharged to "$0.8 \times V_{DD}$" or is charged up to "$0.2 \times V_{DD}$" [17]. In the suggested WRE11T SRAM design, BL can be discharged to the ground and charged to the V_{DD}, relying on the cell's content. Therefore, the worst-case read delay is considered for comparison. The write '1'/'0' delay is defined as a time duration between when the wordline responsible for writing is pulled up to V_{DD} and when the initial Q node's voltage, which is '0'/'1', reaches up to 90%/10% of V_{DD} [18]. In the suggested WRE11T design with the single-ended scheme, the writing '1' to node Q, which accumulates a '0' at the beginning, is more difficult because of the employment of n-type-based transistor M11, which passes a degraded '1', consequently, the writing '1' delay is taken into account for analyses.

The amounts of delays for read and write operations for all the investigated SRAMs at 0.45 V are presented in Table 3.3. The lowest amount of read delay is associated with the traditional 6T design because it employs a fully differential reading structure. The RDFC11T and SPI11T designs have lower reading speed when compared with the suggested WRE11T design because of three serially n-type-based transistors presented in their reading paths. A small difference found in the read delays of the RDFC11T and SPI11T SRAMs is associated with the distinction in the intermediate nodes' parasitic capacitances. The suggested WRE11T design offers 1.59× higher and 1.30×/1.31× lower read delay in contrast to the traditional 6T and RDFC11T/SPI11T SRAMs, respectively. As given in Table 3.3, Owing to the compact and fully differential scheme and the use of one access transistor, the traditional 6T design has the lowest amount of write delay. Though in the writing paths of both the RDFC11T and SPI11T SRAMs exist two serially n-type-based transistors, the bitline in the latter SRAM is kept at '1' for writing '1', which cannot strongly be passed through these types of transistors, consequently, enlarging the write delay. The n-type-based write-access transistor M7 presented in the suggested WRE11T design passes a degraded '1', thereby, prolonging the write delay. The single-ended writing

Table 3.3 Speed performance comparison at 0.45 V

SRAM type	Read delay (ps)	Write delay (ps)
6T [10]	130.28	97.33
RDFC11T [11]	269.89	134.87
SPI11T [7]	270.37	152.18
WRE11T (this work)	207.18	140.25

structure employed in the RDFC11T, SPI11T, and proposed WRE11T SRAMs further increases the write delay. The suggested WRE11T design shows 1.44×/1.04× higher and 1.09× lower write delay in comparison with the 6T/RDFC11T and SPI11T SRAMs, respectively.

3.3.3 Power analysis

An SRAM cell consumes power to execute a read or write operation. This power is a kind of dynamic power [19]. A remarkable part of dynamic power is associated with the activation of the various involved signals and bitlines charge/discharge [8]. An SRAM indicates more dissipated power for a writing execution since its bitlines responsible for writing should be completely discharged to the ground [16]. SRAMs with single-ended operation show a lower amount of dynamic power consumption as compared to SRAMs with fully differential operation due to the bitline activity factor decreasing to half [20]. The consumed amounts of read and write dynamic power components of the investigated designs at 0.45 V are given in Table 3.4. The traditional 6T design has a fully differential scheme and exhibits the highest amounts of read and write dynamic powers. The assertion of multiple involved signals in the SPI11T design compared to the RDFC11T SRAM increases the read dynamic power. The suggested WRE11T SRAM is free from the bitline precharge operation before the beginning of the read cycle, thereby, saving dynamic read power. The suggested design indicates at least a 1.39× improvement in the dynamic read power. The bitline in the RDFC11T SRAM with single-ended scheme is entirely pulled down to the ground for each write operation, consequently, increasing the amount of dynamic write power. The SPI11T design executes the write operation by means of several involved signals, which results in higher amount of dynamic write power in contrast to the suggested design, which employs only one involved signal. As a result, the suggested WRE11T design reduces the amount of dynamic write power by at least 1.23×.

A remarkable part of the overall power is associated with leakage power because, in memory, only the involved SRAM bitcell(s) perform read/write operations, and the other remaining bitcells are in hold mode and just maintain the data [21]. The amounts of leakage power dissipated by the studied designs at 0.45 V are reported in Table 3.4. Because of using more bitlines and the lack of leakage reduction techniques, the leakage power dissipated by the traditional 6T design is the highest compared to the other designs. Since the back-to-back inverters principally contribute to the leakage power dissipated by an SRAM, the SPI11T and suggested WRE11T SRAMs dissipate lesser power during the hold operation than that of the RDFC11T design because of using stacked transistors in the latch core. However, the leakage current flowed via MNR3 presented in the Schmitt-trigger inverter of the SPI11T SRAM increases leakage power dissipation. According to these reasons, the suggested WRE11T design dissipates the least leakage power, offering at least 1.06× improvement.

Table 3.4 A power consumption comparison between the proposed and other studied SRAMs at 0.45 V

SRAM type	Dynamic read power (μW)	Dynamic write power (μW)	Leakage power (μW)
6T [10]	9.62	21.65	1.095
RDFC11T [11]	5.29	17.22	0.826
SPI11T [7]	6.35	13.46	0.767
WRE11T (this work)	3.81	10.95	0.721

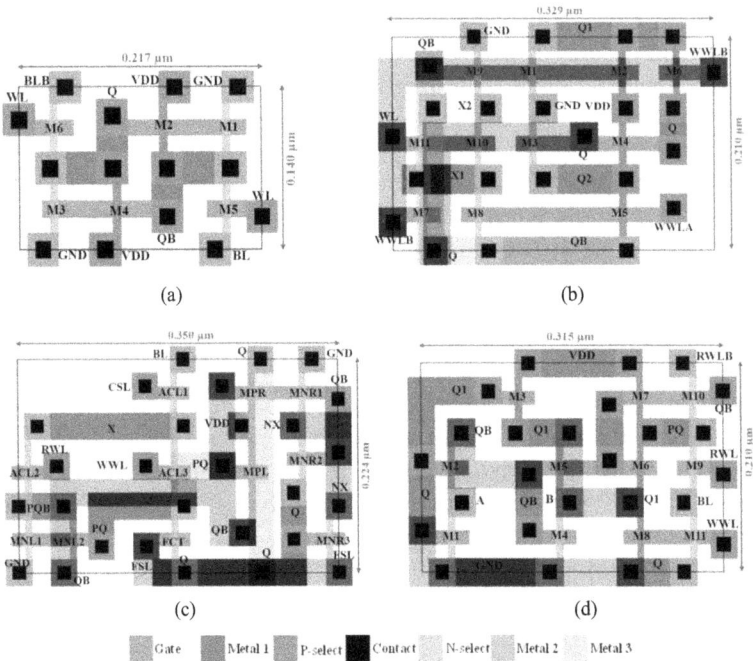

Figure 3.6 Layout of the (a) 6T, (b) RDFC11T, (c) SPI11T, and (d) WRE11T SRAM bitcells.

3.3.4 Area comparison

The layouts of the traditional 6T, RDFC11T, SPI11T, and the suggested WRE11T SRAM bitcells are illustrated in Figure 3.6. All the layouts are drawn utilizing the design rules described for FinFET technology [22]. The layout's height and width for each SRAM bitcell are annotated in Figure 3.6. The 6T SRAM bitcell shows the smallest layout area because it has a simple and compact structure designed with only six transistors. Though the other SRAM bitcells employ 11 transistors, the suggested bitcell occupies a smaller layout area. The suggested design exhibits a 0.06615 μm² layout area, which requires 2.18× higher and 1.04×/1.19× lower area than those of the 6T and RDFC11T/SPI11T SRAM bitcells, respectively.

3.4 CONCLUSION

This chapter presented a highly stable, robust, and ultralow-power design for 11T near-threshold SRAM bitcell. The suggested WRE11T design decreased the consumption of leakage and active power components using only one bitline. These amounts in the suggested design were further improved because of the non-precharge bitline for the read operation (stacked transistors). Moreover, the suggested cell improved the stability in all modes as much as possible using the read-decoupling, feedback-cutting, and multiple stacked transistors. The obtained results showed that at 0.45 V supply voltage, the suggested FinFET-based design enhanced RSNM by at least 1.07× and offered the second-highest values of WSNM, showing 1.49×/1.06× higher and 1.03× lower WSNM in contrast to the 6T/RDFC11T and SPI11T designs, respectively. The read and write delays of the proposed WRE11T SRAM were increased by 1.59× and 1.44×/1.04× in comparison with the 6T and 6T/RDFC11T SRAMs, respectively. However, the suggested WRE11T design offered at least 1.39×/1.23×/1.06× improvement in the read/write/leakage power. When the investigated designs were studied and evaluated under grave global and local manufacturing process variations, the proposed design showed the least variability in the RSNM and WSNM. The suggested WRE11T bitcell's area is 0.06615 µm², indicating a 2.18× higher and 1.19×/1.04× lower area than those of the 6T and SPI11T/RDFC11T SRAM bitcells, respectively.

REFERENCES

1. J. Lv, Z. Wang, M. Huang, and Y. He, "A read-disturb-free and write-ability enhanced 9T SRAM with data-aware write operation," *International Journal of Electronics,* 109, 23–37, 2022.
2. E. Abbasian, "A highly stable low-energy 10T SRAM for near-threshold operation," *IEEE Transactions on Circuits and Systems-I: Regular Papers,* 69(12), 5195–5205, 2022.
3. E. Abbasian, F. Izadinasab, and M. Gholipour, "A reliable low standby power 10T SRAM cell with expanded static noise margins," *IEEE Transactions on Circuits and Systems I: Regular Papers,* 69(4), 1606–1616, 2022.
4. E. Abbasian, M. Gholipour, and S. Birla, "A single-bitline 9T SRAM for low-power near-threshold operation in FinFET technology," *Arabian Journal for Science and Engineering,* 47, 14543–14559, 2022.
5. E. Mahmoodi and M. Gholipour, "Design space exploration of low-power flip-flops in FinFET technology," *Integration,* 75, 52–62, 2020.
6. M. Karamimanesh, E. Abiri, K. Hassanli, M. R. Salehi, and A. Darabi, "A write bit-line free sub-threshold SRAM cell with fully half-select free feature and high reliability for ultra-low power applications," *AEU-International Journal of Electronics and Communications,* 145, 154075, 2021.
7. E. Abbasian, S. Birla, and E. Mojaveri Moslem, "Design and investigation of stability-and power-improved 11T SRAM cell for low-power devices," *International Journal of Circuit Theory and Applications,* 50(11), 3827–3845, 2022.

8. E. Abbasian, S. Birla, and M. Gholipour, "Ultra-low-power and stable 10-nm FinFET 10T sub-threshold SRAM," *Microelectronics Journal, 123,* 105427, 2022.

9. M.-H. Tu, J.-Y. Lin, M.-C. Tsai, S.-J. Jou, and C.-T. Chuang, "Singleended sub-threshold SRAM with asymmetrical write/read-assist," *IEEE Transactions on Circuits and Systems I: Regular Papers,* 57, 3039–3047, 2010.

10. M. R. Jan, C. Anantha, and N. Borivoje, *Digital Integrated Circuits: A Design Perspective,* Prentice Hall: Upper Saddle River, NJ, 2003.

11. S. S. Ensan, M. H. Moaiyeri, and S. Hessabi, "A robust and low-power near-threshold SRAM in 10-nm FinFET technology," *Analog Integrated Circuits and Signal Processing,* 94, 497–506, 2018.

12. Predictive technology model (PTM), Available at https://ptm.asu.edu/ (Accessed 2020).

13. N. Eslami, B. Ebrahimi, E. Shakouri, and D. Najafi, "A single-ended low leakage and low voltage 10T SRAM cell with high yield," *Analog Integrated Circuits and Signal Processing,* 105, 263–274, 2020.

14. P. Sanvale, N. Gupta, V. Neema, A. P. Shah, and S. K. Vishvakarma, "An improved read-assist energy efficient single ended PPN based 10T SRAM cell for wireless sensor network," *Microelectronics Journal,* 92, 104611, 2019.

15. V. Sharma, M. Gopal, P. Singh, S. K. Vishvakarma, and S. S. Chouhan, "A robust, ultra low-power, data-dependent-power-supplied 11T SRAM cell with expanded read/write stabilities for internet-of-things applications," *Analog Integrated Circuits and Signal Processing,* 98, 331–346, 2019.

16. E. Abbasian and M. Gholipour, "Single-ended half-select disturb-free 11T static random access memory cell for reliable and low power applications," *International Journal of Circuit Theory and Applications,* 49, 970–989, 2021.

17. S. S. Ensan, M. H. Moaiyeri, M. Moghaddam, and S. Hessabi, "A low-power single-ended SRAM in FinFET technology," *AEU-International Journal of Electronics and Communications,* 99, 361–368, 2019.

18. E. Abbasian and M. Gholipour, "Robust transmission gate-based 10T sub-threshold SRAM for internet-of-things applications," *Semiconductor Science and Technology,* 37(8), 085013, 2022.

19. E. Abbasian and M. Gholipour, "Improved read/write assist mechanism for 10-transistor static random access memory cell," *International Journal of Circuit Theory and Applications,* 50(10), 3642–3660, 2022.

20. E. Abbasian and M. Gholipour, "Design of a Schmitt-Trigger-based 7T SRAM cell for variation resilient low-energy consumption and reliable internet of things applications," AEU-International Journal of Electronics and Communications, 138, 153899, 2021.

21. E. Abbasian and M. Gholipour, "A low-leakage single-bitline 9T SRAM cell with read-disturbance removal and high writability for low-power biomedical applications," *International Journal of Circuit Theory and Applications* 50(5), 1537–1556, 2022.

22. S. Salahuddin, H. Jiao, and V. Kursun, "A novel 6T SRAM cell with asymmetrically gate underlap engineered FinFETs for enhanced read data stability and write ability," In *International Symposium on Quality Electronic Design (ISQED),* Santa Clara, CA, USA, pp. 353–358, 2013.

Chapter 4

Investigation and optimization of high stability 6T CNTFET SRAM cell with low power

M. Elangovan

Government College of Engineering Srirangam

S. Jayanthi and P. Raja

Sri Manakula Vinayagar Engineering College

4.1 INTRODUCTION: CNT AND CN-FET

MOSFET's electrical characteristics are temperature-dependent and their highest tolerance temperature is 150°C. As compared to MOSFET, the CN-FET can function in a higher temperature range. Temperature has substantially less of an impact on the threshold voltage (V_{th}) of CN-FETs. Henceforth, due to temperature, the CN-FET-based circuits are not easily corrupted [1]. With MOSFETs, the Hox and quantum capacitance have an inverse relationship, whereas, in CN-FETs, the quantum capacitance and Hox have an emphatic correspondence. As a result, the quantum capacitance is decreased by reducing gate oxide thickness, increasing the CN-FET's speed [2,3]. CN-FET's chirality is negatively related to its V_{th}. The high value of V_{th} is achieved by choosing n=0 and the smallest value of "m." The effect on the CN-FET threshold voltage due to temperature fluctuations is adversely related. The increasing temperature reduces the CN-FET's threshold voltage. The temperature effect on the CN-FET threshold voltage is much smaller. Just 4.6% of the threshold voltage of CN-FET [4] increases the temperature shift from 27°C to 227°C. CN-FET has several advantages over silicon MOSFET. The current-carrying limit and CN-FET's I(on)/I(off) are higher than MOSFET. CN-FET transconductance is also higher than MOSFET transconductance. From now on, CN-FET yields more drain current than MOSFET. High strength and low resistance are given by the CN-FET. In the deeply adaptable range, CN-FET's leakage current is minimal. CN-FET's turn "OFF" condition current (Ioff) is straightforward relative to the temperature estimation square temperature rise, henceforth increasing the Ioff. This decreases the I(on)/I(off) value in turn. EG=2acclVccl/D_{CNT} gives the bandgap of CN-FET, where D_{CNT} is the CNT diameter. The saturation and leakage current of the CN-FET is

DOI: 10.1201/9781003459231-4

47

generated by rising the D_{CNT}. That results from drain-induced barrier lowering (DIBL) [5]. The transconductance is the function of the CN-FET's "ON" current (Ion). Relative to temperature, D_{CNT}, and supply voltage, the CN-FET turned-on current is straightforward. As a consequence, transconductance is directly related to variations in temperature, D_{CNT}, and V_{DD} [6]. Thermal energy is obtained by an electron from an increase in temperature. In low-voltage gates, this allows electrons to leap into the conduction band. Due to the increase in temperature, the V_{th} of CN-FET has decreased [7]. Indirectly, the D_{CNT} is proportional to CN-FET's V_{th}. The DIBL effect causes the low gate voltage of the CN-FET to turn "ON." The increasing V_{DD} decreases CN-FET V_{th}. The temperature difference, the D_{CNT}, and the V_{DD} are indirectly connected to the V_{th} of CN-FET [8]. CN-FET's saturation current is a function of its channel length. The saturation current is decreased by the channel length from 10 to 15 nm and the channel length from 15 to 20 nm is fundamentally increased by the saturation current [9]. The shift in Hox affects the CN-FET drain current. The conductivity of the CN-FET is negatively related to the Hox. Whereas, when the Hox is reduced, the V_{th} of CN-FET decreases. Thus increasing the conductivity of the CN-FET. Consequently, the CN-FET's power consumption rises [10]. Different dielectric materials have distinct constant dielectric materials (Kox). Oxide materials like SiO_2, Si_3N_4, Al_2O_3, HfO_2, Ta_2O_5, and TiO_2 have a Kox of 3.9, 7.5, 10, 16, 25, and 55, respectively [11,12]. The capacitance of the CN-FET gate is enhanced by an increase in the dielectric constant. This, in turn, decreases CN-FET's threshold voltage [12]. The oxide material Kox is indirectly proportional to CN-FET's V_{th}. Therefore, with an increase in dielectric constant value, the power dissipation of CN-FET is increased [13,14]. The lowest noise voltage required to change the memory cell state, i.e., change the bit cell's state from 0 to 1 and vice versa is characterized as static noise margin (SNM). An evaluation of a memory cell's stability is done using SNM. SRAM cells' noise resistance to DC noise is exactly proportional to their SNM. By changing the model noise source value from 0 V to V_{DD}, the two inverters 1 and 2's Voltage Transfer Characteristics (VTC) plots and the noise margin are calculated [15,16].

The semiconducting channel of the MOSFET is supplanted by cylindrical carbon nanotube (CNT) structures in the CN-FET. In all respects, such as speed and power utilization, CN-FET performance is excellent in contrast to MOSFET [7]. The CN-FET speed exceeds the MOSFET speed by a factor of 5–8. The power dissipation of the CN-FET is 2–7 times lower than that of the MOSFET. Compared to MOSFET, the power delay product (PDP) of CN-FET is 15–20 times lesser. Good electrical characteristics, such as low bias voltage and high current density, are provided by the CN-FET. Chirality is the term for the

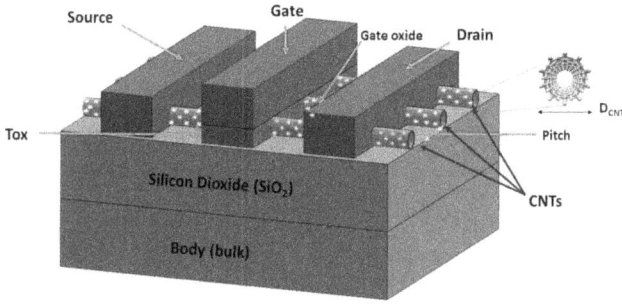

Figure 4.1 CN-FET structure.

angle of the atom structure along the CNT. Depending on its chiral vector, the CNT will behave as a conductor or semiconductor [15,17–19]. If $|n-m|=3i$ or $n=m$, then the CNT behaves as a conductor (where i is a positive integer), else it will function as a semiconductor [20]. For m is zero, a zigzag structure is given for the carbon nanotubes. When n and m are equal, the CNT have an armchair configuration; in every other condition, they have a chiral arrangement [21–24]. The CN-FET structure is shown in Figure 4.1. The WCNT, V_{th}, and D_{CNT} dependency CNT parameters shown are presented in [25]. The CN-FET (W_{CNT}) width is defined by

$$W_{CNT} = (N-1)\,S + D_{CNT} \tag{4.1}$$

N-Number of CNTs, S- spacing between CNTs in parallel, also known as CN-FET pitch value. As shown below, the chirality, V_{th} and D_{CNT} are related to each other.

$$D_{CNT} = \frac{a\sqrt{n^2 + nm + m^2}}{\pi} \tag{4.2}$$

$$V_{th} = \frac{E_g}{2e} = \frac{\sqrt{3}}{3}\frac{aV_\pi}{eD_{CNT}} \tag{4.3}$$

where a- atomic distance between the carbon atoms, $V\pi$ – Energy between carbon bonds ($V\pi$ =3.033 eV) and e-an electron charge [25]. In Figure 4.2, the composition of zigzag, armchair, and chiral carbon nanotubes are shown.

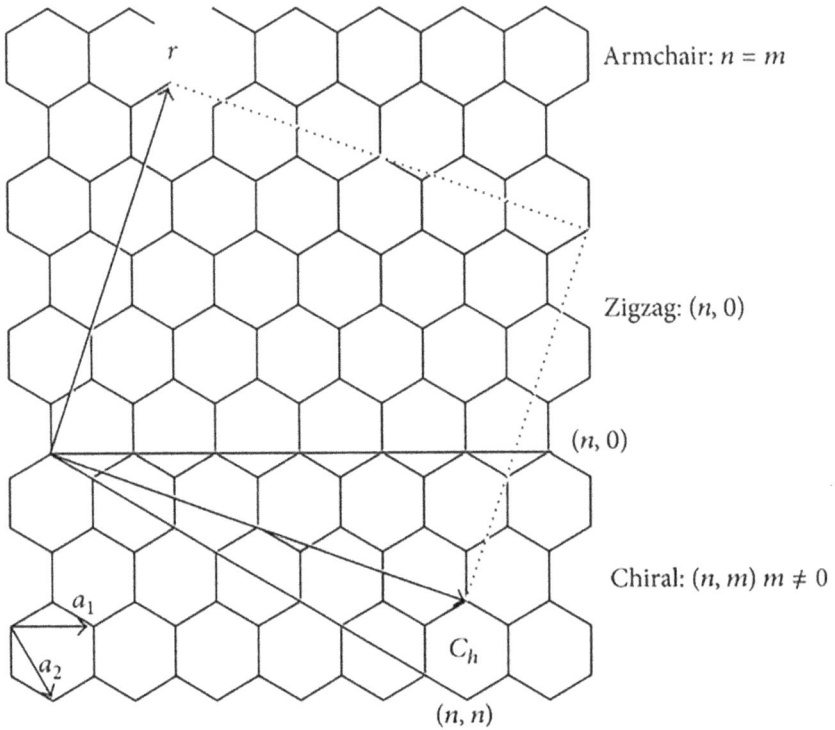

Figure 4.2 Structure of a graphene sheet.

4.2 6T CN-SRAM CELL STRUCTURE AND OPERATION

SRAM is a class of memory that has a bistable flip-flop to store single bit. In contrast to the DRAM cell, the SRAM cell is faster and more expensive. The SRAM cell is usually used for CPUs, and the primary memory of personal computers (PC) uses DRAM [3]. SRAM is the memory of a volatile semiconductor. It retains the data stored in it, as long as the source of power is available. If the V_{DD} is turned "OFF" it loses the data [26]. The T1-T6 CN-FETs create the CN-SRAM cell. This is used to store a single bit of value. The CN-FETs T1-T4 provide a cell for storage. The T5 and T6 CN-FETs form the access transistors. Tasks like reading and writing are carried out by the memory cell via access transistors, which are managed by the WL. Storage nodes for the bit cell are represented by Q and QB. The bit lines (BL and BLB) are employed to transport the data during the writing activity from outside to storage nodes and storage nodes during the reading activity to the outside peripherals. Bit lines are connected to a sense amplifier, which measures the voltage difference between them to determine what data is stored in the storage cell. In Figure 4.3, the structure of the 6T CN-SRAM cell is shown.

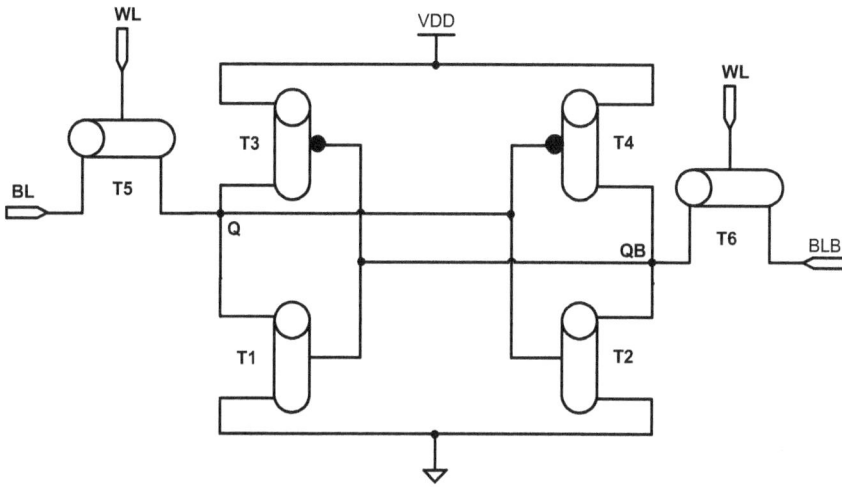

Figure 4.3 6T CN-SRAM cell.

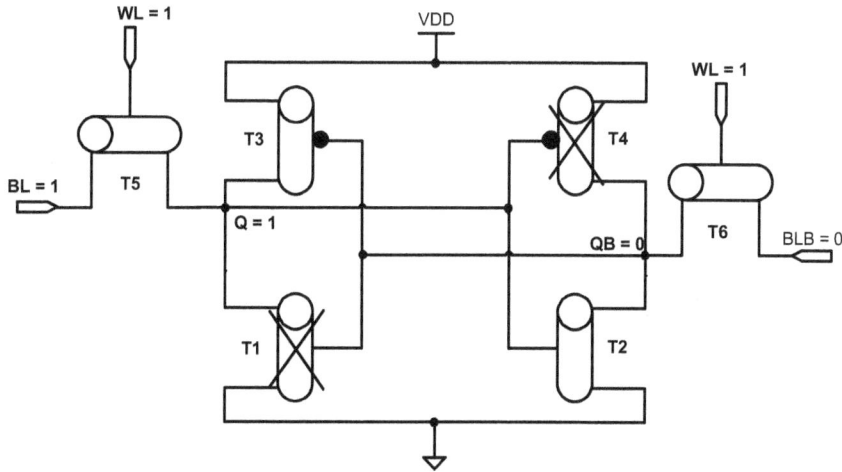

Figure 4.4 6T CN-SRAM cell mode (write "I").

4.2.1 Write mode

Setting BL to high, BLB to low, and WL to high creates the appropriate conditions for performing store "1" in Q (Figure 4.4). As a result, T2, T3, T5, and T6 are turned "ON," whereas T1 and T4 are turned "OFF." Thus V_{DD} is at node Q by the "ON" transistor T3. The node Q of the storage is raised to V_{DD}. Meanwhile, QB is at GND through T2. The QB node of storage is discharged to the ground voltage.

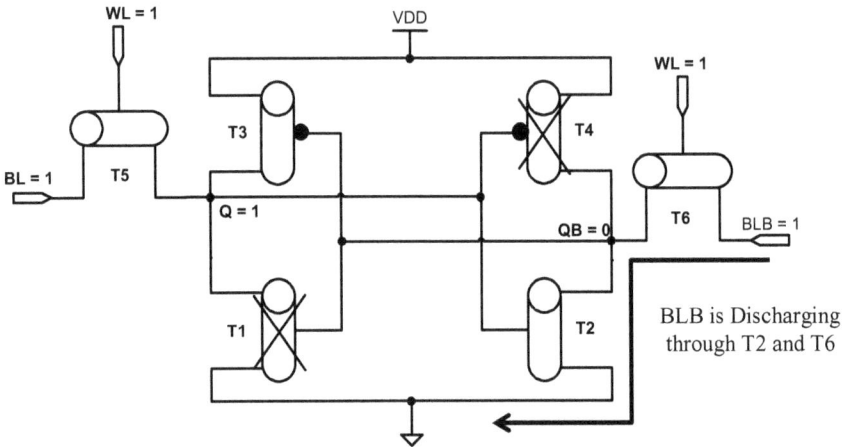

Figure 4.5 6T CN-SRAM cell read mode.

4.2.2 Read mode

Prior to beginning a read operation (Figure 4.5), bit lines must be discon-nected from the storage nodes by momentarily turning the T5 and T6 "OFF." Later, for pre-charging, the bit line and its complement are set to V_{DD}.Bit lines are connected to storage nodes again after a while. The volt-age at BL and BLB varies in accordance with the values in Q and QB. Using the sense amplifier, the potential variations in bit lines are sensed. Finally, the resulting stored value for storage nodes is concluded.

4.2.3 Hold mode

Often called standby mode is the hold mode (Figure 4.6). The stored bits are protected in this mode, as they are in the storage cell. By applying WL=0, the CN-FETs T5 and T6 are switched OFF. The result is that the memory cell's bit lines are detached. In the memory cell, the data is therefore held in its entirety.

4.3 SIMULATION RESULTS AND DISCUSSION

The parameters of the CN-FET include dielectric constant, oxide thickness, pitch value, CNT number, supply voltage, and temperature. By varying the above-mentioned parameters, the power consumption and SNM of the 6T CN-SRAM cell are measured. CN-FET gate oxide's dielectric constant is indirectly related to the memory cell's threshold voltage and noise margin. Based on various materials, the dielectric constant ranges from 3.9 to 55.

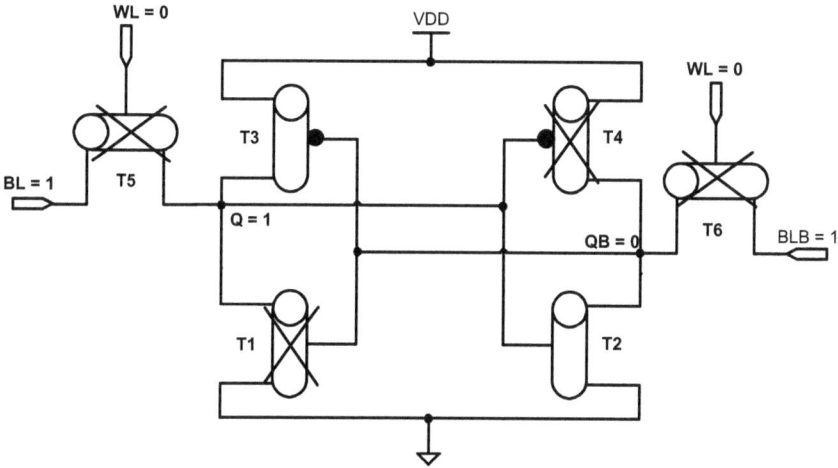

Figure 4.6 6T CN-SRAM cell hold mode.

In comparison with high dielectric constant materials, the materials with low dielectric constant offer a high noise margin. Dielectric constant of the gate oxide does not highly affects the power consumption of the 6T CN-SRAM cell. For different dielectric materials such as SiO_2, Si_3N_4, Al_2O_3, HfO_2, Ta_2O, and TiO_2 during its various modes, the noise margin is calculated for the studied 6T CN-SRAM cell. From the results, it is observed that with the variation in dielectric constant, memory cell stability is negatively increased. Figure 4.7 shows the transients waveform of the proposed SRAM cell.

The simulation's results show that the gate capacitance of the CN-FET grows as the dielectric constant rises (Figure 4.8). This, in turn, lowers the CN-FET threshold voltage. The noise margins of the SRAM cell considered in this work are therefore reduced during its operations. The stability of the cell for SiO_2 with the dielectric constant of 3.9 is 225.75, 255.66, and 288.26 mV, whereas with that of 55 (the highest value), the values of SNM are 106.07, 144.15, and 231.19 mV while reading, writing and in hold condition, respectively.

The various chiral vector values are determined for the noise margin of 6T CN-SRAM cells. The chiral vector nominal values are $m=19$ and $n=0$, respectively. For $n=0$, the nanotubes of carbon give a zigzag shape. As an armchair structure, the value of $n=m$ shapes carbon nanotubes and the carbon nanotubes form the chiral structure for all other cases. In all three situations, the stability is analyzed and the simulation effects are observed. It is noted from the simulation results that the SNM of the read, write, and hold modes is independent of the CNT structure. The SNM only relies on the chiral vector values. The chiral vector value is inversely related to the

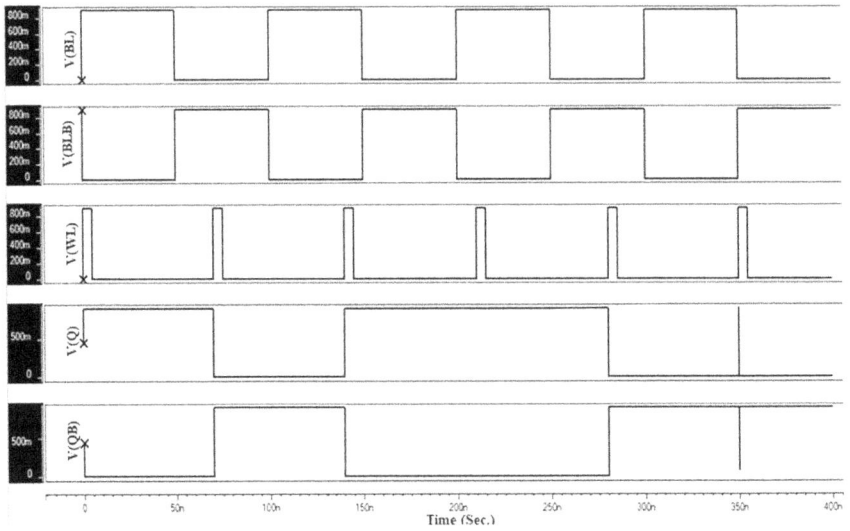

Figure 4.7 Transient response of the memory cell.

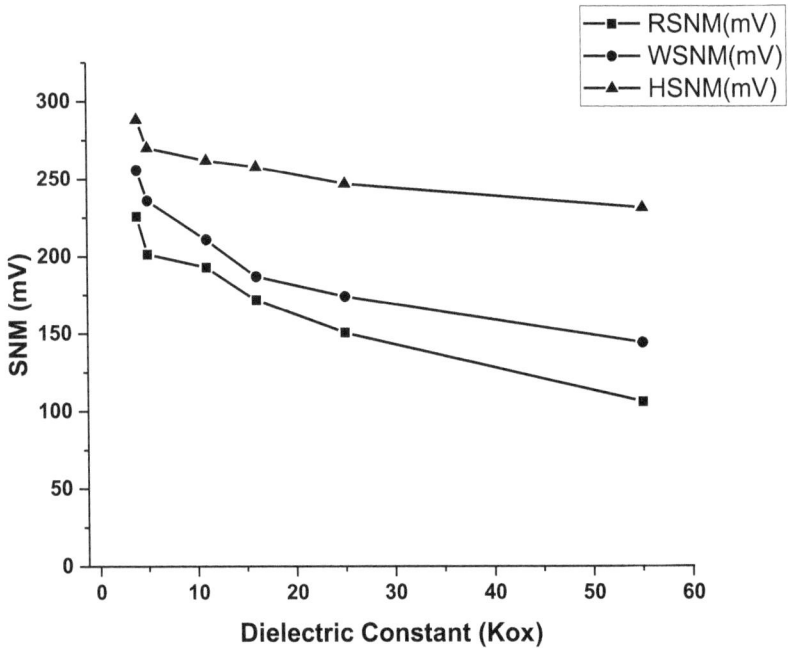

Figure 4.8 SNM comparison of 6T CN-SRAMcell for various Kox values.

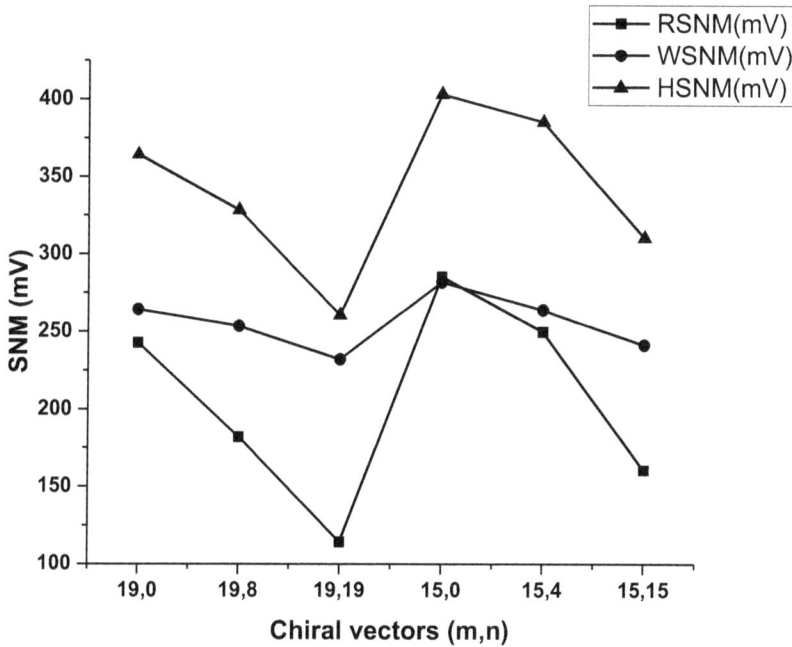

Figure 4.9 SNM comparison of 6T CN-SRAMcell with various chiral values.

threshold voltage. Thus, high stability is given by low chiral values and vice versa. The chiral vector with nominal values has high values of 242.76, 264.18, and 364.14 mV read, write, and hold modes SNM, respectively. In relation to SNM the high value with $m=19$ and $n=19$ give less read, write and hold SNM values compared to nominal values SNM. With three other chiral value set, such as $m=15$, $n=0$, m=15, $n=4$, and $m=15$, $n=15$, the same analysis is performed. For all modes of 6T CN-SRAMcell, the low chiral vector CNT gives high stability. Figure 4.9 shows the dependence of the chiral vector on the SNM.

The SNM of the cell is impacted by the CN-FET gate oxide's thickness. The SNM variation of the SRAM cell is observed to adjust gate oxide's thickness. The CN-FET gate oxide thickness is 20% different from the nominal value $(4 \times 10^{-9} m)$. The fact is that the Hox of the 6T CN-SRAMcell is positively related to the SNM. This is because the V_{th} of the CN-FET is increased as gate oxide thickness is increased. The increase in thickness of oxide layer from $3.2 \times 10^{-9} m$ to 4.8×10^{-9} raises the RSNM, WSNM, and HSNM to 179.52, 184.95 to 198.56 mV, and 253.83 to 266.55 mV, respectively. Figure 4.10 shows the SNM for various oxide thickness values.

The source of power of an SRAM cell ranges between 0.9 and 0.4 V. The memory cell functionality is not reached below 0.4 V. RSNM, WSNM, and HSNM are proportional to the difference in the supply voltage.

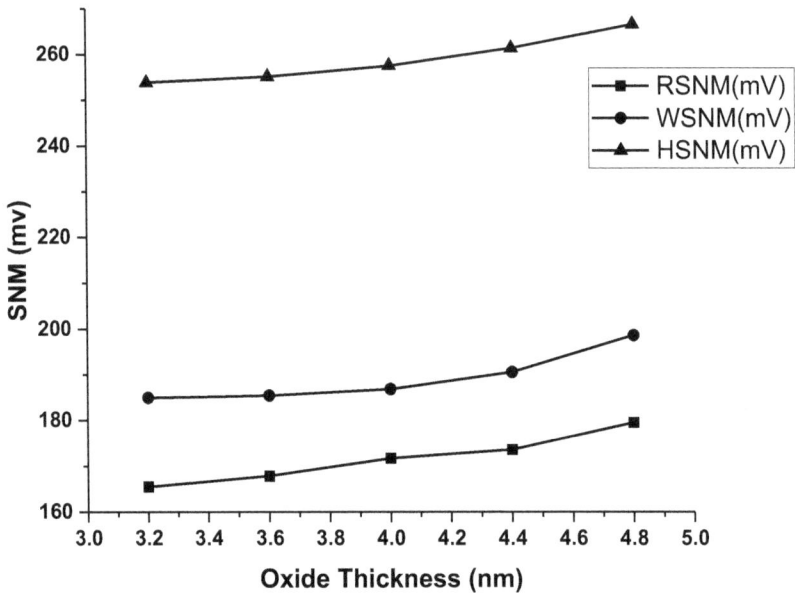

Figure 4.10 SNM comparison of 6T CN-SRAMcell of various oxide thickness values.

The reduction in the voltage of the power supply reduces the stability of the 6T CN-SRAMcell. The results demonstrate that the SNM of the 6T CN-SRAMcell is significantly reduced when the supply to the transistors is varied from 0.9 to 0.4 V. The 6T CN-SRAMcell offers 171.68, 186.83, and 257.52 mV for supply voltage, RSNM, WSNM, and HSNM, respectively. The noise tolerance of the same memory cells during read, write, and hold operating modes at a low V_{DD} of 0.4 V is 84.56, 128.45, and 110.66 mV in that order. Figure 4.11 depicts the SNM levels of 6T CN-FET memory cells with different V_{DD}.

The temperature variation influences the stability of the memory cell. It is found that because of the increase in temperature, the memory cell's SNM is degraded. During writing, reading, and holding operation modes, the temperature of the 6T CN-SRAMcell ranges from 27°C to 125°C. The rise in temperature causes pairs of electrons and holes injections. The low gate voltage is, therefore, necessary to turn the system "ON." This, in essence, decreases the CN-FET's threshold voltage. The RSNM, WSNM, and HSNM are thus decreasing as the temperature is rising. RSNM falls from 171.68 to 153.20 mV, WSNM from 186.83 to 172.02 mV, and HSNM from 257.52 to 243.83 mV when the temperature increases from 27°C to 125°C. Figure 4.12 shows the variance of the SNM due to temperature changes.

The 6T CN-SRAM consumes power, proportional to the CNT diameter used in the CN-FET. It is observed that increasing the chiral vector values increases the nanotube diameter and channel current. Consequently, the

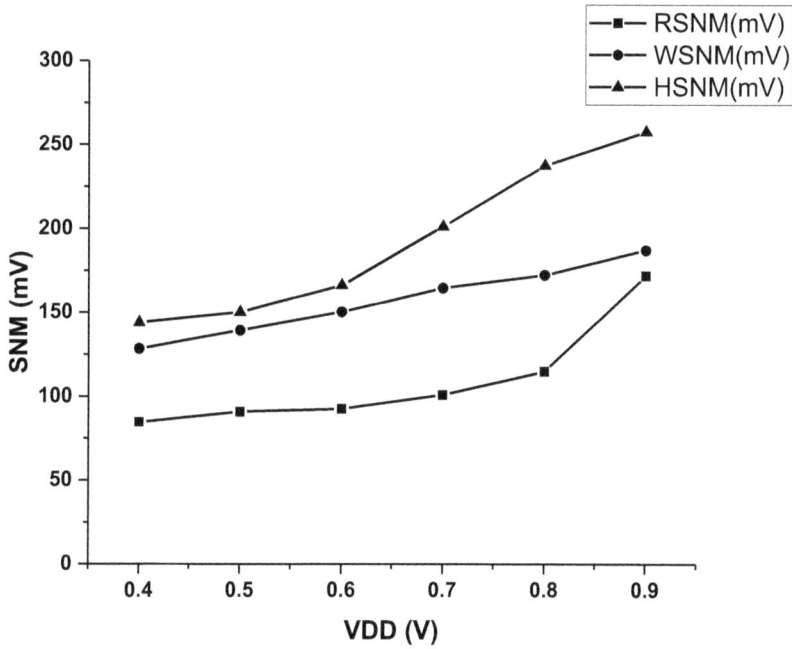

Figure 4.11 SNM comparison of 6T CN-SRAMcell for different supply voltage.

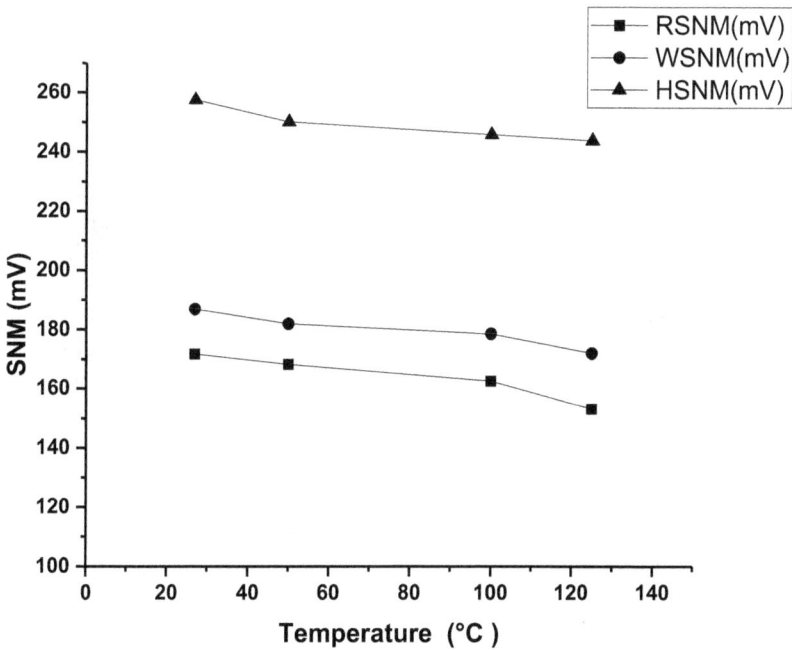

Figure 4.12 SNM comparison of 6T CN-SRAM cell for various tempearture.

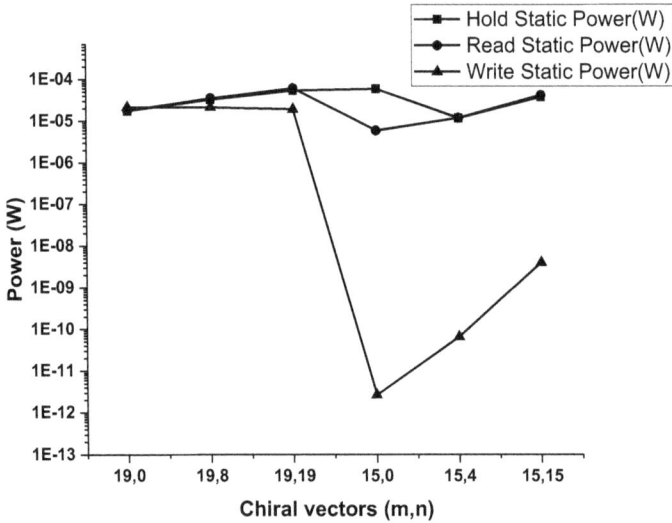

Figure 4.13 Power comparison of 6T CN-SRAMcell for different chiral values.

amount of power consumed for all modes for 6T CN-SRAMcells during its operation. The SRAM cell consumes power as 5.38E-05 W, 6.10E-05 W, and 1.93E-05 W for hold, read, and write modes, respectively, with the maximum value of chiral value as $m=19$ and $n=19$. With minimum chiral values, ($m=15$ and $n=0$) the consumption of power during the hold, read, and write modes is 5.89E-06 W, 5.92E-06 W, and 2.69E-12 W respectively. The power comparison for the different chiral values of 6T CN-SRAMcells is shown in Figure 4.13.

For different dielectric constant values, we observed the power performances of a 6T CN-SRAMcell. The CN-FET gate oxide material's dielectric constant ranges from 3.9 to 55. The power consumed by the cell with the dielectric constant of 3.9 for hold, read, and write modes are 3.64E-06, 3.64E-06, and 2.15E-10.0 respectively. With the maximum value of 55 for the dielectric constant, the power consumed by the 6T CN-SRAMcell during the hold, read, and write mode is found to be 3.66E-05, 4.08E-05, and 2.14E-10.5, respectively. The rise in dielectric constant increases the power consumed by the memory cells. Figure 4.14 depicts the power performances of the 6T CN-SRAMcell with different dielectric constants.

By changing the thickness of gate oxide layer of CN-FET by 20% from its nominal value the power consumed by the memory cell is observed. The oxide thickness of a gate ranges between 3.2E-09 and 4.8E-09 meters. The 6T CN-SRAMcell consumes hold, read and write powers of 1.91E-05, 1.98E-05, and 2.15E-10 watts respectively with Hox=3.2E-09 m. For Hox=4.8E-09 m, hold, read, and write powers of 1.61E-05, 1.65E-05, and 2.15E-10 W, respectively. The threshold voltage of CN-FET is directly related to the gate oxide's thickness. Lowering the gate oxide's thickness

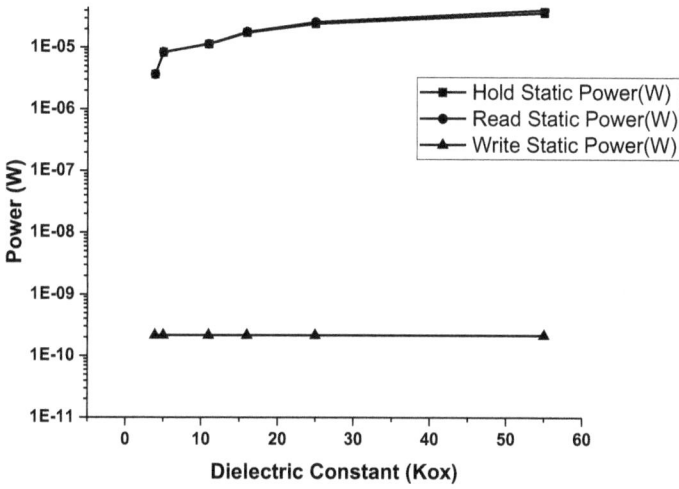

Figure 4.14 Power comparison of 6T CN-SRAMcell for different dielectric constants.

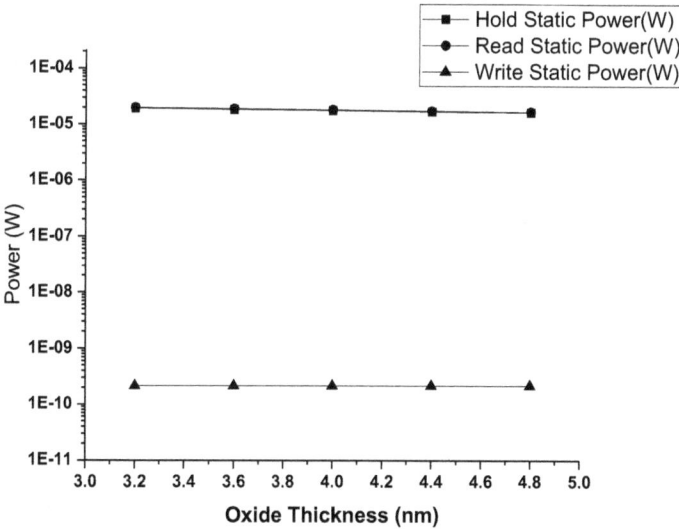

Figure 4.15 Power comparison of 6T CN-SRAMcell for different oxide thickness.

lowers the CN-FET's threshold voltage. Thus, with less gate voltage, the CN-FET is turned "ON" and begins conduction and its power consumption is increased. The memory cell absorbs less power over the high oxide thickness of 4.8E-09 m compared to the low oxide thickness value. Plot of power consumption of SRAMs for distinct oxide thicknesses is shown in Figure 4.15.

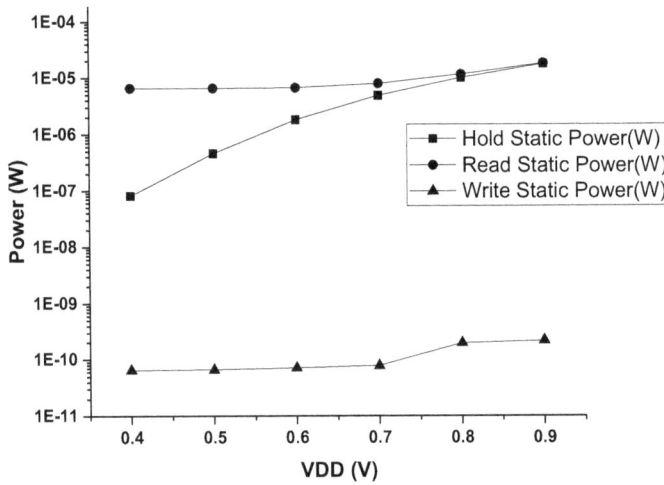

Figure 4.16 Power comparison of 6T CN-SRAMcell for various V_{DD}.

Any SRAM cell's power consumption is related directly to the transistor supply. The 6T CN-SRAM cell supply typically varies from 0.9 to 0.4 V. The reduction in power supply to the transistors lowers the memory cell's power consumption. The power consumption of the hold, read, and write modes are 1.75E-05, 1.80E-05, and 2.15E-10 W for the 0.9 V supply voltage, respectively. If the V_{DD} is reduced to 0.4 V, 8.12E-08, 6.55E-06, and 6.51E-11 W respectively are power consumed by the SRAM memory during hold, read and write modes. Figure 4.16 shows the power variation of the memory cell to the variation of the supply voltage.

For various temperature values, the power consumed by the 6T CN-SRAMcell is determined. By changing the temperature value from 27°C to 125°C, the writing, holding, and reading power consumption of the 6T CN-SRAMcell is observed. The simulation results indicate that the temperature rise increases the memory cell's power dissipation. The power consumed by the 6T CN-SRAMcells for writing, reading, and holding is 9.34E-11, 1.74E-05, and 1.70E-05 W at 27°C, respectively. If the temperature has risen to 125°C, the write, read and hold power values are 5.17E-09, 2.14E-05, and 2.02E-05 W, respectively. Figure 4.17, compares the power required by the 6T CN-FET memory cell for different temperature values.

The spacing between two adjacent CNTs is known as CN-FET pitch and its nominal value is 20e-9 m. The pitch ranges from 10e-9, 50e-9, and 90e-9 m. By varying the pitch, the power consumed by the 6T CN-SRAMcells during the various modes is observed. The results of the simulation show that the rise in pitch also rises the 6T CN-SRAM cell's power. Figure 4.18 compares the power of the 6T CN-SRAM cell at various pitches.

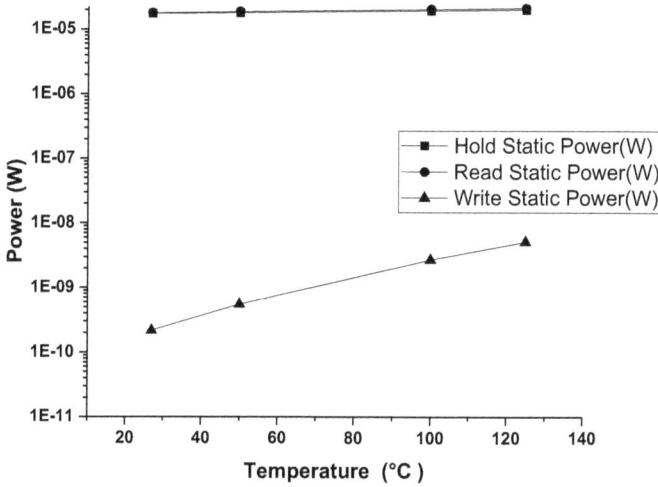

Figure 4.17 Power comparison of 6T CN-SRAMcell for different temperature values.

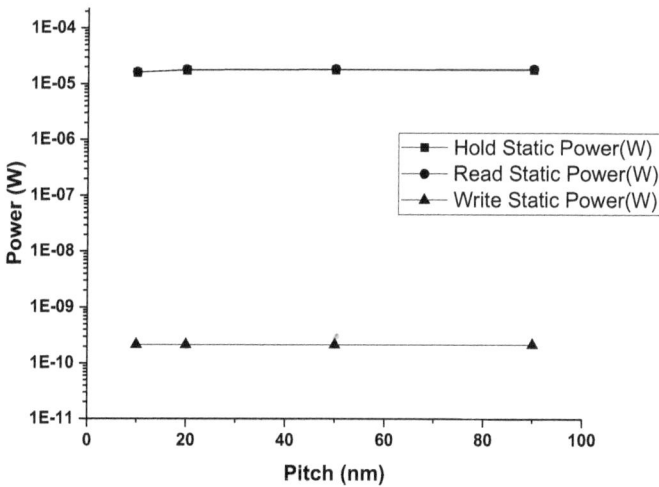

Figure 4.18 Power comparison of 6T CN-SRAMcell for different pitch values.

4.4 PROPOSED METHOD

The $m=19$, $n=0$, Hox=4nm, Kox=16, CNTs=3, and $V_{DD}=0.9$ V are the nominal values of the CN-FET parameters. For nominal values, the 6T CN-FET memory cell has write, read, and hold powers as 2.15E-10, 1.79E-05, and 1.74E-05 W, respectively. For the nominal values, the WNSM, RSNM, and HSNM of the bit cell are 186.83, 171.68, and 257.52 mV

Table 4.1 Proposed optimized CN-FET parameters for 6T
CN-SRAM cell

| CN-FET | CN-FET parameters | | |
	Chiral vectors (m,n)	Number of CNTs	Kox
T1	16,0	6	
T2	16,0	6	
T3	16,0	3	
T4	16,0	3	3.9
T5	19,0	3	
T6	19,0	3	

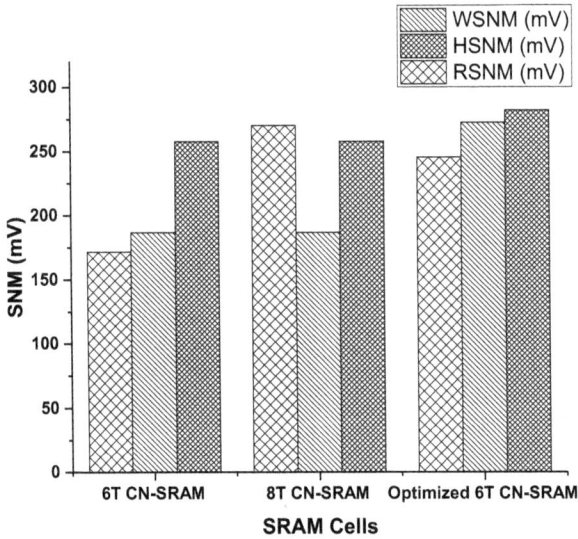

Figure 4.19 SNM comparison of various SRAM cells.

respectively. As mentioned in Table 4.1, the CN-FET parameters are optimized. CN-FET parameter-optimized values enhance the power and noise
performance of the conventional 6T CN-SRAM cell. With optimized
CN-FET parameter values, the traditional 6T CN-SRAMcell consumes
2.97E-11, 1.72E-06, and 2.85E-11 W of power, during its write, read, and
hold modes respectively. For the optimized CN-FET parameter values, the
WNSM, RSNM, and HSNM are 272.39, 245.15, and 281.98 mV respectively. The reduction of the chiral vector increases the CN-FET's threshold
value. This again raises the noise margin and decreases the SRAM cell's power
consumption. Figures 4.19 and 4.20 show the SNM and power comparison

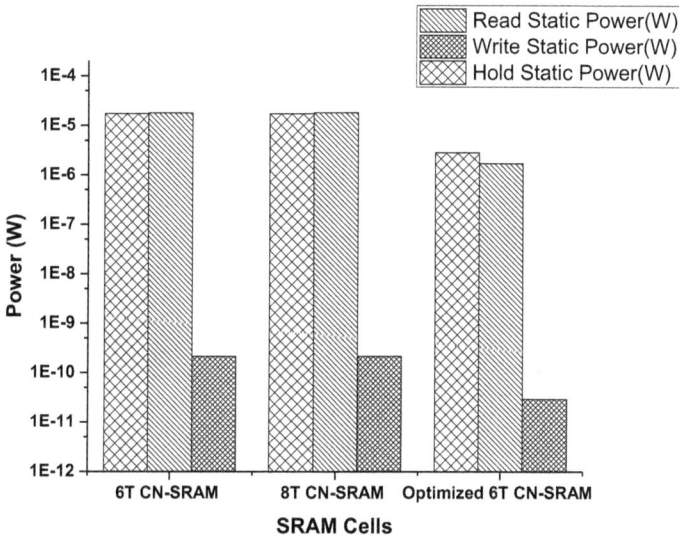

Figure 4.20 Power comparison of various CN-FET SRAM cells.

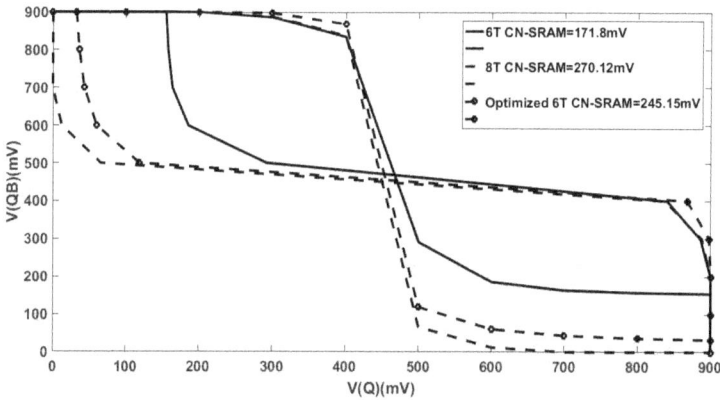

Figure 4.21 RSNM comparison butterfly curve of SRAM cells.

of the 6T CN-SRAM for nominal and optimized CN-FET parameters, respectively.

Figures 4.21–4.23 display the butterfly diagram's comparison outcomes for RSNM, WSNM, and HSNM of various SRAM cells.

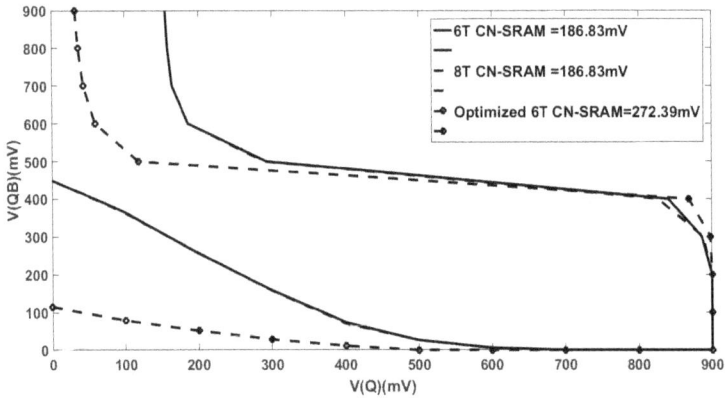

Figure 4.22 WSNM comparison butterfly curve of SRAM cells.

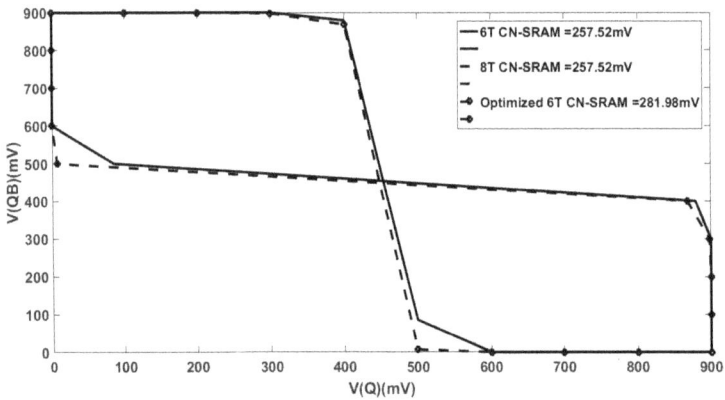

Figure 4.23 HSNM comparison butterfly curve SRAM cells.

4.5 CONCLUSION

The stability and power performances of a 6T CN-SRAM cell for nominal and optimized CN-FET parameters have been reported in this chapter. The optimized parameter values minimize the power of the 6T CN-SRAM by 90.39%, 86.1%, and 99.99% over nominal values during reading, writing, and holding modes of operations. The optimized CN-FET parameters increase the read, write, and hold noise margin of the 6T CN-SRAM cell by 42.8%, 45.8%, and 9.5%, respectively as compared to the nominal case. The observed results show that the stability of the 6T CN-SRAM cell is positively related to the V_{DD}, Hox, and CN-FET pitch values. The RSNM, WSNM, and HSNM are negatively proportional to the temperature and Kox. It is noted that the V_{DD}, temperature, chiral vectors, Kox, and pitch value are positively proportional to the power consumed by the 6T CN-SRAM cell, and they are indirectly proportional to the Hox.

REFERENCES

[1] S. K. Sinha, P. Singh, and S. Chaudhury, "Effect of Temperature and Chiral Vector on Emerging CNTFET Device," 2014, doi: 10.1109/IndiaCom.2014.6828174.

[2] M. Elangovan and K. Gunavathi, "High Stable and Low Power 8T CNTFET SRAM Cell," *J. Circuits, Syst. Comput.*, 29(5), 2020, doi: 10.1142/S0218126620500802.

[3] A. Karimi and A. Rezai, "Improved Device Performance in CNTFET using Genetic Algorithm," *ECS J. Solid State Sci. Technol.*, 6(1), M9–M12, 2017, doi: 10.1149/2.0101701jss.

[4] M. Elangovan and K. Gunavathi, "High Stable and Low Power 10T CNTFET SRAM Cell," *J. Circuits, Syst. Comput.*, 2019, doi: 10.1142/S0218126620501583.

[5] K. R. Agrawal and R. Sonkusare, "PVT Variations of a Behaviorally Modeled Single Walled Carbon Nanotube Field-Effect Transistor (SW-CNTFET)," In *2015 Int. Conf. Nascent Technol. Eng. Field, ICNTE 2015- Proc.*, 2015, doi: 10.1109/ICNTE.2015.7029940.

[6] M. Elangovan and K. Gunavathi, High Stability and Low-Power Dual Supply-Stacked CNTFET SRAM Cell, In *Innovations in Electronics and Communication Engineering: Proceedings of the 6th ICIECE 2017*, vol. 33, pp. 205–210. Springer, Singapore, 2019.

[7] M. S. Benbouza, D. Hocine, Y. Zid, and A. Benbouza, "Energy Optimization Nanotechnology Structures CNTFET GaAs," In *7th Int. IEEE Conf. Renew. Energy Res. Appl. ICRERA 2018*, vol. 5, pp. 522–526, 2018, doi: 10.1109/ICRERA.2018.8566989.

[8] J. Liang, L. Chen, J. Han, and F. Lombardi, "Design and Evaluation of Multiple Valued Logic Gates using Pseudo N-Type Carbon Nanotube FETs," *IEEE Trans. Nanotechnol.*, 13(4), 695–708, 2014, doi: 10.1109/TNANO.2014.2316000.

[9] M. S. Benbouza, D. Hocine, Y. Zid, and A. Benbouza, "New Nanotechnology Structures CNTFET GaAs," In *8th Int. Conf. Renew. Energy Res. Appl. ICRERA 2019*, pp. 799–803, 2019, doi: 10.1109/ICRERA47325.2019.8997103.

[10] C. Venkataiah, "Investigating the Effect of Chirality, Oxide Thickness, Temperature and Channel Length Variation on a Threshold Voltage of MOSFET, GNRFET, and CNTFET," *J. Mech. Contin. Math. Sci.*, 1(3), 232–244, 2019, doi: 10.26782/jmcms.spl.3/2019.09.00018.

[11] S. Lin, Y. Bin Kim, and F. Lombardi, "Design of a CNTFET-based SRAM Cell by Dual-Chirality Selection," *IEEE Trans. Nanotechnol.*, 9(1), 30–37, 2010, doi: 10.1109/TNANO.2009.2025128.

[12] M. S. and A. Rezai, "Improved Device Performance in a CN-FET using La2O3 High- Dielectrics," *J. Comput. Electron.*, 65, 221–227, 2017.

[13] R. Sahoo, S. K. Sahoo, and K. C. Sankisa, "Design of an Efficient CNTFET using Optimum Number of CNT in Channel Region for Logic Gate Implementation," In *2015 Int. Conf. on VLSI Syst., Architec., Technol. Appl. (VLSI-SATA)*, pp. 1–4, 2015, doi: 10.1109/VLSI-SATA.2015.7050473.

[14] A. H. M. Ali, M. H. Ani, and M. A. Mohamed, "Channel Length Effect on the Saturation Current and the Threshold Voltages of CNTFET," *IEEE Int. Conf. Semicond. Electron. Proceedings, ICSE*, pp. 267–269, 2014, doi: 10.1109/SMELEC.2014.6920848.

[15] K. R. Agrawal, S. M. Kottilingel, R. Sonkusare, and S. S. Rathod, "Performance Characteristics of a Single Walled Carbon Nanotube Field Effect Transistor (SWCNT-FET)," In *2014 Int. Conf. Circuits, Syst. Commun. Inf. Technol. Appl. CSCITA 2014*, pp. 30–35, 2014, doi: 10.1109/CSCITA.2014.6839230.

[16] G. Cho, Y. Bin Kim, and F. Lombardi, "Assessment of CNTFET based Circuit Performance and Robustness to PVT Variations," *Midwest Symp. Circuits Syst.*, 1106–1109, 2009, doi: 10.1109/MWSCAS.2009.5235961.

[17] M. Patnala *et al.*, "Low Power-High Speed Performance of 8T Static RAM Cell within GaN TFET, FinFET, and GNRFET Technologies - A Review," *Solid. State. Electron.*, 163(2019), 107665, 2020, doi: 10.1016/j.sse.2019.107665.

[18] P. Dhilleswararao, R. Mahapatra, and P. S. T. N. Srinivas, "High SNM 32nm CNFET based 6T SRAM Cell Design Considering Transistor ratio," In *2014 Int. Conf. Electron. Commun. Syst. ICECS 2014*, 2014, doi: 10.1109/ECS.2014.6892748.

[19] W. Wang and K. Choi, "Novel Curve Fitting Design Methodology for Carbon Nanotube SRAM Cell Optimization," In *2010 IEEE Int. Conf. Electro/Information Technol. EIT2010*, pp. 629–633, 2010, doi: 10.1109/EIT.2010.5612138.

[20] M. T. Ahmadi, J. Karamdel, R. Ismail, C. F. Dee, and B. Y. Majlis, "Modelling of the Current-Voltage Characteristics of a Carbon Nano Tube Field Effect Transistor," *IEEE Int. Conf. Semicond. Electron. Proceedings, ICSE*, 7(2), 576–580, 2008, doi: 10.1109/SMELEC.2008.4770391.

[21] S. R. Shailendra and V. N. Ramakrishnan, "Analysis of Quantum Capacitance on Different Dielectrics and its Dependence on Threshold Voltage of CNTFET," In *2017 Int. Conf. Nextgen Electron. Technol. Silicon to Software, ICNETS2 2017*, pp. 213–217, 2017, doi: 10.1109/ICNETS2.2017.8067933.

[22] S. K. Sinha and S. Chaudhury, "Impact of Temperature Variation on CNTFET Device Characteristics," *CARE 2013-2013 IEEE Int. Conf. Control. Autom. Robot. Embed. Syst. Proc.*, 2013, doi: 10.1109/CARE.2013.6733774.

[23] E. Abiri and A. Darabi, "Design of Low Power and High Read Stability 8T-SRAM Memory based on the Modified Gate Diffusion Input (m-GDI) in 32 nm CNTFET technology," *Microelectronics J.*, 46(120, 1351–1363, 2015, doi: 10.1016/j.mejo.2015.09.016.

[24] I. A. Khan and N. Alam, "CNTFET Based Circuit Design for Improved Performance," *Proc. -2019 Int. Conf. Electr. Electron. Comput. Eng. UPCON 2019*, pp. 1–5, 2019, doi: 10.1109/UPCON47278.2019.8980053.

[25] M. Elangovan and K. Gunavathi, "Stability Analysis of 6T CNTFET SRAM Cell for Single and Multiple CNTs," *2018 4th Int. Conf. Devices, Circuits Syst.*, Coimbatore, India, vol. 2, pp. 63–67, 2018.

[26] A. Karimi, A. Rezai, and M. M. Hajhashemkhani, "Ultra-Low Power Pulse-Triggered CNTFET-Based Flip-Flop," *IEEE Trans. Nanotechnol.*, 18, 756–761, 2019, doi: 10.1109/TNANO.2019.2929233.

Chapter 5

Study of transistor sizing techniques for low-power design in FinFET technology

Ehsan Mahmoodi and Morteza Gholipour
Babol Noshirvani University of Technology

5.1 INTRODUCTION

Nowadays, the industry is focusing on shrinking integrated circuits because of improved performance and smaller circuit areas. It has become increasingly necessary to introduce new transistors as the ascending of MOSFETs has been inadequate. An upcoming technology is the Fin-shaped field effect transistor (FinFET), where the gate bounds the fin-shaped channel (Figure 5.1) [1–3]. Newly developing devices are still being developed such as TMDFETs [4] and GNRFETs [5], FinFET is flagging the scheme for Moore's law because this technology performs well and has a subordinate power consumption than MOSFETs [2,6]. In general, FinFET is classified into SG and IG (shorted and independent gate) modes. In the first one, the gate envelops the fin on three sides (Figure 5.1), whereas in the second one, two independent gates are switched distinctly [1].

This chapter is organized based on precise transistor sizing which explores how flip-flops can be designed through *LE* and *EEC* procedures. Analyses are then conducted on the results in different cases. As a final step, the consequences are presented for having the lowest power and energy consumption with respect to the delay as well as the area for various loads. Following is a brief description of how this paper is organized. A discussion of power optimization methods is provided, which according to in Section 5.2. In Section 5.3, flip-flops and transistor sizing techniques in digital circuits are discussed, and the flip-flop is designed in Section 5.4. Section 5.5 demonstrates the simulation outcomes and eventually the deduction can be seen in Section 5.6.

5.2 POWER OPTIMIZATION APPROACHES

5.2.1 ONOFIC approach

For scheming low-power and fast nanoscale CMOS circuits with low power consumption and low leakage, on/off logic (*ONOFIC*) was proposed in

DOI: 10.1201/9781003459231-5

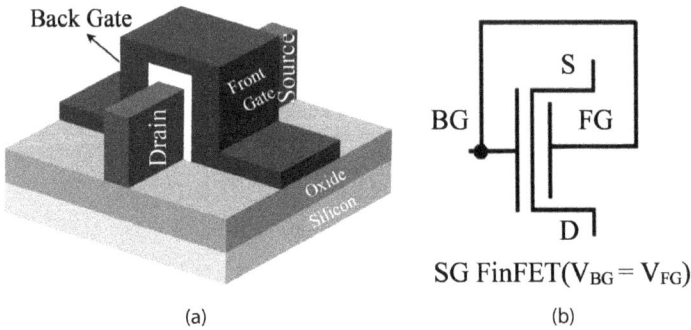

Figure 5.1 (a) FinFET assembly and (b) Representation of shorted-gate mode FinFET.

2014 [7]. With the *ONOFIC* approach, high-performance and low-leakage CMOS circuits can be designed quickly and reliably. The schematic of this structure can be seen in Figure 5.2. When the *ONOFIC* transistors are in the cut-off region, the *ONOFIC* block is *OFF* and the logic block has excessive resistance that can bridle the leakage current exactly, while if they are in the linear, the block is *ON* and provides a good conducting path. Therefore, in both active and standby modes, it gives accurate logic levels and minimizes leakage currents [7].

5.2.2 LECTOR approach

In [8], the LECTOR (Leakage Control Transistor) method is introduced for modeling CMOS-based circuits, significantly reducing the leak flow and dynamic power dissipation. In this method, as shown in Figure 5.2, two transistors are added to control leakage within the logic gate, whose gate terminals are controlled by their respective sources. In addition, one of the LCTs is continuously "near the cutoff voltage," resulting in substantial reductions in leakage current for any mixture of input. In comparison with other methods, *LECTOR* is superior in reducing leakage in both active and inactive conditions of the circuit [8].

5.3 GATE SIZING PROCEDURES

5.3.1 Logical effort

The Logical effort (*LE*) theory offers a way for approximating the smallest lag of digital circuits subjected to linear delay modeling [9]. The *LE* represents the stabilized the delay associated with each gate, and by

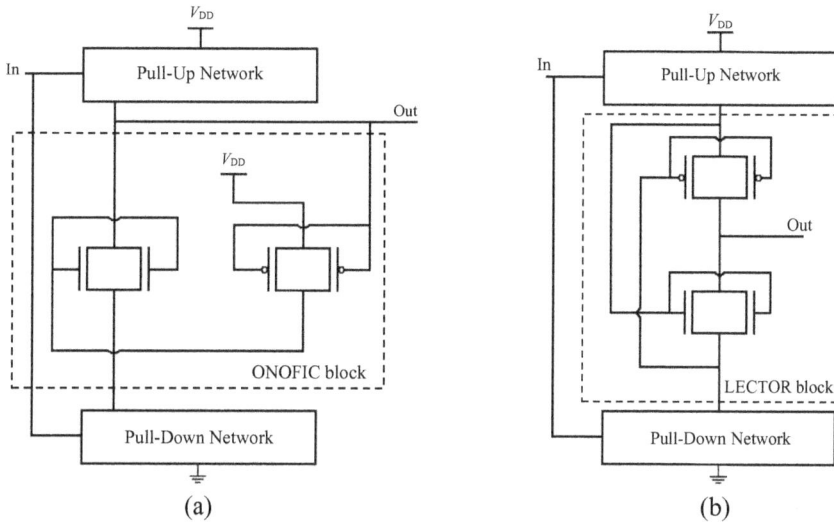

Figure 5.2 (a) ONOFIC schematic and (b) LECTOR schematic.

equating the effort of each stage, the minimum delay can be achieved in multi-stage paths [10,11]. The dimensions of stages separately in a path can be considered as below (starting from the former stage and moving toward the first stage):

$$F = GBH \text{ where } G = \prod g_i, \; BH = \prod h_i, \; H = \frac{C_{\text{out-path}}}{C_{\text{in-path}}} \tag{5.1}$$

$$\hat{f} = g_i h_i = F^{(1/N)} \tag{5.2}$$

$$C_{\text{in}_i} = g_i \frac{c_{\text{out}_i}}{\hat{f}} \text{ where } h_i = \frac{c_{\text{out}_i}}{c_{\text{in}_i}} \tag{5.3}$$

There are four types of stage effort, referred to in the following: logical (g_i), electrical (h_i), branching (b_i), and effort of the path ($\hat{f}$). Meanwhile, N represents the sum of the effort of the phases, and G, H, B, and F are the logical, electrical, branching, and the effort of the path for the entire tracks correspondingly [9].

5.3.2 EEC strategy

The "energy efficient curve" or EEC curvature in a digital circuit consists of the design points under a fixed supply voltage V_{DD} along with the output load C_L and by using it, the conditions of reaching the minimum delay for

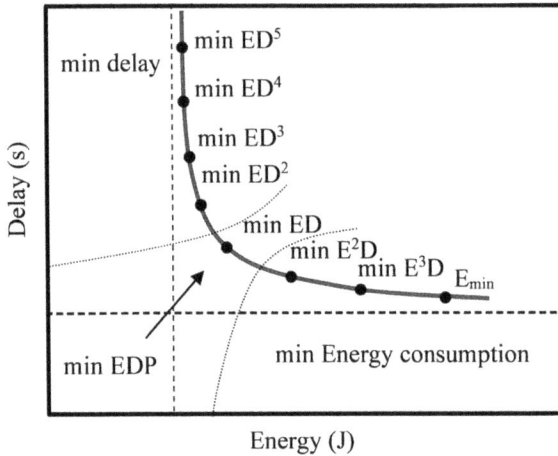

Figure 5.3 EEC hyperbolic curve with adjusting the E^iD^j metrics [11].

any fixed energy consumption can be achieved (or vice versa) [12]. According to the explanations given, the target is to explore entirely in the *E-D* area and find design points that are exactly on the EEC curve (Figure 5.3), which, according to (5.4), can be said to have a hyperbolic shape [13].

$$(E - E_0)(D - D_0) = E_0 D_0 \tag{5.4}$$

the E_0 is the minimum energy that can be estimated by exerting the smallest transistor extents with an accurate process. D_0 is the least reprieve of the circuit which can be gauged by growing the sizes of the transistors properly. For finding some isolated points of the EEC, it is enough to know $E^i D^j$ criteria for a distinct series of amounts i and j, e.g. $[ED^3, ED^2, ED, E^2D, E^3D, E_{min}]$. By appending the finest arguments in the *E-D* dimensions, a hyperbolic fitting can be achieved [10].

5.4 FLIP-FLOP

Flip-flop is one of the essentials in digital circuits and is generally implemented in clock spreading systems, which is one of the major suppliers to chip power [14–16]. It is an essential component of parallel processing and can be utilized to stash status data in consecutive circuits [17,18]. Figure 5.4 shows a Transmission-gate-based flip-flop (TGFF) that is the very common and modest topology of flip-flops [19,20]. In this part, the plans employed to characterize the scheme limits of the flip-flop are described [10].

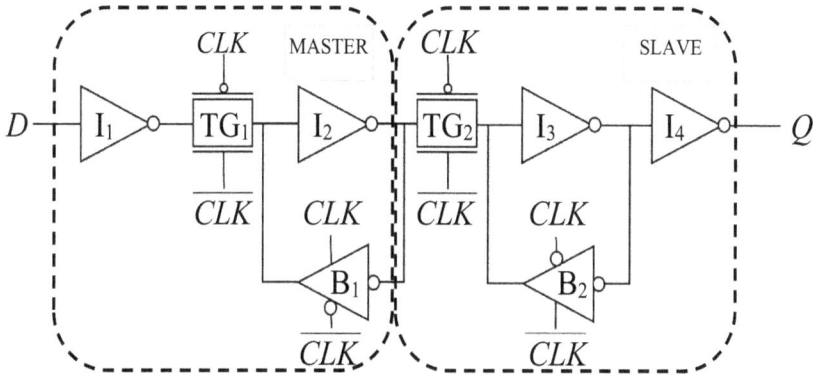

Figure 5.4 Schematic of TGFF, (gate-level of static FF).

5.4.1 TGFF scheme with logical effort approach

In this section, the flip-flop design using the *LE* method is fully explained step by step. This flip-flop generally consists of inverters, transmission gates, and tri-state buffers (a mixture of the TG and the INV), and the proportion of the widths of P-type to N-type of each gate is shown in Figure 5.5. The W_P/W_N in the inverter and TG (transmission gate) are considered 2 and 1 respectively, to preserve reliability by means of standard cell strategy [21,22].

The D to Q route is assumed as the feedback has no effect on the delay. To reduce the number of variables effectively, the INV+TG mixture is presumed as a separate phase. As can be seen in Figure 5.6, by adding TG to the INV the C_{gate} remains constant and only the resistance of the INV+TG is doubled (assuming $R_{INV} = R_{TG}$).

In *LE* method, the factors are found by using the Elmore delay model [18] (All widths are normalized to the minimum width W_{min} in Figure 5.7). All the primary phase devices (INV+TG) are presumed with the equivalent width W_1 (input capacitance of the TGFF). The transistors in the second, third, and fourth stages have the widths W_2, W_3 and, W_4 respectively. In this case, the value of the $C_{parasitic}$ of the devices is the same as the value of the C_{gate} [9]. Table 5.1 illustrates the *LE* factors for diverse phases (respected to each node in the Elmore technique) can be seen in (W_L refers to the load) and the *D*(delay) relations are inscribed as (*C*) *R* multiplication [10,11]. Considering R_{min} (resistance of a device with the smallest size) equal to $1/L_{min}$ the resistance is found as $1/W$. According to [10,11], separating the *Data to Q* path into two chunks performs better, in which two different

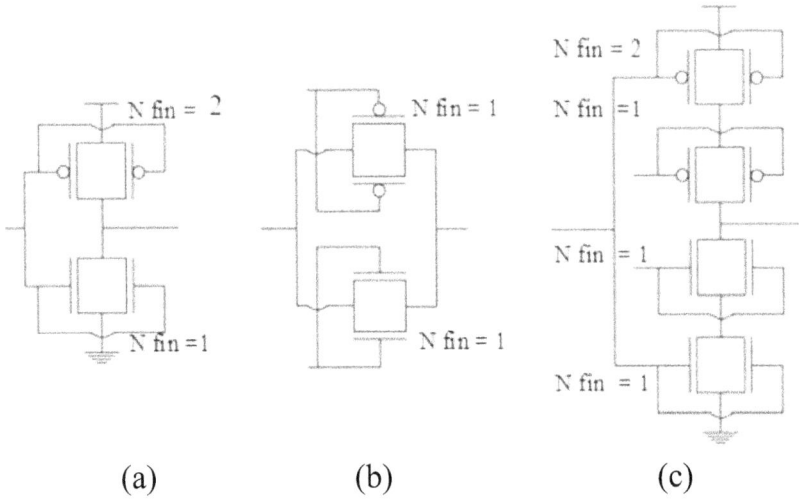

Figure 5.5 Sizing of (a) INV, (b) TG, (c) IN+TG based on LE.

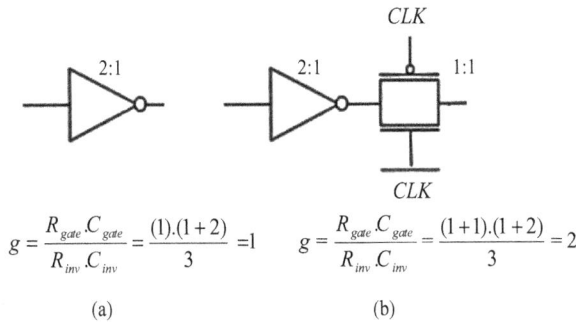

$$g = \frac{R_{gate}.C_{gate}}{R_{inv}.C_{inv}} = \frac{(1).(1+2)}{3} = 1 \qquad g = \frac{R_{gate}.C_{gate}}{R_{inv}.C_{inv}} = \frac{(1+1).(1+2)}{3} = 2$$

(a)   (b)

Figure 5.6 Calculation of g for (a) Inverter and (b) Tri-state buffer.

items related to the factors of the next step are extracted. By applying (5.2) to these two paths, equation (5.5) is obtained. Replacing the results came from Table 5.1 in equations (5.5–5.8):

$$g_{1-1}h_{1-1} = g_{1-2}h_{1-2} = \sqrt{F_1} = \sqrt{G_1B_1H_1} \tag{5.5}$$

$$g_{2-1}h_{2-1} = g_{2-2}h_{2-2} = g_{2-3}h_{2-3} = \sqrt{F_2} = \sqrt{G_2B_2H_2} \tag{5.6}$$

$$G_1 = g_{1-1}g_{1-2} / G2 = g_{2-2}h_{2-2}g_{2-3}h_{2-3} \tag{5.7}$$

$$B_1H_1 = h_{1-1}h_{1-2} / B_2H_2 = h_{2-1}h_{2-2}h_{2-3} \tag{5.8}$$

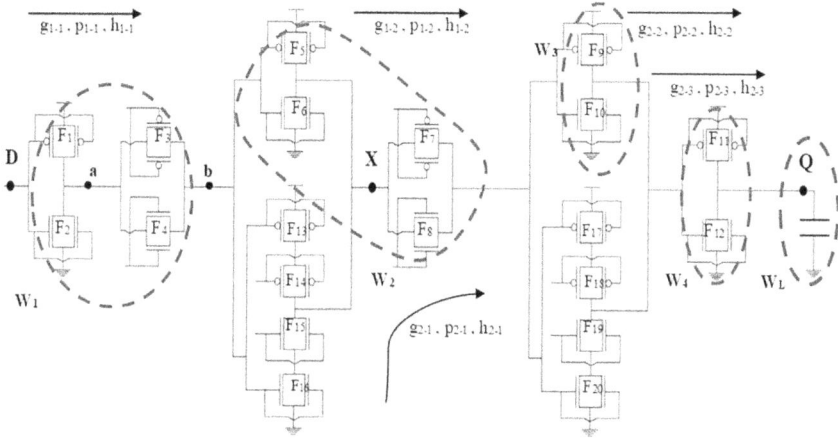

Figure 5.7 TGFF circuit with FinFET devices.

Table 5.1 The LE technique in TGFF (assuming two chunks)

Chunk-stage	Standardized Elmore delay (logical effort) (d)	Logical effort (g)	Electrical effort (h)	Parasitic delay (p)
1-1	$(5W_1)(1/3W_1) + (2W_1 + 2 + 3W_2)(2/3W_1)$	2	$(2 + 3W_2)/3W_1$	3
1-2	$(5W_2 + 2)(1/3W_2) + (2W_2 + 2 + 3W_3)(1/3W_2)$	1	$(4 + 3W_3)/(3W_2)$	7/3
2-1	$(2W_2 + 2 + 3W_3)(2/3W_2)$	2/3	$(2 + 3W_3)/W_2$	4/3
2-2	$(3W_3 + 2 + 3W_4)(1/3W_3)$	1	$(2 + 3W_4)/3W_3$	1
2-3	$(3W_4 + W_L)(1/3W_4)$	1	$(W_L/3W_4)$	1

Since t_{setup} and $t_{CQ\text{-}opt}$ are minimized by the LE method (minimum t_{DQ} is found), equations (5.9–5.11) can be utilized to determine the W_2, W_3 and W_4 variables that minimize t_{setup} and $t_{CQ\text{-}opt}$ given W_1 and W_L values [11].

$$W_2 = (-2 + \sqrt{4 + 2(+3W_3)(3W_1)})/6 \tag{5.9}$$

$$W_3 = (-2 + \sqrt{4 + 2(+3W_4)(3W_2)})/6 \tag{5.10}$$

$$W_4 = (-2 + \sqrt{12(W_3)(3W_L)})/6 \tag{5.11}$$

5.5 SIMULATIONS

In this chapter, HSPICE simulations were performed using the 16 nm PTM–MG technology [23] at the temperature of 25°C, $V_{DD} = 0.85$ V, Level = 72, and the CLK frequency of 1 GHz. The D and CLK are applied to the circuit after passing through a buffer to achieve more sensible outcomes, as shown in Figure 5.8. The FinFET width (W) is equivalent to the sum of the operative width of fins (5.12) [21]. The device parameters are defined as shown in Table 5.2.

$$W = N(2H_{\text{fin}} + T_{\text{fin}}) \tag{5.12}$$

5.5.1 Simulation results for power optimization methods

In this part, the efficiency of FF planned with adding ONOFIC and LECTOR blocks in different loads (W_L) is compared. As can be seen in Table 5.3 the ONOFIC approach has better performance than the LECTOR method.

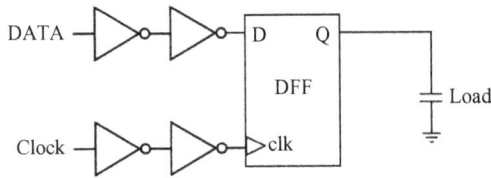

Figure 5.8 Simulation testbench.

Table 5.2 FinFET parameters

Parameter	Value
Length of gate (L_G)	16 nm
Height of FIN (H_{fin})	26 nm
Thickness of FIN (T_{fin})	12 nm
Thickness of oxide (T_{ox})	1.35 nm
Supply voltage (V_{DD})	0.85 V

Table 5.3 Simulation outcomes of TGFF with power optimization methods

W_L	ONOFIC		LECTOR	
	Power (μW)	Delay (ps)	Power (μW)	Delay (ps)
4	2.50	12.43	3.77	12.48
16	3.49	27.78	5.05	27.80
21	4.32	34.13	5.58	34.20
30	5.29	45.29	6.56	45.35
39	6.25	56.82	7.53	56.92

5.5.2 Simulation results for LE-based FF strategy

As described in the previous sections, the transistor sizes of TGFF phases are obtained, and simulations are exerted to estimate delay and energy for different input and output loads. In simulations, different values of W_1 and W_L, which represent input and output capacitances, respectively, are assumed. Table 5.4 demonstrates the sizes W_2, W_3, and W_4 for the LE method derived from equations (5.9–5.11). In order to be used as FinFET sizes, the nearest natural number is considered.

According to [18], t_{CQ} (CLK to Q) can be used to obtain t_{setup} and t_{hold} (Figure 5.9). As can be seen in Table 5.4, the mean power consumption and delay have a definite relationship with W_L, while with the increase of W_1, the delay decreases, and the mean power consumption increases. Conforming to Table 5.4, it can be said that t_{setup} and t_{hold} are only affected by W_1, W_2, and W_3. Also, W_4 and W_L have almost no effect on these two parameters.

5.5.3 Simulation results of the EEC-based FF

Based on the discussion in Section 5.3, the EEC shows the least energy required for a given delay by points in the design space [10,19]. The optimization target is the t_{CQ} delay, which is affected by the stages in the input-to-output path. Since other stages do not affect the target delay, their constant sizes have been chosen. Some of the energy-efficient cases identified by minimizing a few E^iD^j metrics by exploring 1,000 points are illustrated in Figure 5.10. This shows that the search algorithm is effective, as the search structures are similar to the EEC.

The EECs for diverse W_1 and W_L instances are exposed in Figure 5.11. As can be seen in Table 5.5, Although the LE method in each stage has smaller widths, the ED criterion has the best trade-off between E and D.

Table 5.4 TGFF parameters based on logical effort technique

Initials		Widths of each stage			Nearest natural number			Parameters			
W_1	W_L	W_2	W_3	W_4	W_2	W_3	W_4	P (μW)	t_{setup} (ps)	-t_{hold} (ps)	Delay (ps)
I	4	0.6398	0.3708	0.5206	I	I	I	2.41	10.4	4.7	12.52
I	16	0.7045	0.5988	1.5002	I	I	2	4.05	9.14	5.94	19.17
4	16	2.079	1.5210	2.5343	2	2	3	6.60	9.81	7.21	13.93
I	21	0.7247	0.6133	1.8789	I	I	2	4.59	9.14	5.94	22.50
4	21	2.1638	1.7289	3.1614	2	2	3	7.13	9.81	7.21	16.03
I	30	0.7561	0.8180	2.5462	I	I	3	5.93	9.15	5.96	23.30
4	30	2.2905	2.0534	4.2104	2	2	4	8.46	9.80	8.83	17.49
I	39	0.7831	0.9374	3.1734	I	I	3	6.49	9.15	5.96	27.23
4	39	2.3960	2.3358	5.1872	2	2	5	9.80	9.81	8.83	18.46

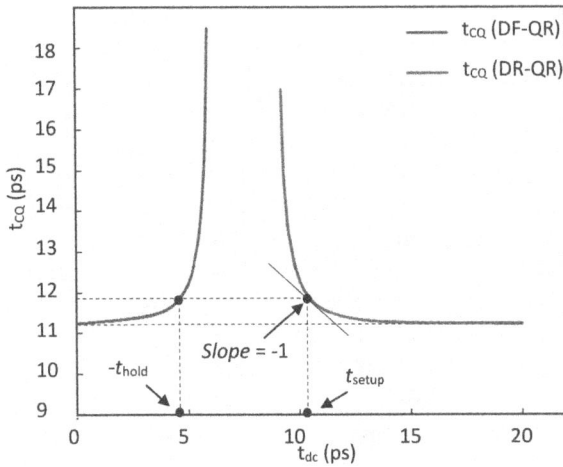

Figure 5.9 Timing parameters of TGFF ($W_I = 1$, $W_L = 4$).

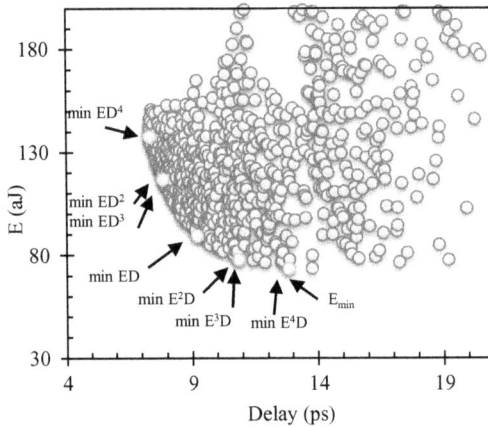

Figure 5.10 E-D design plane for TGFF ($W_I = 1$, $W_L = 16$).

Here the efficiency of the TGFF considered based on the EEC technique is compared (W_1 and W_L are 1 and 16, respectively) under different frequencies, temperatures (±10%) and supply voltages (±10%) that result in Tables 5.6–5.8, respectively. The results indicate that the E^4D scheme has the least power consumption at every frequency, whereas the E^4D and ED show the minimum variations in power consumption with respect to frequency. ED^4-based scheme is also the fastest one, and the E^4D scheme demonstrates more resistance to temperature changes in each parameter. Furthermore, the min delay and P depend on ED^4 and E^4D criteria, respectively, if the supply voltage varies.

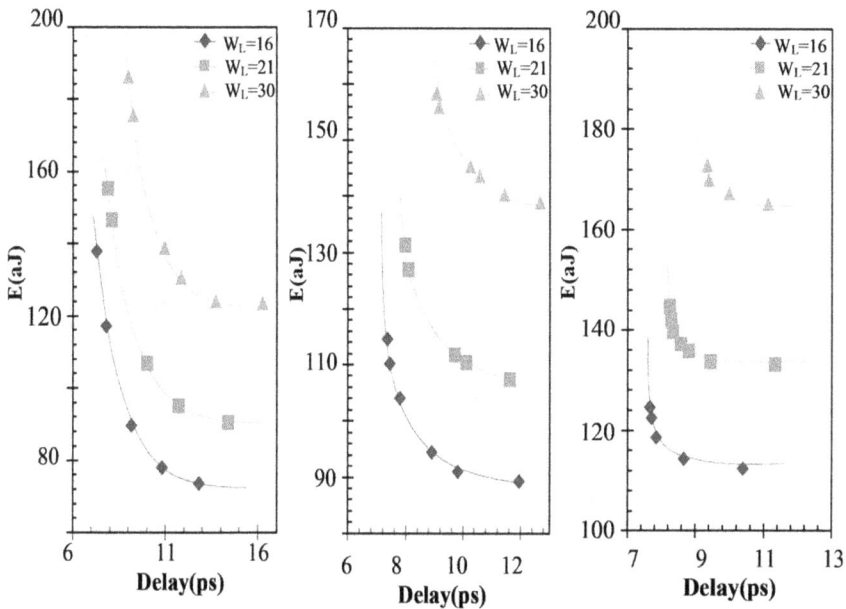

Figure 5.11 EEC in TGFF for W_L = 16, 21, 30 and (a) W_l = 1, (b) W_l = 4, and (c) W_l = 7.

Table 5.5 Measuring the sizes of stages and EDP and comparing them in EEC and LE methods

LE method				EEC metrics											
				ED^4 design				ED design				E4D design			
(W_L, W_l)	W_2	W_3	W_4	EDP (fJ.ps)	W_2	W_3	W_4	EDP (fJ.ps)	W_2	W_3	W_4	EDP (fJ.ps)	W_2	W_3	W_4	EDP (fJ.ps)
(16,1)	1	1	2	1.49	2	2	4	0.945	5	4	6	0.822	10	7	9	1.01
(16,4)	2	2	3	1.28	5	3	5	0.895	9	5	7	0.813	10	7	9	0.842
(16,7)	3	2	3	1.58	7	4	6	0.993	9	5	7	0.931	9	6	9	0.955
(21,1)	1	1	2	2.33	2	2	4	1.30	5	4	7	1.07	10	7	10	1.23
(21,4)	2	2	3	1.84	5	4	7	1.11	10	6	9	1.03	10	7	10	1.05
(21,7)	3	2	4	1.91	7	4	7	1.26	9	6	9	1.16	9	7	10	1.19
(30,1)	1	1	3	3.22	2	2	5	1.70	5	5	9	1.52	10	7	10	1.68
(30,4)	2	2	4	2.59	5	4	8	1.61	10	6	10	1.43	10	7	10	1.44
(30,7)	4	3	5	2.39	8	5	9	1.67	9	6	10	1.60	9	7	10	1.62

5.5.4 Designing for maximum stability

A circuit's stability is measured by its static noise margin (SNM). Due to the fact that both latches are not active at the same time, each latch is considered separately (Figures 5.12 and 5.13) as well as TG transistors and tri-state

Table 5.6 Power analysis of TGFF based on EEC metrics in diverse frequencies

	Design ED⁴		Design ED		Design E⁴D	
f (Hz)	P (μW)	Var. (%)	P (μW)	Var. (%)	P (μW)	Var. (%)
I K	5.6552	-	3.3256	-	1.8684	-
10 K	5.6556	0.007	3.3266	0.030	1.8686	0.010
100 K	5.6567	0.026	3.327	0.042	1.8688	0.021
I M	5.6686	0.237	3.3328	0.216	1.8723	0.209
10 M	5.7868	2.33	3.3905	1.951	1.9071	2.071
100 M	6.9705	23.25	3.9672	19.29	2.2553	20.71
I G	18.82	232.79	9.7433	192.97	5.7504	207.77

Table 5.7 EEC-based TGFFs parameters and distinctions under diverse temperatures

		ED⁴ design			ED design			E⁴D design		
Temp (°C)		Delay (ps)	Power (μW)	EDP (ps)²(μW)	Delay (ps)	Power (μW)	EDP (ps)² (μW)	Delay (ps)	Power (μW)	EDP (ps)²(μW)
	Val	8.48	10.6	762.26	10.31	5.66	601.84	14.11	3.67	730.66
−55	Var. (%)	16.32	−43.67	−24.26	12.31	−41.89	−26.82	10.06	−36.17	−22.68
25	Val	7.31	18.82	1006.49	9.19	9.74	822.49	12.82	5.75	945.02
	Val	6.11	109.0	4069.19	8.22	59.85	4046.92	11.98	32.9	4721.8
125	Var. (%)	−16.18	479.17	304.29	−10.45	514.47	392.03	−6.55	472.17	399.6

Table 5.8 The effect of voltage variations on the TGFF parameters in EEC-based designs

		ED⁴ design			ED design			E⁴D design		
Voltage (V)		Delay (ps)	Power (μW)	EDP (ps)²(μW)	Delay (ps)	Power (μW)	EDP (ps)² (μW)	Delay (ps)	Power (μW)	EDP (ps)²(μW)
0.765	Val	7.89	13.3	829.63	9.86	6.99	681.54	13.66	4.21	786.68
	Var. (%)	7.93	29.33	−17.57	7.29	−28.23	−17.13	6.55	−26.78	−16.75
0.85	Val	7.31	18.82	1006.49	9.19	9.74	822.49	12.82	5.75	945.02
0.935	Val	6.86	26.00	1226.76	8.71	13.34	1011.33	12.2	7.15	1064.2
	Var. (%)	−6.15	38.15	21.88	−5.22	36.96	22.95	−4.83	24.34	12.61

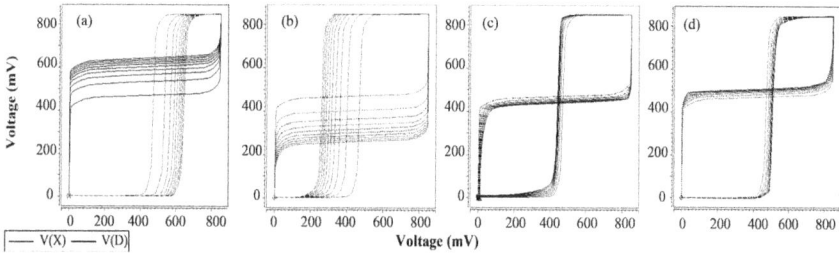

Figure 5.12 Butterfly arcs for various dimensions of (a) F_1 (b) F_2 (c) F_5 (d) F_6, in master latch.

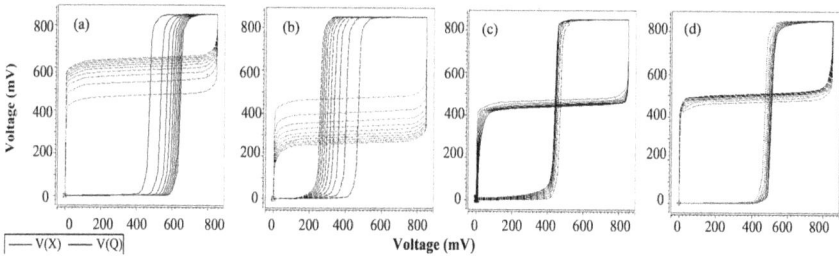

Figure 5.13 Butterfly arcs for various dimensions of (a) F_9 (b) F_{10} (c) F_{11} (d) F_{12}, in slave latch.

Table 5.9 Parameters of the design with the maximum noise margin

Parameter		Val	
		Max SNM (V)	@ Max SNM
Master latch	k_1	1	0.31
	k_2	1	0.31
	k_5	6	0.359
	k_6	1	0.31
Slave latch	k_9	1	0.31
	k_{10}	1	0.31
	k_{11}	6	0.35
	k_{12}	1	0.31

buffers are assumed minimum size since they do not affect stability [10]. The device dimensions are attained as $W_N = k \times W_{N\text{-min}}$ and $W_P = k \times W_{P\text{-min}}$, ($k_i$ is the dimensions constant of the i-th device and $W_{Pmin} = 2 \times W_{N\text{-min}}$). Table 5.9 shows that the max SNM is reached when all devices except F_5 and F_{11} are minimum size, and their sizes are six times that of others.

5.5.5 Introducing power-delay-area and energy-delay-area space

Previously, energy and delay parameters were the only factors considered. While, there are other ones, like area, which play an important role in designing. As part of this section, the A in order to design FFs that have more efficiency is approximated in either the E-D-A or P-D-A space.

In order to estimate the area, a simple but effective approach is used. Previous synthesis tools considered the areas of transistors. This method does the same. According to the design rules provided in [24], the circuit layout can be used for a more accurate estimation of area. The area of this Flip-Flop (Figure 5.4) can be calculated as follows:

$$\text{Area} = [((2W_{I1} + 2W_{I1}) + 2W_{TG1}(2W_{I2} + W_{I2}) + 2W_{TG2} + (2W_{I3} + W_{I3})$$
$$+(2W_{I4} + W_{I4}) + (2W_{B1} + 3W_{B1}) + (2W_{B2} + 3W_{B2}))]W_{\min} \times L_G \quad (5.13)$$

The W_{min} of INV, TG and tri-state buffer are W_{Li}, W_{TGi} and W_{Bi}, respectively. So, the approximate area of the LE-based circuit demonstrated in Figure 5.7 is as follows: $(W_{B1} = W_{B2} = W_B)$:

$$\text{Area} = ((3W_1 + 2W_1) + (3W_2 + 2W_2) + 3(W_3 + W_4) + (10W_B))W_{\min} \times L_G \quad (5.14)$$

For diverse W_1, W_L, TGFF is then reformatted using PDAP (power-delay-area product) and EDAP (energy-delay-area product). Simulations were performed for each pair (W_1, W_L) of these tables, and the minimum PDAPs and EDAPs were reported. From Table 5.10, it is obvious that the min PDAP values occur when W_2 and W_3, the central stages, are the smallest. Though, in the E-D-A space, in larger input and output loads, the min EDAP values have been obtained in the bigger size of the middle stages.

5.5.6 Comparison of presented designs

A comparison of previous parts' designs was conducted in this section on equal terms. The W_1 and W_L are considered 1 and 16, respectively. Based on Table 5.11, it can be seen that the E^4D metric performs optimally in EDP and PDP. Additionally, the ED^4 metric and $ONOIC$ have the least delay, and power, respectively.

5.6 CONCLUSION

The purpose of this chapter was to analyze the efficiency of FinFET-based circuits, and because of that, a TGFF is characterized and assessed in FinFET technology based on its routine parameters. Several methods have

Table 5.10 The results of the design that has the highest noise margin

(W₁, WL)	min PDAP						min EDAP Design					
	$W_2\ W_3\ W_4$	Area (f m²)	Delay (ps)	Power (μW)	PDP (aJ)	PDAP (10^{-30}J.m²)	$W_2\ W_3\ W_4$	Area (f m²)	Delay (ps)	Power (μW)	EDP (10^{-28}J.s)	EDAP (10^{-41}J.s.m²)
(1,16)	1 1 2	24.13	19.2	4.06	77.83	1.88	2 2 3	33.28	13.72	73.53	10.1	3.36
(1,21)	1 1 2	24.13	22.5	4.59	103.34	2.49	2 2 4	35.78	14.39	90.35	13.00	4.65
(1,30)	1 1 3	26.62	23.3	5.94	138.28	3.68	2 2 5	38.27	16.25	123.74	20.1	7.70
(4,16)	1 1 2	36.61	19.37	5.39	104.57	3.83	3 2 4	52.42	11.94	89.17	10.6	5.58
(4,21)	1 1 3	39.1	19.67	6.33	124.45	4.86	3 3 5	57.41	12.33	107.88	13.30	7.64
(4,30)	1 1 3	39.1	23.52	7.28	171.24	6.70	4 3 7	66.56	12.66	139.00	17.59	11.71
(7,16)	1 1 3	51.58	17.72	7.93	140.51	7.25	4 3 5	74.05	10.39	112.41	11.68	8.65
(7,21)	1 1 3	51.58	19.87	8.45	167.98	8.66	4 3 6	76.54	11.33	133.01	15.07	15.34
(7,30)	1 1 3	51.58	23.65	9.41	222.47	11.5	5 4 7	85.70	11.89	165.27	19.65	16.84

Table 5.11 Comparing the efficiency of TGFF in each approach

Method	LE	ED⁴	ED	E⁴D	Min PDAP	Min EDAP	Max SNM	ONOFIC	LECTOR
Power (µW)	4.05	18.8	9.74	5.74	4.06	5.36	5.77	3.49	5.05
Delay (ps)	19.14	7.29	12.31	12.82	19.2	13.72	23.73	27.78	27.80
PDP (aJ)	77.52	137.05	119.9	73.59	77.95	73.53	136.9	96.95	140.39
EDP (10^{-28} J.s)	1.48	1.00	1.48	0.94	1.50	1.01	3.25	2.69	3.90

been used to enterprise and investigate the FF, which can also be considered to design any other circuit. By using the *ONOFIC* approach, the least power consumption is achieved at 3.49 W, while by using ED^4, the least delay of 7.29 seconds is achieved. The *Max SNM* method also achieves a maximum SNM of 0.359 V. Considering the circuit's overall performance, min EDAP and E^4D metric is the best sizing method option, with PDP and EDP as 73.53 aj and 0.94 fJ.ps, respectively.

REFERENCES

[1] A. M. Prateek Mishra and Niraj K. Jha, FinFET circuit design. *Nanoelectronic Circuit Design*, 23–54, 2010.

[2] A. K. Mayur Bhole and Sagar Pawar, "FinFET - benefits, drawbacks and challenges," *International Journal of Engineering Sciences & Research Technology*, 2, 3219–3222, 2013.

[3] E. Abbasian, M. Gholipour, and S. Birla, "A single-bitline 9T SRAM for low-power near-threshold operation in FinFET technology," *Arabian Journal for Science and Engineering*, 47(11), 14543–14559, 2022.

[4] Y.-Y. C. Morteza Gholipour and Deming Chen, "Compact modeling to device- and circuit-level evaluation of flexible TMD field-effect transistors," *IEEE Transactions on Computer-Aided Design of Integrated Circuits and Systems*, 37(4), 820–831, 2017, doi: 10.1109/TCAD.2017.2729460.

[5] Y.-Y. C. Morteza Gholipour, Amit Sangai, Nasser Masoumi, and Deming Chen, "Analytical SPICE-compatible model of Schottky-Barrier-type GNRFETs with performance analysis," *IEEE Transactions on Very Large Scale Integration (VLSI) Systems*, 24(2), 650–663, 2016, doi: 10.1109/TVLSI.2015.2406734.

[6] E. Abbasian, "A highly stable low-energy 10T SRAM for near-threshold operation," *IEEE Transactions on Circuits and Systems I: Regular Papers*, 69(12), 5195–5205, 2022, doi: 10.1109/TCSI.2022.3207992.

[7] M. P. V.K. Sharma and B. Raj, "ONOFIC approach: low power high speed nanoscale VLSI circuits design," *International Journal of Electronics*, 101(1), 61–73, 2014/01/02 2014, doi: 10.1080/00207217.2013.769186.

[8] N. R. N. Hanchate, "LECTOR: a technique for leakage reduction in CMOS circuits," *IEEE Transactions on Very Large Scale Integration (VLSI) Systems*, 12(2), 196–205, 2004, doi: 10.1109/TVLSI.2003.821547.

[9] R. F. S. Ivan Sutherland, Bob Sproull, and David Harris, *Logical Effort: Designing Fast CMOS Circuits*. Morgan Kaufmann, Massachusetts, United States, 1999.

[10] M. G. Ehsan Mahmoodi, "Design space exploration of low-power flip-flops in FinFET technology," *Integration*, 75, 52–62, 2020, doi: 10.1016/j.vlsi.2020.06.006.

[11] E. C. Massimo Alioto, Gaetano Palumbo, *Flip-Flop Design in Nanometer CMOS*. Springer, 2016.

[12] E. C. Massimo Alioto and Gaetano Palumbo, "General strategies to design nanometer flip-flops in the energy-delay space," *IEEE Transactions on Circuits and Systems I: Regular Papers*, 57(7), 1583–1596, 2009.

[13] A. J. M. Paul I Pénzes, "Energy-delay efficiency of VLSI computations," In *Proceedings of the 12th ACM Great Lakes symposium on VLSI*, New York, USA, pp. 104–111, 2002.

[14] Y. Li, L. Chen, I. Nofal, M. Chen, H. Wang, R. Liu, Q. Chen, M. Krstic, S. Shi, G. Guo, S. H. Baeg, S.-J. Wen, and R. Wong, "Modeling and analysis of single-event transient sensitivity of a 65 nm clock tree," *Microelectronics Reliability*, 87, 24–32, 2018, doi: 10.1016/j.microrel.2018.05.016.

[15] B. A. V. Anandi, "Design of a D flip flop for optimization of power dissipation using GDI technique," In *2017 2nd IEEE International Conference on Recent Trends in Electronics, Information & Communication Technology (RTEICT)*, pp. 172–177, IEEE, 2017, doi: 10.1109/RTEICT.2017.8256580.

[16] D. Z. P. Taehee Lee and Joon-Sung Yang, "Clock network optimization with multibit flip-flop generation considering multicorner multimode timing constraint," *IEEE Transactions on Computer-Aided Design of Integrated Circuits and Systems*, 37(1), 245–256, 2017, doi: 10.1109/TCAD.2017.2698025.

[17] T. K. D. Peiyi Zhao and M.A. Bayoumi, "High-performance and low-power conditional discharge flip-flop," *IEEE Transactions on Very Large Scale Integration (VLSI) Systems*, 12(5), 477–484, 2004, doi: 10.1109/TVLSI.2004.826192.

[18] D. M. H. Neil and H. E. Weste, *CMOS VLSI Design: A Circuits and Systems Perspective*. Pearson Education India, 2015.

[19] J. W. T. Dejan Markovic and Vivek K. De, *Transmission-Gate based Flip-Flop*. Washington, DC: Patent and Trademark Office, U.S. Patent No. 6,642,765, 2003.

[20] B. N. Dejan Markovic and Robert Brodersen, "Analysis and design of low-energy flip-flops," In *Proceedings of the 2001 International Symposium on Low Power Electronics and Design*, pp. 52–55, 2001, doi: 10.1145/383082.383093.

[21] V. V. Guo Xinfei, Gonzalez-Guerrero Patricia, Mosanu Sergiu, and Stan Mircea R, "Back to the future: digital circuit design in the finfet era," *Journal of Low Power Electronics*, 13(3), 338–355, 2017, doi: 10.1166/jolpe.2017.1489.

[22] R. N. A. Shiva Taghipour, "Aging comparative analysis of high-performance FinFET and CMOS flip-flops," *Microelectronics Reliability*, 69, 52–59, 2017, doi: 10.1016/j.microrel.2016.12.012.

[23] Predictive Technology Model (PTM), 2012. Available on: https://ptm.asu.edu/.

[24] L. T. Clark, V. Vashishtha, L. Shifren, A. Gujja, S. Sinha, B. Cline, C. Ramamurthy, and G. Yeric, "ASAP7: A 7-nm finFET predictive process design kit," *Microelectronics Journal*, 53, 105–115, 2016, doi: 10.1016/j.mejo.2016.04.006.

Chapter 6

Emerging nano devices for low-power applications

A research point of view

G. Boopathi Raja
Velalar College of Engineering and Technology

6.1 INTRODUCTION

The metal-oxide-semiconductor field effect transistor (MOSFET) has evolved over the past 40 years into the fundamental component of practically all computing hardware. Their popularity has steadily increased as a result of the feature size, which has now shrunk to 0.1 μm. The constraints of manufacturing processes and the principles of quantum physics, however, may soon stop further feature size reduction. As a result, various transistor substitutes are being researched for use in ultra-dense circuits. These novel gadgets, whose sizes are on the scale of tens of nanometers, are known as nanodevices, and the field in which they are studied is known as nanotechnology.

An overview of the research on electronic switching devices at the nanoscale scale is given in this publication. Future ultra-density integrated electronic computers are projected to be made using such components. Werst first outlines the challenges posed by the downscaling of FET devices before going into the new solutions: (1) Transistors made of carbon nanotube devices based on the quantum effect and one electron and (2) molecular electronic devices. Each class of device's fundamental functioning principle is covered. Each new device's current state of the art is described, along with any unresolved research issues [1–4]. A potential timeline for their widespread deployment is then provided.

Based on the working theories and fabrication processes, the devices are divided into three major categories [5,6]:

- Carbon nanotube transistors
- Solid-state quantum effect devices
- Molecular electronic devices

The first class of devices resembles ordinary MOSFETs, but they differ in terms of their size and the substance they are constructed of (carbon nanotubes). Both the second and third classes make use of quantum

DOI: 10.1201/9781003459231-6

effects, although they are constructed in distinct ways. The manufacturing processes used for the solid-state devices are similar to those used for MOSFETs. These devices make use of quantum mechanical processes, but they also benefit from decades of advancements in MOSFET manufacturing technique [7].

A novel method called molecular electronics is needed, as are new components and a new mode of operation. The fact that molecules naturally form at nanoscale dimensions provides the impetus for such rapid change. Molecules may be manufactured similarly, inexpensively, and readily, in contrast to nanostructures formed from bulk substances. The creation of molecular switches and the exact assembly of those switches into the structures required for accurate computing pose two major obstacles.

In this work, researchers first provide a general overview of the issues raised by further reduction of the existing MOS technology. Then, they discuss the devices in a carbon nanotube, their working principles, and their performance compared with other devices. They concluded by providing a timeline for when these devices may be implemented in mass-produced integrated circuits.

The total power consumed by an IC is divided into two categories: static and dynamic. Leakage current and DC current sources are frequently the sources of static power. The use of frequency-dependent dynamic current commonly reigns supreme in terms of total power. It is the consequence of the crowbar action of switching transistors coupled between the supply and capacitive nodes, which charge and discharge alternately [8,9].

6.1.1 CMOS inverter

The common CMOS inverter's capacitive charge characteristics are shown in Figure 6.1. The power consumption is proportional to the clocking frequency as a result of the dynamic charging. The crowbar effects are depicted in Figure 6.2. During a logic transition, there is a short current flow that occurs whenever both transistors become active. This current flow consumes energy.

The main goal of a low-power designer is to minimize supply voltage, clocking frequencies, DC source values, circuit complexity, and switching node capacitance to reduce current consumption [10]. Analog circuitry is the finest type of architecture for implementing circuit functions that are constantly in use. To achieve low power, an analog designer typically:

- Innovative electronic circuit methods for complicated functions.
- Deployment in low-threshold CMOS technology.
- Holding CMOS transistors in the weak inversion (sub-threshold) region.
- Circuits for voltage and/or current control

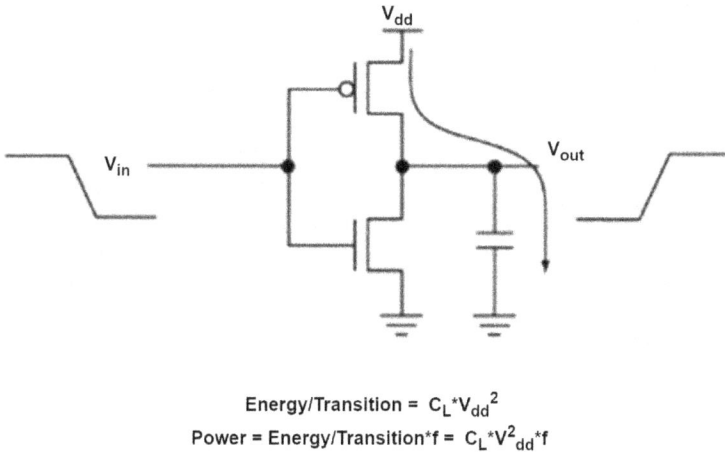

$$\text{Energy/Transition} = C_L{}^*V_{dd}{}^2$$
$$\text{Power} = \text{Energy/Transition}{}^*f = C_L{}^*V^2{}_{dd}{}^*f$$

Figure 6.1 Capacitive charge characteristics of CMOS inverter.

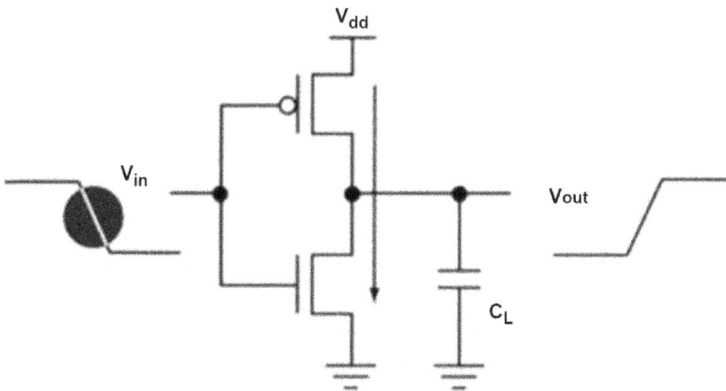

Figure 6.2 Crowbar effects in CMOS inverter.

1. In a typical CMOS inverter, the power consumption is approximately equal to the clocking frequency due to dynamic charging.
2. In a typical CMOS inverter, power is used due to the brief current flow that happens when both transistors activate during a logic change.

6.1.2 Weak inversion

Weak inversion design allows for minimum switching and leakage currents at very low supply voltages (Vt+200 mV). For low-frequency applications, it performs effectively. A weakly inverted channel of insignificant minority

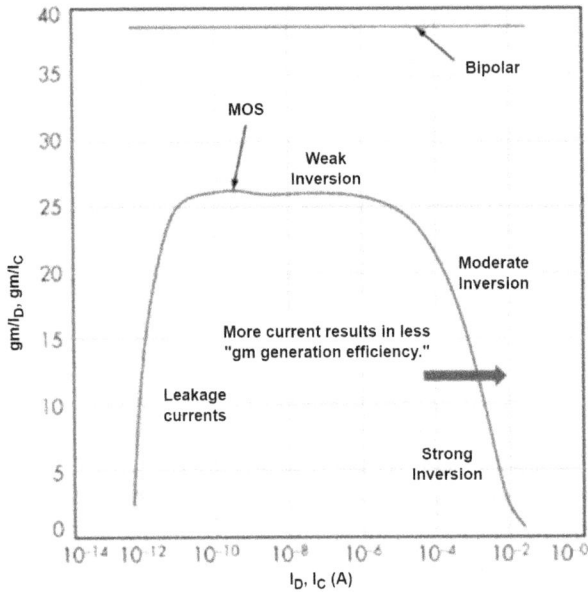

Figure 6.3 Transconductance, gm versus current.

carriers forms beneath the gate of the MOS transistor during sub-threshold operation under conditions of considerable electron energy band bending, generally up to 100 mV below the projected threshold voltage, as shown in Figure 6.3 [11].

The MOSFET weak inversion area resembles the more desired properties of a bipolar transistor, according to the gm/Ic ratio graph of a MOS and bipolar transistor.

Additionally, the drain current (I_D), which mimics bipolar characteristics, has a diffusion component that dominates a drift component due to gate-induced band bending within the channel in the absence of large minority carriers. MOS transistors operating in weak inversion have a transconductance (gm) that is dependent on temperature and current but independent of the process spread.

By constructing large area and long-channel devices, surface state effects may be statistically averaged away. Long-channel device design reduces threshold voltage sensitivity to changes in drain-source voltage. A CMOS inverter that is operating in the weak inversion region may have its output entirely flip among supply rails with an input switching window of 30–40 mV while it is under dc bias. This is possible because of the way the inverter works. In addition, the MOS output impedance is around 5 M when it is operating in the saturated zone.

When developing for low-current consumption, engineers make full use of these sub-threshold features. Furthermore, when the strength and frequency are increased, so is the capacity to identify minuscule signals amid the noise. Therefore, low-power operation limits the least observable signal level and necessitates specialized design expertise for reducing circuit noise.

6.1.3 Designer's approach

The Swiss watch industry was the first to develop techniques for ultra-low-power design to maximize the benefits of utilizing analogue circuitry for low-power continuous-time applications. These applications include the electronic circuit for maintaining quartz-stabilized oscillators. Ultra-low-power design techniques were initially developed in Switzerland.

For improved precision, event-triggered processes with relatively extended idle times are commonly accomplished in the digital space with only a power management function. A typical strategy used by a digital designer might be:

- Implementing multi-level thresholds
- Efficient formulation of high-level conceptualization to logic gate level
- Multiple-frequency clock source
- New methods of logic reduction
- Optimized voltage domains
- Customized level-shifting cells
- Power management (power cycling, clock gating)

6.2 NANOSCALE MOSFETs

A transistor seems to be a switch that regulates the flow of current throughout its channel in a digital circuit according to the condition of the device's gate terminal. The circumstance that enables or disables the gadget establishes its kind. If a voltage is applied to the gate of a device to regulate the current that is flowing through the channel, then that device is known as a Field Effect Transistor (FET), and if a current is flowing through the gate, then that device is known as a Bipolar Junction Transistor (BJT) [9,12]. The most widely used transistor in modern circuits is a MOSFET, a variation of the FET. This section will examine the device's operation to explore the issues with MOSFET miniaturization below 0.1 μm.

6.2.1 Structure and operation of a MOSFET

The source, Drain, and Gate are the three terminals of a MOSFET [13]. Figure 6.4 depicts the molecular structure of a MOSFET. It is supported by

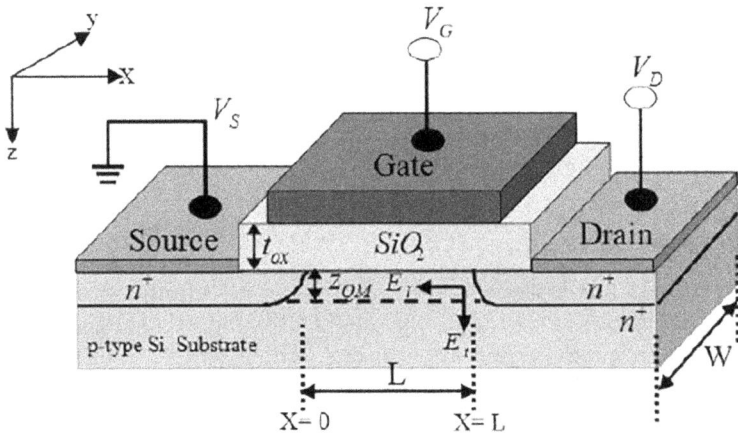

Figure 6.4 Structure of MOSFET.

a silicon substrate that is doped and crystallized. Because dopant impurities, including boron or arsenic, can increase the amount of mobile positive or negative charges in silicon, which is a poor conductor due to its purity. Positively doped (p-type) silicon has electron vacancies termed as holes that serve as positive charge carriers, whereas negatively doped (n-type) silicon has surplus electrons.

An n-type MOSFET, as seen in Figure 6.4, has two severely doped n-type source and drain regions sandwiched between a light-doped p-type channel. The gate terminal consists of an electrode made of metal that is separated from the channel by an oxide barrier that acts as an insulator. The voltage at the gate of the MOSFET has the effect of altering the electric field in the channel of the device, which in turn affects the ability of the device to conduct current [12]. Few negative charge carriers are present in the channel and there is very little current flow when the voltage on the gate is low. However, if the gate is kept at a high voltage, more carriers were drawn to the area below the gate, and as a result, the channel conducts freely, increasing the device's current output. As a result, the MOSFET functions as a two-state device that switches between the on (high channel conductivity) as well as off (low channel conductivity) state in response to the voltage applied to the gate terminal.

Over the past three decades, integrated circuit technology has advanced using a straightforward scaling concept. Simply reducing the size of the circuit's dimensions, such as the wire's length and breadth and the transistors' sizes, will result in smaller MOSFETs. The feature size is a metric that affects how big the circuit is in comparison to circuits from earlier generations. The chip must be manufactured with this minimum conductor (wire) width. Every length is a multiple of this shortest unit. Designers would seek to reduce the feature size to accommodate more transistors on a device of a given area. Over the years, the feature size has steadily shrunk till it has reached 0:1um. Dense chips such as the Intel Pentium are now possible because of this.

6.2.2 Problems with nanoscale MOSFETs

The industry and other research organizations strive to further reduce feature size and advance MOSFET technology despite significant obstacles. Since the industry has long relied on MOSFET-based CMOS technology, switching to a new one requires a significant expenditure. Although 25 nm-sized functioning transistors have been created, there are still issues with large-scale circuit design that need to be resolved [13–15].

6.2.2.1 High electric fields

As a result of scaling down, the electric field intensity across the gate oxide is increased since the power supply voltage cannot be reduced in proportion to the channel length. The oxide field has a maximum value of 5 MV/cm for 0.1 μm channel length devices, whereas the silicon field has a value of more than 1 MV/cm [15]. As the channel size shrinks to nanometer-sized proportions, these numbers will continue to rise. The high fields cause the device's performance to suffer from greater leakage currents. In the worst situations, the field results in an avalanche collapse of the barrier, allowing the electrons to conduct freely and causing damage to the device through current surges.

6.2.2.2 Power supply and threshold voltage

To maintain the active power and electric field within acceptable bounds, one would like to proportionally lower the supply voltage when the MOSFET channel is scaled down. The threshold voltage, however, cannot be significantly decreased. The reason for this is that the power used by the device in its steady state, or the quiescent state power, should be managed. Since the leakage current through the device is the primary source of power consumption under these conditions, the threshold voltage is kept high in order to reduce the amount of leakage current. However, the greater threshold voltage produces considerably finer lines between on and off states, increasing the likelihood that the device would reach an indeterminate state, which is neither on nor off. Voltage scaling is a serious issue due to inductive effects and noise margin.

6.2.2.3 Heat dissipation

Heat is produced by transistors in their resistive components as a type of energy loss. If this heat is not effectively dispersed, hot spots may develop on the circuit. The device's performance degrades, malfunctions, or even gets destroyed because of these hot spots that lead the material to overheat.

6.2.2.4 Interconnect delays

The resistance rises as the wire width narrows, lengthening the delay. Shrinkage would significantly lengthen connectivity delays compared to gate delays. Scaling is done to make chips faster and denser, in addition to increasing chip density. Due to significant connectivity delays, the devices might not be significantly quicker.

6.2.2.5 Vanishing bulk properties

An optical filter is used during the manufacturing process to dope the substrate. Since it becomes harder to distinguish the finer doping zones as feature size drops, the bulk may not be evenly doped at such tiny scales. This might not produce transistors where they should, which would cause the circuit to malfunction.

6.2.2.6 Shrinkage of gate oxide layer

An oxide layer of 30 A is necessary for 0.1um CMOS devices running at 1.5 V [15]. This is about equivalent to 10 layers of silicon atoms. Quantum mechanical tunneling occurs when there is a thin oxide layer, causing leakage and through gate. The possibility of further lowering oxide thickness is diminished as a result.

Due to ineffective doping techniques and the emergence of quantum phenomena, the aforementioned challenges exist. To overcome these challenges and create systems that would take the quantum effects into consideration. The silicon on insulator (SOI) device, which has a substrate made of a partly depleted insulator, is an attempt in this direction [16]. Another strategy is to produce MOSFETs using silicon germanium (SiGe) rather than standard silicon. Beyond the scope of this investigation [17,18], further discussion of these devices is not necessary. The choice of silicon as a material causes the issue rather than the fundamental design of the transistor. Because of this, several research teams have employed extremely small carbon tubes to create semiconductor switches that are simultaneously smaller and quicker than silicon MOSFETs.

6.3 CATALYST FOR LOW-POWER, LOW-VOLTAGE DEVELOPMENT

The Swiss watch industry, an important source of wealth for the region's economy, discovered in the 1960s that consumer preferences were shifting from mechanical to electronic devices, with prices falling in tandem. The Swiss Watch Industry Federation (FH) then expands its study of electronic timepieces [8].

The problem was to lower the power consumption of the electronic quartz clock because it had previously been developed. To create the electronic quartz stabilized wristwatch, the Swiss Laboratory for Watch Research (LSRH) established the Centre ElectroniqueHorloger (CEH) joint research laboratory in LSRH building in Neuchatel in 1962.

By running CMOS transistors in their previously undiscovered sub-threshold zone, Eric Vittoz, a CEH employee in Neuchatel, created a break-through in low-power design. At ESSCIRC'76, he delivered his first paper, and his patent enabling amplitude regulation of the 32-kHz oscillator cir-cuit for low operational current quickly established the gold standard for quartz stabilized watches as shown in Figure 6.5.

At ESSCIRC'76, Eric Vittoz demonstrated these weak inversion designs as well as the MOS transistor weak inversion model.

Prior to then, it was thought that MOS transistors' sub-threshold region, which manifested as a dc leakage current throughout CMOS digital cir-cuits, was an unfavorable feature. Consequently, it hadn't been thoroughly examined. Figure 6.6 shows the CMOS inverter with drain junction leak-age and sub-threshold current path.

This circuit demonstrates how it is simple to mix up the leakage cur-rents in the drain junction area of a MOS transistor with those in the sub-threshold region.

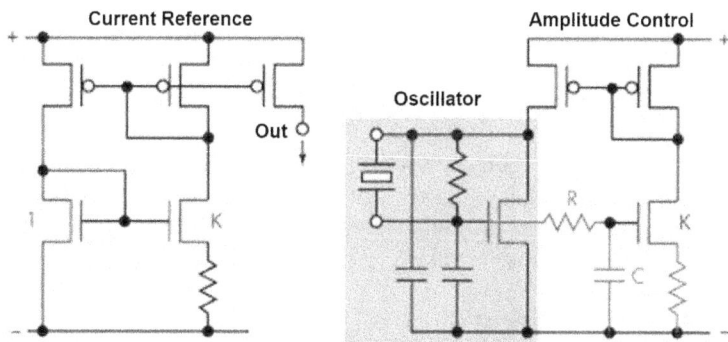

Figure 6.5 Amplitude regulation of the 32-kHz oscillator circuit.

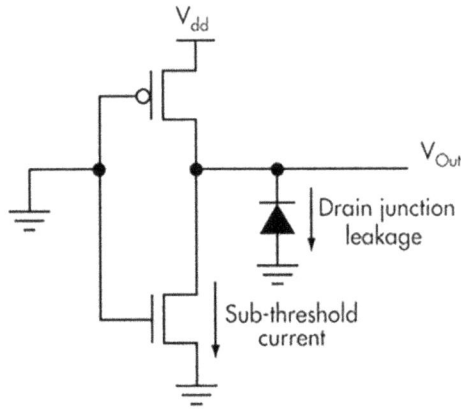

Figure 6.6 CMOS inverter with leakage path.

To create EM Microelectronics (Swatch Group), Ebauches Electronique SA acquired a CMOS licence from Hughes Aircraft. EM Microelectronics will design and produce ICs for the watch sector. It contributed to the circumstances needed for the establishment of an industrial-scale low-power, low-voltage competence by delivering its first CMOS circuit in 1975.

Based on Eric Vittoz's work, these circumstances led the Swiss semiconductor company Faselec to transition from bipolar to CMOS technology for the construction of ultra-low-power integrated watch circuits.

In the vicinity of this time, Jean Hoerni, a Swiss national who had previously worked at Intersil and was the inventor of the planar fabrication technique for silicon technology, established Eurosil electronic GmbH in Munich with the intention of applying the planar technique to low-power CMOS applications. EM Microelectronics and Faselec were the two firms that were engaged in this incident. By controlling the voltage, it was possible to circumvent the patents held by Vittoz, and this method was subsequently improved upon by the addition of a separate amplitude regulation circuit that made use of the back gate bias effect.

When CEH and CSEM merged in 1983, the result was the formation of the Swiss Centre for Electronics and Microtechnology (CSEM). It was from this institution that the ultra-low-power fabless semiconductor business Xemics, which was recently acquired by Semtech Inc., was spun off in 1997. Xemicsspecialises in battery-powered sensors and wireless transceiver applications.

The primary recipient of this wireless know-how is a significant Swiss medical OEM engaged in the manufacture of cutting-edge personal hearing instruments. Since then, Swiss low-voltage, low-current circuit prowess has propelled the semiconductor sector for battery-powered wireless consumer goods.

6.3.1 Industrial cross-fertilization

Watches were a trendsetter in bringing important technology to commercial maturity and popular awareness. The watch and microelectronics industries have benefited from technological cross-fertilization. The power budget of a watch circuit is divided between the stepper motor and oscillator block. Back-EMF sensing methods are used in an incorporated stepper motor driver created in the 1980s therefore for watch industry to reduce current consumption. Hard-disk stepper motor devices for laptops and tablets now benefit from this adaptive motor driving technology, which increases battery life. Semiconductor firms in the watch, clock, and calculator industries started focusing on RFID tagging technology in the late 1980s. Early in the 1990s, the American Veterinary Industry received its first prototypes attributable to subsequent groundbreaking work.

A cost-effective entry point into a market controlled by Gemplus (France), a company lacking experience in low-power design, arose during this period as smart-card patents started to expire. This made it possible to enter the contactless RF smart-card market at the same time as radio-controlled precision timekeeping watches. Both of these goods need to be designed using ultra-low-power RF methods.

Low-power design strategies also started to benefit formerly huge and bulky mobile phones. The 32-kHz quartz crystal watch circuit serves as the foundation for the real-time clock (RTC) that triggers sleep mode when the device is idle. The process of miniaturization of these devices, as well as the amount of time they can run on a single charge, has benefited simultaneously from developments in battery technology.

The growing use of RF front-end circuits in CMOS technologies is a reflection of the Swiss semiconductor industry's transition from bipolar to CMOS in the late 1970s. The low-threshold voltage alternatives available with modern conventional CMOS processes are equivalent to the groundbreaking low-threshold technology used by the watch industry during the early 1980s. This innovation has helped Bluetooth and ZigBee produce single-chip, inexpensive and low-power communication systems.

Silicon-based MEMS and other space-saving technologies are crucial for adding new functionalities and improving the shrinking of mechanical watch components and have enabled the creation of new product categories including mobile phones, RFID transponders, and, more recently, implantable medical devices.

As a result, better, less noticeable hearing aid devices (MEMS-based hearing aids) which are more affordable and have a longer battery life have been able to be miniaturized. Based on MEMS technology, the next generation of active implanted drug delivery devices will distribute medications to afflicted regions in a regulated manner.

Conventional semiconductor technologies are becoming an increasingly acceptable option for specialized low-power applications within the industry that is still in the process of emerging for implanted medical devices. An ultrasonic pacemaker with a transceiver implant was developed by Cambridge Consultants for EBR Systems. This device eliminates the need for invasive surgery and pacing leads by transmitting electrical current to an electrode that is surgically placed in the patient without the need for wires.

Low-power needs are linked with energy-collecting techniques. Techniques for converting kinetic energy and wirelessly transmitting power, which were initially developed for the watch and RFID sectors, respectively, are gaining popularity as potential alternate power sources for implanted technology.

In-body micro-generators are being developed as part of the 2-year SIMM (self-energizing implantable medical microsystem) initiative to provide power for implanted medical devices such as electrical stimulators, pacemakers, and body area network applications.

Mammals have a natural battery located deep inside their inner ears, a chamber loaded with ions that generate an electrical potential to power brain messages. Researchers from the Massachusetts Institute of Technology (MIT) have discovered that this chamber represents an additional potential source of power for electronic devices that are surgically implanted.

Patrick Mercier, a former graduate student at MIT who is currently an assistant professor at the University of San Diego, worked at the Microsystems Technology Laboratories of MIT (MTL) on the formation of an embedded radio transmitter circuit that is powered by this biological battery without resulting in any hearing loss. Mercier is now an assistant professor at the University of San Diego.

The application of technology that is used for microsystems to the rapidly emerging field of nanotechnology is making things that were unthinkable only a few short years ago more likely to become a reality. The use of manufactured nanorobots to perform cellular repairs is being studied in research labs (nano-medicine). Low-power design determines which energy sources are best for self-powered nanosystems.

Therefore, the effective creation of high-tech items for the future world depends on low-power design methodologies.

6.4 CARBON NANOTUBE FIELD EFFECT TRANSISTORS (CNTFET)

A carbon nanotube device is analogous to a MOSFET due to the fact that a gate is used to control the amount of current that passes through the device by varying the field that is present across a channel. The method by which electrons are transported from the source to the drain is novel in this case. These devices use a small tubular structure called a carbon nanotube in place of a channel whose field may be regulated by a gate electrode.

Depending on whether the tube is straight or twisted, it might be considered conducting or semiconducting [19–22]. Compared to silicon MOSFETs, these devices are far more compact and smaller. The function of a carbon nanotube in a CNTFET is discussed in this section along with its fundamental physics.

6.4.1 Basic physics of the carbon nanotube

Graphene is a material which may be rolled up into a cylindrical tube to generate carbon nanotubes. Graphene is composed of a single layer of graphite atoms that are organized in a hexagonal pattern similar to that of chicken wire mesh [22]. Figure 6.7 depicts the structure of single-walled and multi-walled CNTs. The graphene molecules fall under the category of fullerenes, which are close-casted molecules with exclusively hexagonal and pentagonal interatomic bonding networks, as a result of their hexagonal structure. They have excellent elastic and tensile qualities because of their hexagonal configuration. Because of their durability, the tubes bounce back to their original form after being bent or pressed. They are helpful in circuits since they can be cooled off more quickly. They also transmit heat extremely well. They also possess peculiar qualities about electrical conduction. Depending on how they are rolled, they may be made to function either as a metal or a semiconductor.

To better grasp carbon nanotubes, a few fundamental terminologies are defined. The nanotube in Figure 6.8 that is unrolled is an example. The figure defines the two unit cell vectors, designated as a1 and a2, respectively. A chiral vector is a vector that is normal toward the circumferential vector throughout the direction where it is being rolled. This vector is known as

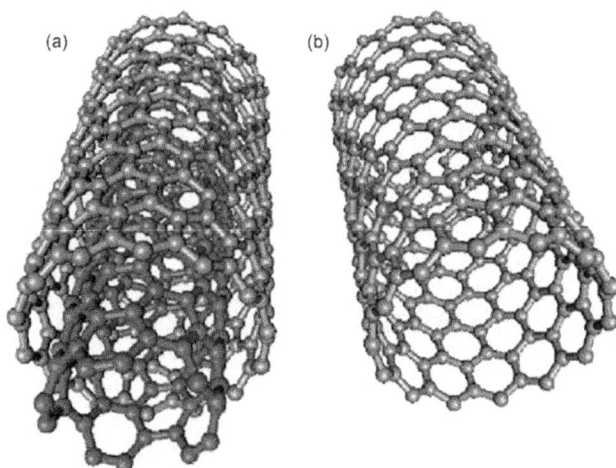

(a)   (b)

Figure 6.7 Structure of single-walled and multi-walled CNTs.

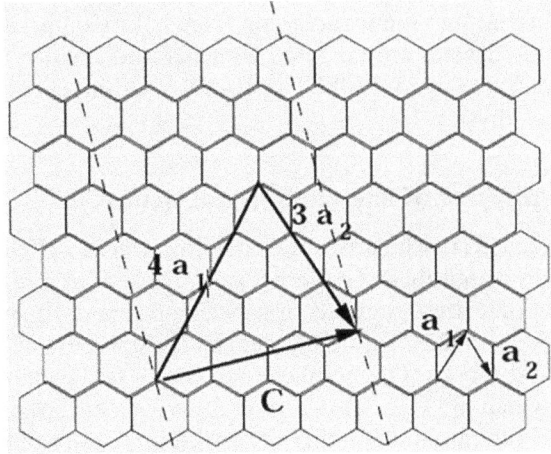

Figure 6.8 Chirality of a (4,3) carbon nanotube.

the direction in which the object is being rolled. The chiral vector is the horizontal vector that extends from one of the open ends of the tube to the other when it has been rolled [15]. The chiral vector may be represented graphically in terms of the unit vectors, as seen in Figure 6.8.

$$\vec{C} = n\vec{a_1} + m\vec{a_2}$$

where n and m are integers. These numbers characterize the nanotube as $(n;m)$. When rolled along the chiral vector, for instance, the tube in Figure 6.8 would resemble a (4; 3) tube. A straightforward rule of thumb has been developed to distinguish between metal and semiconductor tubes:

- The tube is metallic if n-m is divided by 3.
- A semiconductor is present in the tube if n-m is not divided by 3.

Researchers may categorize the tubes among three groups based on their chirality. The tubes in Figure 6.9 are an example. According to (n,m), a carbon nanotube can be categorized as:

- Zig-zag if either $n=0$ or $m=0$
- Armchair if $n=m$
- Chiral if $n \ne m$

When the need for metallic properties is paired with the armchair type, then the armchair type will always be metallic. However, the other two kinds can be classified as metallic or semiconducting according to the chiral condition they are in.

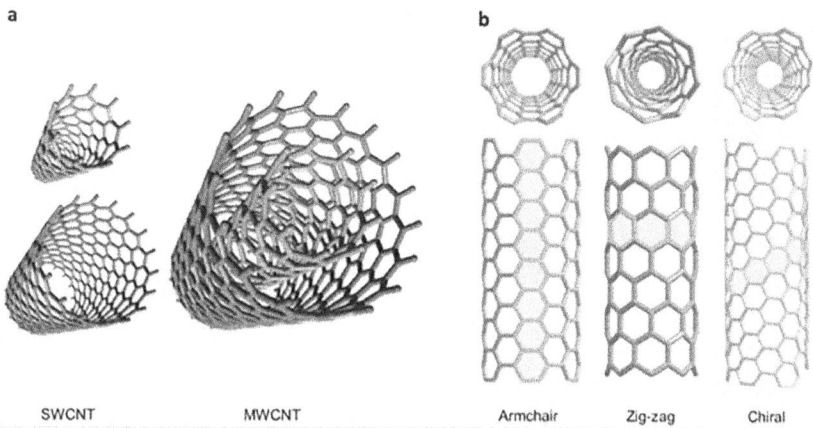

Figure 6.9 Different kinds of tubes: zigzag, armchair, and chiral [23].

From the tubes, it is possible to make single-walled nanotubes (SWNT) as well as multi-walled nanotubes (MWNT) [20]. SWNTs are bundled one on top of the other to form MWNTs. As we discuss in the next section, CNTFETs may be produced using either variety of tubes.

6.4.2 Basics of CNTFETs

A CNTFET and a MOSFET are similar from the outside. The source, drain, and gate are the three terminals on both. A carbon nanotube that is positioned between the source and the drain is what really creates the channel [22]. The gate regulates the field that crosses the nanotube, which in turn regulates the current flowing from the source to the drain. In a CNTFET, a nanotube replaces the channel of a traditional MOSFET.

The gold electrodes used in the initial generation of CNTFETs were placed on top of the tube to create the channel. The insulator-separated gate has situated on either side of or beneath the tube. This design had a flaw in that the tube was open to the air, which meant that it could only function as a p-type transistor due to a tube feature. The device's size expanded as a result of the thick gate oxide that was required to provide insulation.

When exposed to air, the CNTs become inadvertently p-type devices. This is because the oxygen in the air causes a change in the Fermi level at the contacts, bringing it closer to the valence band. As a direct consequence of this, holes experience a lower level of resistance than electrons do, and as a consequence, they may tunnel through the device at a faster rate. An undoped CNTFET that is then subjected to air will result in the production of a p-type device, also known as a conductor of holes.

Table 6.1 Physical characteristics of graphene in comparison to other materials [24]

Parameters/ Materials	Current density (A/cm²)	Electron mobility (cm²/V.s)	Melting point (K)	Thermal conductivity (10³/W/ mK)	Mean free path (nm)
Carbon nanotube	>10⁹	>10,000	3,800	1.75–5.8	>10³
Graphene	>10⁸	>10,000	3,800	3–5	1×10³
Si	-	1,400	1,687	0.15	20–30
Cu	>10⁷	-	1,357	0.385	40

A significant advancement was the second generation of CNTFETs. The gate electrode is positioned above the tube to provide a barrier between it and the air in the surrounding environment. This also resulted in a reduction in the circuit's total capacitance. Table 6.1 shows the comparison of the physical characteristics of graphene with other materials such as graphene, Si, and Cu.

An n-type device is required in addition to a p-type to create standard CMOS circuits. There are two methods for doing this: doping and annealing. The procedure of annealing involves briefly burning the tube to 450°C in a nitrogen atmosphere. This technique increases the Fermi level to the conduction band and decreases the electron barrier by pushing out the oxygen that the tube had absorbed. This is accomplished by forcing out the oxygen. The tube, which functions as an n-type device, conducts electrons more readily than holes as a result.

Doping is a different method of changing a p-type tube to an n-type tube. In this method, a potassium-based electron donor is used to dope the tube. Once the tube has been sufficiently doped, the additional electrons weaken the barrier strength, allowing the electrons to tunnel through it and cause the device to behave as an n-type device.

Even though we were successful in converting the p-type tubes to the n-type, the tubes will revert to their p-type behavior if they are exposed to air. Therefore, when they are changed to n-type, the tubes must be covered. The gate acts as a natural cover for the tube, which is another factor contributing to the success of the second-generation CNTFETs.

6.4.3 Fabrication

CNTFETs were first imagined and developed in laboratories more than 10 years ago, but the major barrier was the absence of a mass manufacturing method that would make it possible to manufacture integrated circuits. The primary issue was that the tubes could not be automatically positioned at precise positions, and there was no way to predict whether a tube would indeed be metallic or semiconducting when it was positioned at a certain spot. This was a substantial obstacle until the Avouris group at IBM came up with a method that they named constructive destruction [25]. The employment of metallic and semiconducting tubes that are wrapped one over the other in MWNTs. After the MWNTs have been arranged in the appropriate manner, the tubes that are not wanted are removed. If a tube must transform into metallic, the metallic tube is preserved, and the semiconducting tube is removed by the use of chemical depositions. The metallic tube in the MWNT is damaged if a semiconducting tube is desired at a certain place. In this way, there is no longer a difficulty with the kind and placement of tubes in a circuit.

6.4.4 Performance comparison with other devices

Table 6.2 shows the performance comparison like Delay, Leakage Power, total power consumption, and Power Delay Product of various devices such as CNTFET and GNRFET.

Table 6.3 shows the read and write delays of 6T SRAM cell under various devices such as GNRFET, FinFET, and CNTFET. The power dissipated by a single-bit 6T SRAM cell based on GNRFET is the lowest, at roughly 23.46 W, compared to SRAM cells based on CNT-FET and finFET, which both dissipate 284.3 and 30.6 W, respectively.

Table 6.2 Comparative analysis for STI with CNTFET and GNRFET [26]

Parameters\Devices	GNRFET	CNTFET
Delay (ps)	11.2	13.5
Leakage power (nW)	0.078	0.0621
Total power (nW)	117	239
PDP (e-18)	1.3	3.23

Table 6.3 Read and write delays of 6T SRAM cell under various devices [26]

Operation\Devices	GNRFET	FinFET	CNTFET
Write 1	2.31E-10	9.88E-10	3.71E-10
Write 0	1.43E-11	9.74E-09	3.71E-10
Read 1	4.60E-09	1.99E-08	1.99E-09
Read 0	2.42E-09	6.99E-09	9.10E-09

6.4.5 Summary: CNTFET

- The CNTFETs are electronic devices with a channel formed of a carbon nanotube that operates on a similar concept to the MOSFET. Depending on how they are rolled, carbon nanotubes can be produced to be metallic or semiconducting. Transistors make use of semiconducting tubes. Both p and n-type devices have now been created, and according to IBM, a method for mass-producing them has been discovered [26]. Due to this, they are strong candidates to take the position of MOSFET. However, there are still a few issues:
- Scaling up further is problematic.
- Multi-level interconnects using various metal layers, for example, are currently not possible with carbon nanotubes.
- The manufacturing level of the new fabrication technique is not yet reached.

Despite these issues, the industry is anxiously anticipating the creation of pioneering carbon nanotube chips.

6.5 LOW-POWER MARKET ANALYSIS

The market for industrial semiconductors is estimated to be worth $30 billion worldwide. The market for medical electronics accounts for more than 10% of that market and is predicted to expand at a 10% annual pace.

New applications for patient-implanted sensing devices in the medical field provide affordable healthcare solutions by eliminating the need for routine clinic visits. The consequent massive expansion in the industry for personal battery-powered medical equipment is fueling the production of micro-powered (wireless) implanted devices for activities such as cerebral stimulation and heart pacing. These implanted devices will be used for activities such as stimulating the brain and pacing the heart.

In recognition of the potential for reduced medical expenditures as well as increased patient quality of life, money from the European Union has been used to generate a wide range of medical implants that will assist patients who have certain nervous system impairments. These implants will help patients with certain conditions that affect the nervous system.

The current RFID industry is valued about $12 billion and consists of transponder/tagging technologies, contactless smart cards, active and passive RFID, as well as related readers. Critical issues, such as tag yield in contrast to price, frequency acceptability, specification creep, and required performance criteria, are being solved to make it possible for the RFID industry to develop rapidly.

According to IDTechEx, the value of the whole industry, which includes both systems and services, will skyrocket to $26 billion by the year 2016. This includes a lot of newly emerging industries, including the one for active RFID real-time positioning systems, which will generate more than $6 billion in revenue in 2016.

6.6 CONCLUSION

The newest market segment to gain from the semiconductor industry's quest for lower power, smaller size, more complicated functionality, and longer battery life is the medical profession. Rapid advancements in wireless sensors that continuously monitor variables such as blood pressure, heart rate, and blood sugar have the potential to significantly enhance healthcare and transform medicine.

For successful therapy to begin, medical issues must be accurately and promptly identified. Technology for low-power implanted medical devices will make it possible for everyone to adopt accurate, real-time patient monitoring and targeted treatment administration.

REFERENCES

[1] BoopathiRaja, G., & Madheswaran, M. (2013). Design and performance comparison of 6-T SRAM cell in 32nm CMOS, FinFET and CNTFET technologies. *International Journal of Computer Applications*, 70(21), 1–6.
[2] Raja, G. B., & Madheswaran, M. (2016). Performance comparison of GNRFET based 6T SRAM cell with CMOS FINFET and CNTFET technology. *International Journal of Innovative Research in Science and Engineering*, 2(5), 197–204.
[3] Raja, G. B., & Madheswaran, M. (2013). Design and analysis of 5-T SRAM cell in 32nm CMOS and CNTFET technologies. *International Journal of Electronics and Electrical Engineering*, 1(4), 256–261.
[4] Raja, G. B., & Madheswaran, D. M. (2013). Design of improved majority logic fault detector/corrector based on efficient ldpc codes. Published in ijareeie in vol. 2.
[5] Raja, G. B. (2023). *Performance Review of Static Memory Cells Based on CMOS, FinFET, CNTFET and GNRFET Design. Nanoscale Semiconductors*. Boca Raton: CRC Press, pp. 123–140.
[6] Raja, G. B. (2022). Impact of nanoelectronics in the semiconductor field: Past, present and future. In Birla, S., Singh, N., & Kumar Shukla, N. (Eds.). *Nanotechnology: Device Design and Applications Nanotechnology* (1st ed., pp. 75–91). Boca Raton: CRC Press.
[7] Raja, G. B., & Madheswaran, M. (2013). Logic fault detection and correction in SRAM based memory applications. In *2013 International Conference on Communication and Signal Processing* (pp. 215–220). Melmaruvathur, India: IEEE. doi: 10.1109/iccsp.2013.6577046

[8] Vittoz, E. (2008). The electronic watch and low power circuits. *IEEE Solid-State Circuits Society Newsletter 13*(3), 7–23.

[9] Weste, N. H., & Eshraghian, K. (1985). *Principles of CMOS VLSI Design: A Systems Perspective.* Boston: Addison-Wesley Longman Publishing Co., Inc.

[10] MIT News, Medical devices powered by the ear itself, November 7, 2012.

[11] Aguirre, P. & Silveria, F. (2012). *Optimizing Analog IC Design When Every Nano Amp Counts.*

[12] Raja, T., Agrawal, V. D., & Bushnell, M. L. (2004). A tutorial on the emerging nanotechnology devices. In *17th International Conference on VLSI Design. Proceedings* (pp. 343–360). Mumbai, India: IEEE. doi: 10.1109/ICVD.2004.1260946

[13] Asai, S. & Wada, Y. (1997). Technology challenges for integration near and below 0.1/spl mu/m. *Proceedings of the IEEE, 85*(4), 505–520.

[14] Nagata, M. (1992). Limitations, innovations, and challenges of circuits and devices into a half micrometer and beyond. *IEICE Transactions on Electronics, 75*(4), 363–370.

[15] Taur, Y., Buchanan, D. A., Chen, W., Frank, D. J., Ismail, K. E., Lo, S. H., ... Wong, H. S. (1997). CMOS scaling into the nanometer regime. *Proceedings of the IEEE, 85*(4), 486–504.

[16] Colinge, J. P. (1998). Silicon-on-insulator technology: Past achievements and future prospects. *MRS Bulletin, 23*(12), 16–19.

[17] Cressler, J. D., & Niu, G. (2003). *Silicon-Germanium Heterojunction Bipolar Transistors.* Norwood, MA: Wrtech House.

[18] Frensley, W. R. (1987). Gallium arsenide transistors. *Scientific American, 257*(2), 80–87.

[19] Ahmad, S. (2002). Special issue on nanotechnology-Guest editorial. *IETE Technical Review, 19*(5), 234–236.

[20] Dresselhaus, M. S., Dresselhaus, G., & Eklund, P. C. (1996). *Science of Fullerenes and Carbon Nanotubes: Their Properties and Applications.* USA: Elsevier Science.

[21] Ebbesen, T. W. (1996). *Carbon Nanotubes: Preparation and Properties.* Boca Raton: CRC Press.

[22] Kanwal, A. (2003). *A Review of Carbon Nanotube Field Effect Transistors.* University Samuel Ginn College of Engineering-in Alabama. https://www. eng. auburn. edu/~ vagrawal/TALKS/nanotube_v3, 1

[23] Negri, V., Pacheco-Torres, J., Calle, D., & López-Larrubia, P. (2020). Carbon nanotubes in biomedicine. *Surface-Modified Nanobiomaterials for Electrochemical and Biomedicine Applications, 378*, 15. https://doi.org/10.1007/s41061-019-0278-8.

[24] Kausar, A. (2018). Advances in polymer/graphene nanocomposite for biosensor application. *NanoWorld Journal, 4*(2), 23–28.

[25] Avouris, P. (2003). *IBM Research: Building Carbon Nanotube Transistors.* https://domino.watson.ibm.com/Comm/bios.nsf/pages/transistors.html.

[26] Sandhie, Z. T., Ahmed, F. U., & Chowdhury, M. (2020). GNRFET based ternary logic-prospects and potential implementation. In *2020 IEEE 11th Latin American Symposium on Circuits & Systems (LASCAS)* (pp. 1–4). San Jose, Costa Rica: IEEE. doi: 10.1109/LASCAS45839.2020.9069028

Chapter 7

A review of the current research on graphene and its promising future

Sujit Kumar
Dayananda Sagar College of Engineering

H.K. Yashaswini, V. Trupti,
Hannah Jessie Rani, and Raghu N
Jain (Deemed-To-Be-University)

7.1 INTRODUCTION

Sir Benjamin Collins Brodie described the thermally shortened graphite oxide layer in 1859. Graphene was not discovered until 1916, and its existence was not even considered till 1947. Later, P. R. Wallace pondered the possible presence of graphene. In 1948, Ruess and Vogt showed the first carefully measured view of graphene from a single layer using electron microscopy [1]. In the same year, the quest to "isolate graphene" started, and in 2004, Geim and Constantine "twitched" graphene layers from graphite, an achievement that would go down in history and earn them the 2010 Nobel Prize [1].

Hanns-Peter Boehm coined the term graphene by combining the words "graphite" with the suffix '-ene'. It is a 2-D hexagonal lattice carbon with a single layer of atoms. Graphene is distinguished by a distinct set of features that distinguishes it from other carbon allotropes. In terms of strength-to-weight ratio, it is around one hundred times as strong as the strongest steel. Materials science and condensed-matter physics are rapidly advancing in this area.

2-D material possesses outstanding crystal and electrical characteristics, yet it is very rare. Graphene's unique electronic spectrum has enabled a new paradigm in condensed-matter physics that allows testing to reveal graphene's spectacles, which are undetectable in high-energy physics [2].

7.2 PRESENT AND APPROACHES

Graphene's atomic thinness and extreme hardness, along with its extreme lightness and pliability, make it a revolutionary material. Graphene can absorb up to 3 wt.% hydrogen due to its high surface area, strong gas adsorption capabilities, and atomic structure.

DOI: 10.1201/9781003459231-7

High electrical and thermal conductivity makes graphene development molecular-proof. Silicon can't equal electron speed. Potential applications in materials science are highlighted by their transparency, and electrical and optical conductivity [3]. Layer 0.34 graphite diffraction signal is shown as a solid graphene sheet. In theory, this might be used for multi-walled carbon nanotubes as well. We examine graphene melt problems using a transmission electron microscope and two-dimensional dendric crystallization. It is the most expensive material on Earth, too. The following are several ways for processing graphene [4].

7.2.1 Graphene extracted from rice husks

Graphene is extracted from rice husks using a calcination and chemical activation method. Crystalline nanoscale graphene and individual sheets of graphene, frequently with one, two, or more layers, make up the bulk of the sample. Monolayers of graphene with topological defects within and on the edges of the hexagonal lattice are the most common source of the grooved kind. Both forms of graphene have atomically flat surfaces and edges, and a microscope view of graphene is presented in Figure 7.1.

7.2.2 Growth of epitaxial graphene

Epitaxial growth, whereby a graphene seed grows with respect to the atomic structure, is used in this method. Single-layer crystal growth is always of interest when a considerable amount of film materials must be grown. Homoepitaxy is epitaxial growth on the same substrate. The graphene

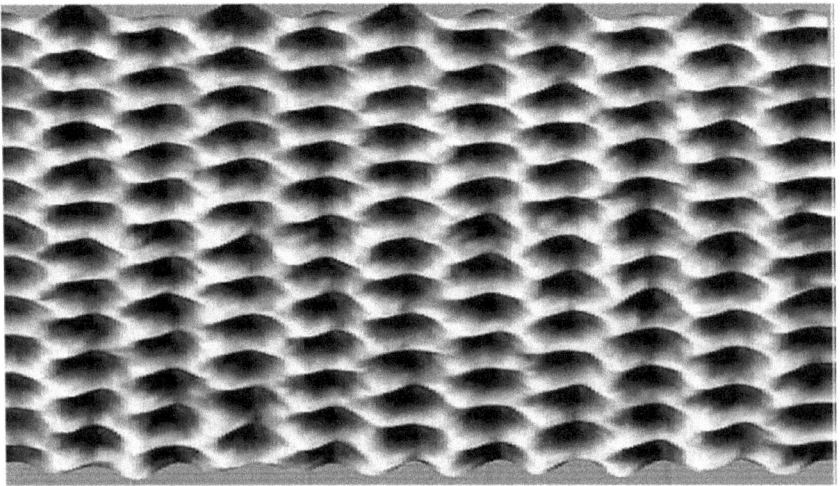

Figure 7.1 Graphene microscopy [5].

substrate may change the sample's structured thickness layer, making it hard to make any inference about its genuine qualities [5].

7.2.3 Saving on silicon carbide

Silicon carbide is converted to graphene by heating it to temperatures over 1100° at low pressures. The size and shape of the wafer play a role in the epitaxial growth process. Polarity in silicon or carbon is utilized to control film thickness, mobility, and carrier density. It is mostly concerned with electronic applications. Silicon carbide is used to make graphene. This approach can only create a tiny structure of graphene.

7.2.4 An ethanol solution containing sodium and reduced hydrazine

Density functional theory was used to study hydrazine and heat-treated graphene oxide deoxygenation processes. Researchers reduced graphene oxide paper to a single layer by immersing it in pure hydrazine [6]. In a recent publication, researchers explained the steps necessary to reduce ethanol using sodium metal, then pyrolyze the resulting ethoxide to produce graphene in very small quantities. Rinsing the sodium salts out of the body is another reason [7].

7.2.5 Deposition of chemical vapors and colorant production

Sheets of graphene larger than $1\,cm^2$ may be made by CVD on thin nickel layers, a technique that yields outstanding quality. This sheeting can be applied to a variety of substrates with success, suggesting a wide range of potential electronic applications [8]. AM is a kind of additive manufacturing that also goes by the names fast prototyping and direct digital manufacturing (DDM). Material is deposited in successive layers using precise geometric shapes designed in computer-aided design software. Objects in three dimensions are made by successively adding increasingly thinner and more precise layers of material. The neighboring layer bonds are linked together by melting completely or partly. Metal powder, thermoplastics, ceramics, composites, and glass can all be utilized to layer the material in the same or different ways.

7.2.6 Adding graphene with lasers to colorant industrial components

Fast prototyping using laser-induced graphene (LIG), a conductive layer produced using CO_2 laser direct scanning on ULTEM. Graphene, a derivative of carbon, was used to create this conductive layer because of its porous nature. Increased laser boundaries wrote an interdigital electrode capacitor on AM components. Electrochemistry favors LIG capacitors. This allows electronic printing to be integrated with 3-D printing technology and explores material-deposition-system and fabrication compatibility utilizing multiple printing methods [9].

7.2.7 Low graphene loading and nanocomposites

Rapid prototyping and nanotechnology may enable new factories to make 3-D twisted materials with improved characteristics and diverse applications. A novel mixing technique combines low graphene oxide (GO) loadings with hydroxyl and carboxyl surface functional groups to produce thermoplastic materials additive manufacturing-friendly. Interfacial adhesion is improved when graphene nanosheets are dispersed in a thermoplastic solution, which is what GO does. Nanocomposites for static and dynamic loading need just 0.06% by weight of GO to improve thermoplastic strength, toughness, stiffness, and strain to failure. GO reduces phase separation and graphene clumping in ABS resin. GO is also used as a filler material. Acetone evaporation can be used to lower strain-to-failure, increase fracture strength and stiffness, and decrease toughness.

7.2.8 Translucent and supple conductive films made from graphene for screens and electrodes

Transfer printing and solution-based technologies enabled large-area graphene circuits. Chhowalla proposed a reliable technique for producing reduced graphene oxide (RGO) thin films [9]. Thus, optoelectronic qualities may be adjusted over numerous orders of magnitude, benefiting transparent and flexible semiconductors or semi-metals. Thinner films behave like ambipolar transistors like graphene, whereas thicker films behave like semi-metals like graphite. The suggested deposition process transformed graphene's fundamental features into technological gadgets. Hong et al. [10] designed large-scale transparent electrodes. Two approaches were employed to generate patterns in nickel films and transport them to other substrates using CVD. Transferred graphene sheets were transparent and had low-sheet resistance.

Low-temperature graphene monolayers on SiO_2 substrates displayed half-integer quantum Hall effect. Thus, CVD graphene resembles hand-cut graphene. Using graphene's excellent mechanical characteristics, the scientists demonstrated how flexible, stretchy, and foldable electronics may employ highly conductive and transparent electrodes. Colombo et al. employed copper foils to manufacture homogeneous, high-quality large-area graphene sheets [11]. CVD and methane were utilized to generate centimeter-thick graphene sheets on copper substrates with a substantial surface area.

Only 5% of the graphene layers in the films were multilayers. Graphene layers crossed copper surface steps and grain boundaries. Because carbon was hard to dissolve in copper, this process of growth seemed to stop on its own. The authors also came up with a way to transfer graphene films to any surface. The silicon–silicon dioxide dual-gated field-effect transistor demonstrated electrons could travel $4,050 \, cm^2$ per volt per second at ambient temperature.

Photonics and optoelectronics might use graphene's mobility, optical transparency, flexibility, strength, and environmental resilience. Ferrari et al. [12] described graphene photonics and optoelectronics. In this review, graphene-based and GO-based transparent conducting films were used to make many photonic and optoelectronic devices and equipment, including window electrodes of inorganic, organic, and dye-sensitized solar cells, organic light-emitting diodes, light-emitting electrochemical cells, touch screens, and flexible smds.

Hong et al. [13] made heaters that work well, are flexible and clear, and are made from large-scale graphene films. CVD on Cu foils yielded low-sheet resistance graphene sheets with 89% optical transmittance. Low-voltage transparent heaters benefit from transparent, low-sheet resistance graphene sheets. Graphene-based heaters outperformed transparent indium tin oxide heaters in heat dispersion and temperature profiles (ITO). Transparent, flexible graphene-based heaters are likely to be utilized widely, especially in smart windows that can be heated.

In [14], De and Coleman calculated the DC/op ratio, DC/op=0.7, 4.5, and 11. Transparent conducting graphene film figures of excellence. These parameters limited graphene flakes, stacks, and films. Transparent electrodes needed higher graphene flake networks. Graphite has excessively low resistance for current-driven transparent conductors with a conductivity ratio of 11. Substrate-induced doping may increase 2D DC conductivity enough to make graphene transparent. Choi and Hong [15] showed that large-scale CVD graphene sheets on Cu foils may be used to make high-performance, flexible, and transparent heaters. After repeated transfers and chemical doping procedures, graphene sheets had 43/sq sheet resistance and 89% transparency, making them perfect for low-voltage transparent heaters. Graphene-based heaters outperformed ITO-based transparent heaters in heat dispersion and temperature profiles. Also, the authors confirmed that mechanical strains of up to 4% did not have a big effect on how well the heater worked. Thus, flexible, transparent graphene-based heaters might be employed in automotive defogging/deicing systems and smart windows. Kim et al. [16] created high-performance organic field-effect transistor graphene electrodes.

The authors regulated graphene electrode's work function by functionalizing SiO_2 substrates to improve these devices (SAMs). NH_2-terminated SAMs are substantially n-doped graphene, whereas CH_3-terminated SAMs mitigated SiO_2-induced p-doping. This altered graphene electrode's work functions. This research can be used for numerous graphene-based electrical and optoelectronic devices. Lee et al. constructed flexible organic light-emitting diodes using graphene electrodes with high work functions and low-sheet resistances [17]. This allowed high-luminous efficiency LEDs.

P(VPF-TrFE) as a doping layer between graphene layers may considerably reduce sheet resistance. Ahn et al. [18] produced a flexible, clear acoustic actuator and nanogenerator utilizing a multilayer graphene/P(VPF-TrFE)/graphene sheet. The acoustic actuator worked at many frequencies. ZnO and PZT nanogenerators produced comparable voltage and current. The authors dynamically rolled graphene/P(VDF-TrFE)/graphene multilayer devices.

Cho et al. created high-performance graphene sheets quickly using hydrogen-free rapid thermal chemical vapor deposition (RT-CVD) and roll-to-roll etching [19]. This method could be used to make graphene films in large quantities that meet industrial standards for size, uniformity, and reliability. The methods for transferring graphene films were also worked out. Transmission electron microscopy, Raman spectroscopy, chemical grain boundary investigations, and electrical device testing have characterized RT-CVD graphene's remarkable uniformity and stability. Modern smartphones use capacitive multi-touch devices with RT-CVD films. In photonics and optoelectronics, graphene sheets have superior mechanical and optical properties. They conduct less than transparent ITO electrodes. Graphene translucent thin-film devices are less effective. Ahn et al. [20] electrostatically doped graphene sheets with ferroelectric polymers to solve this issue. These graphene sheets' ferroelectric polarization made thin organic solar panels (OSCs). Chemically doped graphene-based OSCs were less stable and less efficient. OSCs using a few micrometers of ultrathin ferroelectric film as a substrate were bendable and durable. Roll them into a 7-mm cylinder.

Gradeak et al. [21] found another technique to develop flexible graphene-based OSCs. These scientists found that heating the MoO3 layer that blocks electrons and placing the ZnO layer that transports electrons directly on graphene increases efficiency. Flexible graphene-polyethylene naphthalate OSCs were also shown. Flexible OSCs with graphene anodes and cathodes converted electricity at record-high efficiencies of 6.1% and 7.1%, respectively. This innovation enabled flexible OSCs with graphene electrodes to be produced easily and repeatedly. Ahn and Hong [22] found graphene promising for transparent, flexible electrodes in a brief overview of the tests above. Graphene electrodes operate in numerous photonic and optoelectronic devices. Ahn and Hong predicted graphene will first be used in flat, simple structures such as touch displays, smart windows, electromagnetic interference shields, lighting, and transparent heaters. Flexible displays and microelectronic devices would come a few years later.

7.2.9 Conductive inks for electronics that can be printed

Noh et al. [23] pointed out that printed electronics has been promising to change the electronics industry for at least 10 years by making it possible to make very large quantities of low-cost electronic circuits and sensors. They also called for high-mobility printed semiconductors, low-sintering temperature conducting inks, and higher resolution and uniformity printing methods.

Jang inkjet printer patterned graphene sheets, which have high line resolution and conductivity [24]. Control sheet resistance with GO ink and print layers. Later, Ferrari with his team [25] showed that inkjet printing is a good way to make graphene devices over a large area. The graphene-based ink was made by exfoliating graphite in N-methyl pyrrolidone while it was in a liquid phase. The ink formed thin-film transistors and transparent and conductive patterns with 80% transmittance and 30 k/cm^2 sheet resistance. This result made it possible for graphene devices to be printed, flexible, and see-through on any substrate.

Inkjet printing may produce various high-resolution graphene designs, as pointed out by Ostling et al. [26]. Most electronic devices with good performance could be made with a few printing passes and a simple baking step. Inkjet printing conductive inks may be created from graphene, according to Torrisi and Coleman [27]. First, liquid phase exfoliation in water and organic solvents may produce several pure graphene nanosheets, which are hundreds of nanometers broad and less than one nanometer thick. This stable ink can be made in normal conditions, is easy to make the same from batch to batch, and has good rheological properties for printing and coating.

In the experiments of Casiraghi et al. [28], nanosheet-based inks were used to make some heterostructures. The scientists noted that stacking 2D materials may provide the material with novel electrical and optical features. Graphene, hBN, and tungsten disulfide are 2D materials (MoS2). These heterostructures may be produced from chemically peeled 2D crystals, the scientists revealed. Devices may be made cheap and scalable.

Hersam et al. [29] have shown how graphene gravure printing can swiftly create conductive patterns on flexible surfaces, which is important for making large-area flexible electronics. The authors chose the right inks and set the printing parameters so that patterns could be made with a resolution of 30 m or less. A step of mild annealing made conductive lines that worked well and were uniform. This is a good way to use graphene in printed and flexible electronics with a large surface area.

Coleman and his colleagues successfully printed liquid-exfoliated graphene and MoS2 nanosheets using an inkjet printer [30]. Nanosheets with precisely determined diameters and concentrations of up to 6 mg/mL/L may be used in the production of inks. At a temperature of 70°C, graphene traces were printed without the need for any chemicals or heating. Percolation effects were seen in thinner traces, whereas traces with a thickness of more than 160 nm, independent of thickness, had a conductivity of 3,000 S/mL. The authors also showed how to use solvent-exfoliated, size-selected MoS2 nanosheets to print semiconducting traces. Graphene interdigitated array electrodes, and these kinds of traces can be used to make photodetectors that are all printed.

Magdasi and Kamyshny [31] reviewed conductive nanomaterials and printed electronics research. The scientists employed nanoparticle metal, carbon nanotube, and graphene inks for inkjet printing. The review included the fundamental features of conductive nanomaterials used in printed electronics, how to maintain them stable in dispersions, how to generate conductive inks, and how to get conductive patterns by various sintering procedures. Conductive nanomaterials employed in electrical devices such as transparent electrodes, solar cell metallization, RFID antennae, light-emitting devices, etc. were briefly examined.

Hersam and colleagues showed how intense pulsed light annealing of graphene inks can quickly process inkjet-printed graphics on various substrates [32]. And 20 mg/mL/L graphene ink produced 25,000 S/mL in one printing run. This shows that using graphene in flexible and high-performance printed and flexible electronics is feasible and effective.

In another recent research [33], Park et al. employed electrodynamic technology to directly print RGO on flat or extremely curved surfaces with good resolution. The authors demonstrated complicated geometric designs using high-resolution electrodynamic inkjet printing of RGO. Flat and highly curved substrates may be used (with a radius of curvature of less than 60 mm). RGO patterns and all-printed RGO transistors identified fraudulent coins. This has great security and electronics possibilities.

7.2.10 Biomedical engineering

7.2.10.1 Vector-based gene transfer

Graphene's use in medication delivery, gene delivery, and protein delivery has been foreseen. It has been shown that the cationic polymer polyethyleneimine (PEI) may serve as a functional and procedural scaffold for the use of graphene in gene transfer. PEI has been examined as a potential non-viral gene vector due to strong electrostatic connection that it shares with the negatively charged phosphates found in DNA and RNA. The effects of chemical modifications on transfection efficiency, cell selectivity, and cytotoxicity are also investigated. The limited scope of PEI's biological uses may be attributed to the material's poor biocompatibility and high toxicity. GO may have several uses in the field of biomedical engineering. Some of these uses include molecular imaging, biological imaging, cancer treatment, and drug/gene delivery [34].

7.2.10.2 Bioengineering of biological tissues

Because of its structure and features, nanomaterials based on graphene are developing into "2-D unique materials" [35]. Its chemical analogues serve as a paradigm for studying quantum phenomena and represent a groundbreaking new class of nanomaterials [36]. Regenerative medicine is an interdisciplinary subject that focuses on the creation of biological replacements for damaged or diseased tissues or organs. Wound repair, regenerative medicine, tissue engineering, and stem cell research have all recently benefited from graphene-based materials. Being strong, elastic, and adaptable on flat surfaces, it may also affect an extensive variety of functions. Bimolecular sensors, the learning of cellular twitching and signaling, and electrospun, biodegradable films and hydrogels are all possible applications [37–45].

7.2.10.3 Scientific method in bioimaging

The use of superparamagnetic iron oxide nanoparticles (IONP) for targeted medication administration and bio-imaging presentations is commonplace. Ocular absorption from the near-infrared to the visible range is guided by a magnetic field for confined photothermal elimination of cancer cells. Graphene nanocomposites that are biocompatible and have several functions hold considerable promise for use in cancer diagnosis and treatment [46].

7.2.11 Microelectronics

7.2.11.1 Metal-air batteries based on graphene

Metal serves as anode, oxygen-rich air serves as cathode, and aqueous solution serves as an electrolyte in metal–air batteries. Because oxygen is required in the cathode, the cathode material in batteries uses a graphene rod because it is porous and allows air to flow through it. Not having to worry about running out of juice first and foremost would be a huge benefit of such a battery. Metal–air batteries, which are an alternative to lithium–ion batteries used in most EVs (and other devices), may be refueled with only water [47]. Graphene nanosheets (GNs) that have been heat-treated to remove metals still reduce air oxygen and have better cycling performance than GNSs in rechargeable Li–air battery air cathodes. Cost-effective, long-lasting, and highly-active electrocatalysts are in high demand across several energy-related industries [48].

7.2.11.2 Luminescent diodes made from graphene

The heat generated by LED lighting may be dissipated using graphene. LED filament lights with a graphene layer are called graphene LED bulbs. Graphene aids in the dissipation of heat from LED bulbs, allowing for more luminosity at lower wattages. They will be more effective than their conventional equivalents, using less energy and so costing less to operate. The lifespan of these bulbs is so long that it will be extended by 10%. First, a flexible substrate, such as a flexible printed circuit board (PCB), is needed for making LED light bulbs with graphene filaments. After applying graphene-based heat dissipation ink to the substrate's backside, the PCB is cut into multiple graphene filaments. The flexible substrate has LED chips on the front and copper wires on both sides for electrical circuits and heat conduction. The flexible substrates back receive graphene-based heat dissipation ink before or after molding the LED chips/phosphor. Furthermore, the graphene filaments are permanently arranged in an arch [49].

7.2.11.3 Antennas made of graphene

Graphene's potential as an electromagnetic passive and antenna holds great promise for enabling further downsizing [50]. There has been a lot of progress in antenna applications [51], thanks to the EM characteristics at different frequency bands. Sandwiching nanoscale antennas between two monolayers results in a light detector that does the same with electromagnetic radiation [52]. By incorporating graphene inlays onto antennas, a data transfer rate of one terabit per second is achievable. As a result, information may be sent instantly between close-by gadgets, and it is limited to transmissions within a radius of 1 m.

7.2.11.4 Technologies integrated into clothing

Recent advances in sensor technology have led to novel applications for graphene-woven fabrics (GWFs). With the GWFs adhered on a composite sheet of polymer and medical tape, we have a wearable, flexible strain sensor. The evolution of wearable electronics relies on the improvement of input device technologies in a flexible and conformal form factor [53]. Currently, the most popular applications for flexible and wearable electronics are in the fields of healthcare and monitoring. Supercapacitors and sensors made from graphene are making significant strides in the field of wearable electronics.

7.2.11.5 Superior capacitors based on graphene

Wearable electronic sensors use 2D graphene nanoparticles due to their inherent flexibility. A graphene-based superconductor made of 2D graphene nanosheets (2D GNs) and highly conductive polyoxometalate measures pulse rate (POM). POM/2D GNs nanocomposites stabilize cycling by evenly covering gene nanoparticles with POM. Strengthening polyoxymethylene (two-dimensional) grafted nanostructures (GNs) over a conductive adhesive substrate reactive to external pulse beats creates these wearable electronic sensors. This makes it suitable for use in health-tracking technology for humans [54].

7.2.11.6 Sensor based on graphene fabric

The greater electrochemical property of materials like nanocomposites has prompted a large reaction in their use as strain sensors [55]. Conducted cotton fabric was created using the synthesis and deposition of GO [56]. It was previously suggested [57] that flexible touch sensors be manufactured and used across both the mechanical and electrical domains. For medical monitoring, flexible electronic clothing and skin-adhering gadgets have smart fabric sensors that are both portable and rubbery in texture. Its powerful skin-adhering feature allows for effective observation of an extensive variety of human activities in both parched and damp environments. Graphene-coated fabric (GCF) sensors include octopus life patterns that respond to stress and pressure. Helpful for keeping tabs on things like heart rates, vocal vibrations, and motion detection [58].

7.2.12 Manufactured printing methods and equipment

7.2.12.1 Functional inks based on graphene

Possibilities for the future are bright when graphene-based ink is combined with supercapacitors [59]. Lithium-Ion batteries may be printed using a solid-state gel polymer electrolyte and a composite functional ink based on GO [60]. Graphene patterns with no coffee rings have been successfully printed using the inkjet printing technique [61]. Graphene has been demonstrated to have excellent thermal stability and corrosion resistance, making it an attractive contender for use in electronics that must perform under high temperatures. Graphene provides several benefits over conventional functional inks, including being non-toxic, environmentally friendly, quick-drying, recyclable, and affordable. Superior chemical stability and inertness of graphene inks provide corrosion protection for a wide variety of substrates.

7.2.13 Engineered textiles

7.2.13.1 Fabric made from graphene

GOs have reactive dye-like hydroxyl and carbonyl groups readily accessible [62]. As a result of their porous nature, graphene sheets with intricate

geometries have exceptional electrochemical characteristics [63]. Graphene
sheets grew into a three-dimensional network to boost polyaniline's conductivity [64]. Graphene-based materials are being studied to provide
wearable electronics with advanced sensing, temperature control, chemical, mechanical, and radiation protection, and energy storage capabilities.
Our hypothesis was that graphene sheets would have an extra, unforeseen
use, such as repelling mosquitoes. Recent studies have shown that graphene
oxide nanosheet coatings in a dry condition significantly reduce skin bit-
ing. The primary reason is that the mosquitoes' chemosensory receptors
are being disrupted by the graphene coatings, so they cannot detect the
chemical attractants on the skin. Because of its excellent puncture resistance, a graphene sheet may withstand being submerged in water or human
sweat without being damaged. Mosquitoes are unable to detect blood or
bite because multilayer graphene, an ultrathin but robust substance, may
block chemical impulses [65].

7.2.14 Space

Spray-applied metallic plasma coatings are ideal for use in the aircraft sector for repairing engine components [66]. Carbon materials, like any other
aerospace material, are employed for extended periods and systems. High
transition temperature, thermal stability, chemical resistance, etc. are desir-
able [67]. Since it is extremely conductive at ambient temperature, multilayered graphene may be utilized to produce ultrathin microwave coatings
for aeronautical applications. Catalytic chemical vapor deposition is used to
create a nanometrically thin Cu catalyst sheet that is sandwiched between
a dielectric (SiO_2) substrate and the Ka-band-stacked graphene. Since the
manufactured graphene absorbs about 35%–43% of the incoming power, it
is very effective at blocking electromagnetic (EM) radiation. The absorption
of electromagnetic signals by this multilayered graphene is much smaller
than the surface depth of typical metals, resulting in a relatively strong electromagnetic shrinkage property, as shown in the Ka-band [68].

7.2.15 Encounters

Processing graphene for integration with other goods is the industry's
biggest issue. Classifying various forms of graphene may aid in the commercialization process. Future price reductions are possible if commercial
production processes are established. New research and development efforts
are underway to maximize this material's potential applications throughout industries, from electronics to biology [69]. Intriguing and promising
of showcasing improved technologies, the obstacles are [70]. Graphene's
chemical production may benefit greatly from wet chemical modification
[71]. Commercialization of manufacturing techniques will be affected by

the most important elements [72]. There will be a significant increase in the use of enhanced electrodes in electrochemical energy storage systems in the future years [73–77].

7.2.16 Consensus and potential next steps

Graphene has several interesting uses and is already being put to work in many significant fields. Almost 50 businesses are actively marketing their use of graphene. Graphene's commercial potential has increased in recent years. Graphene's lack of a bandgap prevents it from being used as an on/off switch, even though it is an excellent conductor. The problem is more severe for nanoelectric devices [78]. Recently, there have also been some attempts made to address this limitation. Growing graphene on a Ni thin film has been shown to provide high-quality graphene, an area where studies are currently being conducted. Graphene might potentially replace silicon in electrical circuits, although this will need study and development in the future. Graphene's potential uses in energy storage devices, biomedicine, membranes, and composition structure are still being investigated. Graphene has the potential to replace silicon in a variety of electrical applications [79]. This means that in the not-too-distant future, it will be easy to find gadgets that are fast, responsive, adaptable to different situations and have the features you need [80].

Another promising area for graphene's future use is as an intermediate material in manufacturing, which promises to open a massive market and boost demand. Graphene and graphene-based nanostructures have several uses beyond those already mentioned, including anti-corrosion technologies [81], nanocellulose, GO-based lightweight anisotropic foams for thermal insulation and fire protection [82], aeronautical applications of graphene [83], etc. As anticipated by Novoselov and Geim, the subject of graphene science and technology has been shown to be a "cornucopia of possible applications." [84].

REFERENCES

[1] Andre K. Geim, *Phys. Scr.* T146 (2012) 014003.
[2] Andre K. Geim, Philip Kim, *Sci. Am.* (2017) Retrieved August 25, 2017.
[3] Kuen Soo Kim, Yue Zhao, Houk Jang, Sang Yoon Lee, Jong Min Kim, Kwang S. Kim, Jong-Hyun Ahn, Philip Kim, Jae-Young Choi, Byung Hee Hong, *Nature* 457 (2009) 706–710, Retrieved August 25, 2017.
[4] Bor Jang. U.S. Patent Application 11/442,903, filed September 28, 2006.
[5] K.S. Novoselov et al., *PNAS* 102 (2005) 10451–10453.
[6] Laura Mgrdichian. (2008) PhysOrg.com. Retrieved August 25, 2017.
[7] Researchers Discover Method for Mass Production of Nanomaterial Grapheme. (2008) PhysOrg.com. Retrieved August 25, 2017.

[8] Mohammad Choucair, *Nat. Nanotechnol.* 4 (2008) 30–33.

[9] Lishi Jiao, Zhong Yang Chua, Seung Ki Moon, Jie Song, Guijun Bi, Hongyu Zheng, Byunghoon Lee Jamyeong Koo, *Nanomaterials* 9 (1) (2019) 90.

[10] Brennan E. Yamamoto, A. Zachary Trimble, Brenden Minei, Mehrdad N. Ghasemi Nejhad, *J. Thermoplast. Compos. Mater.* 32 (3) (2019) 383–408.

[11] Sumit Goenka, Vinayak Sant, Shilpa Sant, *J. Control. Release* 173 (2014) 75–88.

[12] Shin, Su Ryon, Yi-Chen Li, Hae Lin Jang, Parastoo Khoshakhlagh, Mohsen Akbari, Yu. Amir Nasajpour, Shrike Zhang, Ali Tamayol, Ali Khademhosseini, *Adv. Drug Deliv. Rev.* 105 (2016) 255–274.

[13] Roberto Gonzalez-Rodriguez, Elizabeth Campbell, Anton Naumov, *PLoS One* 14 (6) (2019) e0217072.

[14] Eunjoo Yoo, Haoshen Zhou, *ACS Nano* 5 (4) (2011) 3020–3026.

[15] Min Zeng, Yiling Liu, Feipeng Zhao, Kaiqi Nie, Na. Han, Xinxia Wang, Wenjing Huang, Xuening Song, Jun Zhong, Yanguang Li, *Adv. Funct. Mater.* 26 (24) (2016) 4397–4404.

[16] Chung-Ping Lai, U.S. Patent 9,933,121, issued April 3, 2018.

[17] Julien Perruisseau-Carrier, In: *2012 Loughborough Antennas & Propagation Conference (LAPC)*, pp. 1–4. Loughborough, UK: IEEE, 2012.

[18] Juan Sebastian Gomez-Diaz, In: *2012 International Symposium on Antennas and Propagation (ISAP)*, pp. 239–242. Nagoya Congress Center in Nagoya, Japan: IEEE, 2012.

[19] Zheyu Fang, Zheng Liu, Yumin Wang, Pulickel M. Ajayan, Peter Nordlander, Naomi J. Halas, *Nano Lett.* 12 (7) (2012) 3808–3813.

[20] Yan Wang, Li Wang, Tingting Yang, Xiao Li, Xiaobei Zang, Miao Zhu, Kunlin Wang, Wu. Dehai, Hongwei Zhu, *Adv. Funct. Mater.* 24 (29) (2014) 4666–4670.

[21] Minpyo Kang, Jejung Kim, Bongkyun Jang, Youngcheol Chae, Jae-Hyun Kim, Jong-Hyun Ahn, *ACS Nano* 11 (8) (2017) 7950–7957.

[22] Taiping Xie, Li Zhang, Yuan Wang, Yajing Wang, Xinxing Wang, *Ceram. Int.* 45 (2) (2019) 2516–2520.

[23] Wen Long Gu, Yong Nan Zhao, In: *Advanced Materials Research*, vol. 331, pp. 93–96, Switzerland: Trans Tech Publications, 2011.

[24] Seyed Hamed Aboutalebi, Rouhollah Jalili, Dorna Esrafilzadeh, Maryam Salari, Zahra Gholamvand, Sima Aminorroaya Yamini, Konstantin Konstantinov, et al., *ACS Nano* 8 (3) (2014) 2456–2466.

[25] Fu. Shao, Shao-Wei Bian, Quan Zhu, Mei-Xia Guo, Si Liu, Yi-Hang Peng, *Chem. An Asian J.* 11 (13) (2016) 1906–1912.

[26] Sungwoo Chun, Wonkyeong Son, Da Wan Kim, Jihyun Lee, Hyeongho Min, Hachul Jung, Dahye Kwon, et al., *ACS Appl. Mater. Interfaces* 11 (18) (2019) 16951–16957.

[27] Yanfei Xu, Matthias Georg Schwab, Andrew James Strudwick, Ingolf Hennig, Xinliang Feng, Zhongshuai Wu, Klaus Müllen, *Adv. Energy Mater.* 3 (8) (2013) 1035–1040.

[28] Kun Fu, Yibo Wang, Chaoyi Yan, Yonggang Yao, Yanan Chen, Jiaqi Dai, Steven Lacey, et al., *Adv. Mater.* 28 (13) (2016) 2587–2594.

[29] Jiantong Li, Fei Ye, Sam Vaziri, Mamoun Muhammed, Max C. Lemme, Mikael Östling, *Adv. Mater.* 25 (29) (2013) 3985–3992.

[30] Xiao Li, Rujing Zhang, Yu Wenjian, Kunlin Wang, Jinquan Wei, Wu Dehai, Anyuan Cao, et al., *Sci. Rep.* 2 (2012) 870.
[31] Jiesheng Ren, Chaoxia Wang, Xuan Zhang, Tian Carey, Kunlin Chen, Yunjie Yin, Felice Torrisi, *Carbon* 111 (2017) 622–630.
[32] Xiao Lee, Tingting Yang, Xiao Li, Rujing Zhang, Miao Zhu, Hongze Zhang, Dan Xie, et al., *Appl. Phys. Lett.* 102 (16) (2013) 163117.
[33] Cintia J. Castilho, Dong Li, Muchun Liu, Yue Liu, Huajian Gao, Robert H. Hurt, *Proc. Nat. Acad. Sci.* 116 (37) (2019) 18304–18309.
[34] David Ward, Ankur Gupta, Shashank Saraf, Cheng Zhang, Tamil Selvan Sakthivel, Swetha Barkam, Arvind Agarwal Sudipta Seal, *Carbon* 105 (2016) 529–543.
[35] Ayesha Kausar, Irum Rafique, Bakhtiar Muhammad, *Polym.-Plas. Technol. Eng.* 56 (13) (2017) 1438–1456.
[36] P. Kuzhir, N. Volynets, S. Maksimenko, T. Kaplas, Yu Svirko, *J. Nanosci. Nanotechnol.* 13 (8) (2013) 5864–5867.
[37] Dimitrios Bitounis, Hanene Ali-Boucetta, Byung Hee Hong, Dal-Hee Min, Kostas Kostarelos, *Adv. Mater.* 25 (16) (2013) 2258–2268.
[38] Liang Ma, Jinlan Wang, Feng Ding, *Chem. Phys. Chem.* 14 (1) (2013) 47–54.
[39] Siegfried Eigler, Andreas Hirsch, *Angewandte Chemie Int. Ed.* 53 (30) (2014) 7720–7738.
[40] Amaia Zurutuza, Claudio Marinelli, *Nat. Nanotechnol.* 9 (10) (2014) 730.
[41] Wei Lv, Zhengjie Li, Yaqian Deng, Quan-Hong Yang, Feiyu Kang, *Energy Storage Mater.* 2 (2016) 107–138.
[42] Kostas Kostarelos, Kostya S. Novoselov, *Nat. Nanotechnol.* 9 (10) (2014) 744.
[43] D. Cohen-Tanugi, J. C. Grossman *J. Chem. Phys.* 141 (2014) 074704.
[44] M. S. H. Boutilier, C. Sun, S. C. O'Hern, H. Au, N. G. Hadjiconstantinou, R. Karnik, *ACS Nano* 8 (2014) 841.
[45] S. C. O'Hern, M. S. H. Boutilier, J.-C. Idrobo, Y. Song, J. Kong, T. Laoui, M. Atieh, R. Karnik, *Nano Lett.* 14 (2014) 1234.
[46] C. Sun, M. S. H. Boutilier, H. Au, P. Poesio, B. Bai, R. Karnik, N. G. Hadjiconstantinou, *Langmuir* 30 (2014) 675.
[47] R. R. Nair, H. A. Wu, P. N. Jayaram, I. V. Grigorieva, A. K. Geim, *Science* 335 (2012) 442.
[48] H. W. Kim et al., *Science* 342 (2013) 91.
[49 H. Li, Z. Song, X. Zhang, Y. Huang, S. Li, Y. Mao, H. J. Ploehn, Y. Bao, M. Yu, *Science* 342 (2013) 95.
[50] R. K. Joshi, P. Carbone, F. C. Wang, V. G. Kravets, Y. Su, I. V. Grigorieva, H. A. Wu, A. K. Geim, R. R. Nair, *Science* 343 (2014) 752.
[51] K. Celebi, J. Buchheim, R. M. Wyss, A. Droudian, P. Gasser, I. Shorubalko, J.-I. Kye, C. Lee, H. G. Park, *Science* 344 (2014) 289.
[52] S. Hu et al., *Nature* 516 (2014) 227.
[53] D.-Y. Koh, R. P. Lively, *Nat. Nanotechnol.* 10 (2015) 385.
[54] S P. Surwade, S. N. Smirnov, I. V. Vlassiouk, R. R. Unocic, G. M. Veith, S. Dai, S. M. Mahurin, *Nat. Nanotechnol.* 10 (2015) 459.
[55] L. Wang, L. W. Drahushuk, L. Cantley, S. P. Koenig, X. Liu, J. Pellegrino, M. S. Strano, J. S. Bunch, *Nat. Nanotechnol.* 10 (2015) 785.
[56] C. A. Merchant et al., *Nano Lett.* 10 (2010) 2915.

[57] G. F. Schneider, S. W. Kowalczyk, V. E. Calado, G. Pandroud, H. W. Zandbergen, L. M. K. Vandersypen, C. Dekker, *Nano Lett.* 10 (2010) 3163.

[58] S. Garaj, W. Hubbard, A. Reina, J. Kong, D. Branton, J. A. Golovchenko *Nature* 467 (2010) 190.

[59] S. Liu, T. H. Zeng, M. Hofmann, E. Burcombe, J. Wei, R. Jiang, J. Kong, Y. Chen, *ACS Nano* 5 (2011) 6971.

[60] O. N. Ruiz, K. A. S. Fernando, B. Wang, N. A. Brown, P. G. Luo, N. D. McNamara, M. Vangsness, Y.-P. Sun, C. E. Bunker, *ACS Nano* 5 (2011) 8100.

[61] B. M. Venkatesan, R. Bashir, *Nat. Nanotechnol.* 6 (2011) 615.

[62] J. Hong, N. J. Shah, A. C. Drake, O. C. DeMuth, J. B. Lee, J. Chen, P. T. Hammond, *ACS Nano* 6 (2012) 81.

[63] V. G. Kravets et al., *Nat. Mater.* 12 (2013) 304.

[64] K. Kostarelos, K. S. Novoselov, *Science* 344 (2014) 261.

[65] A. Servant, V. Leon, D. Jasim, L. Methven, P. Limousin, E. V. Fernandez-Pacheco, K. Prato, K. Kostarelos, *Adv. Healthc. Mater.* 3 (2014) 1334.

[66] M. Drndić, *Nat. Nanotechnol.* 9 (2014) 743.

[67] K. Kostarelos, K. S. Novoselov, *Nat. Nanotechnol.* 9 (2014) 744.

[68] D. Wang et al., *ACS Nano* 4 (2010) 1587.

[69] D. Yu, L. Dai, *J. Phys. Chem. Lett.* 1 (2010) 467.

[70] Y. Zhu et al., *Science* 332 (2011) 1537.

[71] Y. Cao, X. Li, I. A. Aksay, J. Lemmon, Z. Nie, Z. Yang, J. Liu, *Phys. Chem. Chem. Phys.* 13 (2011) 7660.

[72] H. Wang, Y. Yang, Y. Liang, J. T. Robinson, Y. Li, A. Jackson, Y. Cui, H. Dai, *Nano Lett.* 11 (2011) 2644.

[73] J. Xiao et al., *Nano Lett.* 11 (2011) 5071.

[74] L. Xiao, W. Wang, D. Choi, Z. Nie, J. Yu, L. V. Saraf, Z. Yang, J. Liu, *Adv. Mater.* 23 (2011) 3155.

[75] F. Du, D. Yu, L. Dai, S. Ganguli, V. Varshney, A. K. Roy, *Chem. Mater.* 23 (2011) 4810.

[76] H. R. Byon, S. W. Lee, S. Chen, P. T. Hammond, Y. Shao-Horn, *Carbon* 49 (2011) 457

[77] M. F. El-Kady, V. Strong, S. Dublin, R. B. Kaner, *Science* 335 (2012) 1326.

[78] A. Zhamu, G. Chen, C. Liu, D. Neff, X. Wang, B. Z. Jang, *Energy Environ. Sci.* 5 (2012) 5701.

[79] J. Luo, X. Zhao, J. Wu, H. D. Jang, H. H. Kung, J. Huang, *J. Phys. Chem. Lett.* 3 (2012) 1824.

[80] X. Yang, C. Cheng, Y. Wang, L. Qiu, D. Li, *Science* 341 (2013) 534.

[81] M. F. El-Kady, R. B. Kaner, *Nat. Commun.* 4 (2013) 1475.

[82] Z. Fan, J. Yan, G. Ning, T. Wei, L. Zhi, F. Wei, *Carbon* 60 (2013) 558.

[83] W.-W. Liu, Y.-Q. Feng, J.-T. Chen, Q.-J. Xue, X.-B. Yan, *Adv. Funct. Mater.* 23 (2013) 4111.

[84] Y. Meng, Y. Zhao, C. Hu, H. Cheng, Y. Hu, Z. Zhang, G. Shi, L. Qu, *Adv. Mater.* 25 (2013) 2326.

Chapter 8

Analysis and optimization for low-power **SRAM** cells

Deepika Sharma and Shilpi Birla

Manipal University Jaipur

8.1 INTRODUCTION

To minimize the silicon area of the devices and achieve high speed and performance, they are now being considerably shrunk down. Supply voltage (V_{DD}) and device size are the only factors that design engineers have control over, making them the most important ones. To reduce static power dissipation, the V_{DD} is usually scaled down. The threshold voltage (V_{th}) should be scaled down as well, though, to get good performance. With a drop in V_{th}, the subthreshold leakage current grows exponentially, increasing static power dissipation. Gate and subthreshold current are the two principal factors that contribute to static power dissipation [1]. The amount of leakage power depends on the contribution made by the leakage currents in each transistor [2]. Sources of leakage include subthreshold and gate leakage. Band-to-band tunneling leakage currents are so insignificantly small that they can be ignored by modern technology [3]. A transistor's gate leakage current hits subthreshold levels when the gate oxide thickness is 3 nm and lower. A transistor's gate leakage current grows exponentially as the gate oxide thickness drops. Gate oxide voltage also grows exponentially with increasing gate oxide voltage.

8.2 MEMORY ARRAY

SRAM is an essential part of many digital systems, including high-performance computers and smartphone processors. Digital logic is dominated by dynamic power, which was formerly manageable by lowering supply voltage. The V_{DD} for digital circuitry has reached about 1V [4,5]. As the SNM reduces, the transistor mismatch increases with decreasing V_{DD} [6]. Furthermore, for reliable read-and-write operation, the cell noise margin is tightly restricted. In addition, stochastic process fluctuations drastically reduce the noise margin when the device size is scaled. Due to the sizing of SRAM cells on a nanometer scale, the changes in electrical parameters, such as V_{th} and sheet resistance, are steadily reduced by modifications in

DOI: 10.1201/9781003459231-8

process factors, such as oxide thickness, diffusion depths, and impurity concentration densities [6]. Supply voltage, threshold voltage, and transistor-sizing ratios all significantly affect the bit yield for SRAM in the context of all these parameters [7]. In light of this, choosing the best cell design for SRAM is difficult. When two identical transistors are put closely together, the transistor mismatch results in significant changes in the electrical parameters of the transistors, such as the V_{th}, body, and current factor, which reduces the predictability and controllability of the design. The SRAM cells' stability is greatly impacted by the rise in unpredictability and fluctuation [8].

8.3 RANDOM ACCESS MEMORY

It is constructed of ICs that enable random access to the stored data for devices [9]. The information or instructions stored in random access memory (RAM) are considered volatile because they will be lost if the supply power is turned off. The concept behind the term "random" is that any piece of data, no matter where it is physically located or how closely it relates to the previous piece of data, can be returned at a predetermined time. The data transfer time is longer than the movement time, and the retrieval time varies according to the physical location of the data in these devices. The fact that recovery times are rapid and dependable makes RAM superior to storage kinds requiring physical movement. Since there is no physical movement required, the time it takes to access the data is brief, and it is constant because it is not dependent on where the physical head is at any particular time. A RAM chip's access time is the same for every bit of data it contains. The disadvantages include the cost of actually transferring media and data loss in the event of a power outage. Since RAM is dependable and quick, it is employed as "primary memory" or major storage. Memory sticks, which are easily upgradable modules, are available. These are easily deleted when they break or the system has to increase its memory due to current requirements. Even though there is a little amount of RAM built into the CPU, it is more commonly referred to as "cache" memory than RAM.

8.4 STATIC RANDOM ACCESS MEMORY

Simply put, the term "static" means that the memory retains its data as long as the power is present. Writing to or reading from memory locations in any order, regardless of the memory location that was most recently accessed, is known as random access. Two cross-coupled inverters with four transistors each store a bit in static random access memory (SRAM). The stable states "0" and "1" stand in for the two states of this storage cell. The bits stored inside the SRAM are accessed using the access transistors when it is in an active state. As a result, six MOSFETs are frequently required to store one memory bit. The bit lines should be linked to the cell depending on the output of the two access transistors, which are controlled by the WWL. Both read-and-write operations involve sending data via word line. Both the bit lines are complementary to each other to increase the noise margin.

8.5 DYNAMIC RANDOM ACCESS MEMORY

In dynamic random access memory (DRAM), every bit of data is kept in a capacitor inside an integrated circuit. The circuit needs to be updated regularly to store the data because the capacitors leak charge. DRAM only needs one transistor and one capacitor for each bit, as opposed to SRAM, which needs six transistors per bit. This criterion allows DRAM to achieve extremely high densities. DRAM has a volatile nature; hence, it loses data when the supply is OFF. To avoid leakage from ever corrupting the memory cells' contents, the periodic refresh method should examine the contents of the memory cells before writing them. Typically, updates should occur every 1–4 ms. Every row of the circuit is renewed using technology meant to refresh memories for a longer period.

8.6 CONVENTIONAL 6T SRAM CELL

Figure 8.1 depicts the basic SRAM cell. It has two cross-coupled inverters, inv1 (M1 and M2) and inv2 (M3 and M4), as well as two access transistors (M5 and M6). The bit lines (BL and BLB) are connected to the source/drain terminals of both pass transistors, and the word line (WWL) is connected to the gate terminals. The word line was switched on before selecting the desired bit cells in a particular specified row. The selected bit cells are read from and written to using the bit lines pair, in comparison. The SRAM cell will hold both the stored data (Q) and its complement (QB).

The logical gate inverter's purpose is to produce the opposite of the input. This can be linked to other logic gates in a chain to create extremely complex logic circuits. The OR operator and the inverters can be used to build any type of logical circuit. On the other hand, a power inverter is a tool that can be used to convert DC electricity to AC [9]. This enables using a battery—which produces DC—to power common appliances, which need AC.

8.6.1 Modes of operation

In SRAM simulation, there are three different modes: standby, read operation, and write operation. These are covered individually as follows:

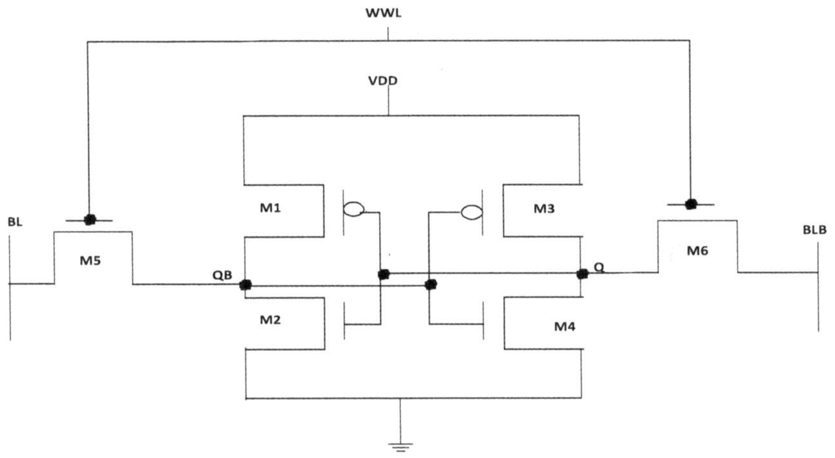

Figure 8.1 Conventional 6T SRAM cell [9].

8.6.1.1 Standby operation

Since the word line is not asserted in this operation, the M5 and M6 transistors are turned off, and the cell is thus severed from the bit line. As long as the supply is there, the cross-coupled inverters will be storing data of either type. As long as the power supply is on, the two connected inverters will continue to reinforce one another, and the data will remain in the hold state.

8.6.1.2 Read operation

Assume that the bit cell contains the logic '1'. The WWL is then asserted high to begin the read mode after both bit lines have been pre-charged to half of the V_{DD}. Bit lines receive the values stored at storage nodes, with one BL remaining at its initial charge and the other two-bit lines being discharged through M6 and M2 to a logic 0. Through M5 and M3, the value at BL is drawn up toward V_{DD}.

8.6.1.3 Write operation

The word line is asserted in this mode of operation (WWL=1). Let's assume that the bit cell in Figure 8.1 has a stored value ($Q=1$) and that data "0" needs to be placed into the bit cell to overwrite the previous value. By asserting 0 V to the BL and raising the BLB to supply voltage, this write-0 process can be carried out (V_{DD}). The new state of an SRAM cell can easily overrule the previous state. When constructing SRAM cells for cache memory, some essential factors must be taken into consideration for the SRAM to function properly. The stability of read-and-write operations depends on significant design issues.

8.7 PERFORMANCE PARAMETERS OF SRAM CELLS

The stability and delays related to various procedures are connected to several performance factors.

8.7.1 SRAM power

Average power is a metric that is becoming more crucial. When a bit is stored inside a cell, standby power is the amount of energy used by the cell. Leakage of the charge is to cause this. Since the SRAM cell is made up of two cross-coupled inverters, there are two leakage current channels between ground to source when data is stored in them. The cell's standby power dissipation is brought on by this. Leakage in the design is also caused by reverse saturation current, subthreshold current, latch-up, and other factors. In general, MOSFETs constantly have very little internal leakage (for silicon, it is of the order of nA). Therefore, the primary reason for standby dissipation in the memories is this leakage.

The propagation delay in SRAM is influenced by both the wire delays and the column height. Thus, segmentation is used to shorten the delay. An increase in one results in a decrease in the other since a device's power delay product is constant; consequently, both outcomes are possible. Greater power dissipation comes at the expense of a smaller reduction in delay when the FinFET device in the SRAM cell is larger. To reduce power dissipation, it is required to lower leakage currents, which enables a longer channel length or higher transistor threshold voltages. These two performance indicators are trade-offs since longer channels have larger delays [10].

8.7.2 Delay

This delay shows the speed of the write and read operations. This is the difference in time between the bit line and the associated storage node. The write access time is determined by measuring the interval between the asserting of the WWL signal and the cell flip. The write or read process proceeds more swiftly as the interval gets shorter.

8.7.3 Power delay product

As the name implies, this parameter in a certain circuit is the result of the power and delay parameters. For separate write and read operations, distinct power delay products are obtained. For a device, the power delay product (PDP) is always constant. This parameter is a constant for any device. When one parameter is increased, the other value is decreased.

8.7.4 Static noise margin

Another important consideration when analyzing circuits is the static noise margin (SNM), which is a stability statistic. SNM [11] is the maximum noise voltage required on the STV nodes of an SRAM to switch a cell's state. SNM can be represented visually as the greatest square that can be calculated from the back-to-back linked inverter pairs two-voltage transfer characteristic (VTC) curves. Use these steps to calculate SNM: Perform a DC analysis at the Q and QB, and isolate the cross-coupled inverters' feedback. There are determined to be two VTC curves. After one of them is inverted, the other is merged with it. The resulting graph, referred to as the butterfly curve, is the combined graph. The SNM is the side of the greatest square that can be changed inside the butterfly curve. The SNM is chosen from the side of the smaller curve if two squares are possible. SNM can be broken down into three categories: write, read, and hold.

8.7.4.1 Hold SNM

It is defined as the minimum voltage at the storage node needed to disrupt the data that is being held in a hold mode. By using butterfly curves, the stability has been evaluated. The word line is asserted low in the hold state. The inverter's feedback is divided. Each inverter's input is swept from 0 to V_{DD}.

8.7.4.2 Write SNM

The WSNM of a bit cell indicates how easy or difficult it is to write data into the cell [12–14]. The WSNM is calculated using the word-line voltage sweep method [15]. The WBL is set to the proper voltage to allow data stored at nodes to be flipped. A supply voltage and a ground voltage are alternately applied to the write word line. Writing into the bit cell is made simpler by the wide write margin.

8.7.4.3 Read SNM

RSNM is a statistic that assesses a bit cell's read stability. The length of the longest side of the largest square fastened into the butterfly lobes can be used to calculate its size. In the read state, WL, BL, and BLB are raised. The feedback from the inverters is divided. The input of each inverter is swept from 0 to V_{DD}.

8.7.5 Power dissipation

The efficiency of the SRAM cell in portable devices is measured by its power dissipation. Because of its low SCEs and low leakage current in FinFET devices, this cell has a basic advantage in that it dissipates less power and has faster access times. An increase in power dissipation in the SRAM cell results from a large driving current, which also reduces access time [10].

8.7.6 Power consumption

A major amount of an application processor's overall power is provided by large embedded SRAM arrays. The power consumption of an SRAM array has very long idle times and very short active times. The biggest issue with huge arrays is standby power consumption. Thus, for low-power VLSI applications, minimizing leakage in huge memory arrays has become essential. The cell area is negatively impacted by using longer channel lengths. As a result, greater channel lengths are only sometimes employed (for instance, on access transistors, which also increase cell stability) [16]. Using higher transistor threshold voltages has a negative impact on access time because of the reduced read current [17]. The RSNM and WSNM, however, are improved by them. High V_{th} NMOS pull-down devices tend to raise the inverter trip point while high threshold PMOS loads tend to lower it. Raising the NMOS transistors threshold voltage typically has a beneficial impact since the pull-down devices may generate more current than the PMOS load. The trip voltage is more likely to be affected by raising the V_{th} of the NMOS transistors [18], which leads to wider RSNM and WSNM because the pull-down devices may drive more current than the PMOS load.

8.7.7 Cell area

A memory array's functionality and density are two of its most crucial features. The choice of sufficiently large SNMs, which are determined by the V_{DD}, device sizing, and, to a lesser extent, by the choice of transistor V_{th}, ensures functionality for huge memory arrays. Increased transistor size improves noise margins while decreasing cell area reduces density [19].

8.8 POWER REDUCTION TECHNIQUES IN SRAM CELL

High-performance ICs now dissipate much more power due to the scalability of technology. The battery life of portable gadgets is decreased by higher power dissipation. Furthermore, because of the accompanying higher power density, the increasing power dissipation limits the ability of technology to continue scaling. If the word line is set to low then the transistors M5 and M6 remained OFF and disconnected the cell from both bit lines. Until they are linked to the supply, the two cross-coupled inverters made up of M1-M2 and M3-M4 will continue to reinforce one another. SRAM hence uses a significant amount of power in this mode. High power dissipation prevents high-performance circuit design paths. Circuit designers must reduce power dissipation since batteries only store a certain amount of energy [20]. As a result, many methods have been suggested to minimize the circuit's power consumption by minimizing leakage current.

8.8.1 Multi-threshold CMOS (MTCMOS)

This technique helps in reducing the delay or increasing power by using different threshold transistors. The multiple threshold voltage is the gate voltage at which an inversion layer begins to form at the intersection of the transistors substrate (body) and an insulating layer (oxide). Because they flip more quickly, low V_{th} devices are vital for lowering clock durations on critical delay lines. A larger static power value is present in low V_{th} devices. In order to reduce static power dissipation on non-critical channels, high V_{th} devices are used; however, they have a delayed cost. The static leakage limit of most high V_{th} devices is ten times greater than that of low V_{th} devices [21].

According to this method, transistors with low V_{th} are disconnected from the power source by placing a high V_{th} sleep transistor at the upper and lower of the circuit. The hold state causes the high Vth transistors to disconnect and the low Vth logic to lose power when the sleep signal is asserted. When the system is in standby state, this results in a very little subthreshold current flowing from supply to ground. One significant drawback of this approach is how difficult it is to size and section sleep transistors for large circuits.

8.8.2 Dual threshold technique

This technique uses the approach of a part of the circuit operating at a higher threshold while the other component operates at a lower threshold. This technique can be applied in two different ways: first, using symmetric cells, and second, using asymmetric cells. Since M5 and M6 are the driving transistors in Figure 8.1, a low threshold might be used there, and the shorter switching time caused by these thin channel devices allows for quick access to the cell.

The remaining transistors, M1, M2, M3, and M4, have minimal subthreshold leakage current because they have thick channels with high thresholds. Low-power operation in SRAM is thus made possible by the reduction in leakage current. Transistors whose thresholds are contributing to leakage should be made high depending on the value that has been stored. Because M2 and M6 have high thresholds, in this case, read access time is degraded. As a result, BL discharge takes a long time. Transistor M4 should have a high threshold because PMOS is not significantly impacted by read access time [22]. Making the M2 high threshold is the second improvement because a low threshold is unnecessary.

8.8.3 Gated V_{DD} technique

With a drop in threshold voltage, leakage current grows exponentially. The primary idea behind including the second transistor in the supply voltage or ground circuit is that it will function as follows: it will turn on during active portions and off during inactive portions. The stacking effect is the primary cause of the decreased leakage current [22].

In gated-V_{DD} using an NMOS transistor, an extra NMOS transistor is connected between SRAM and the ground path. Similarly, SRAMs PMOS transistors and the supply path can be connected using gated-V_{DD} and PMOS transistors. When the cell performs read-and-write operations, the gated-V_{DD} transistor turns on, and when the cell is in a hold state, it turns off. The width should be sufficient to carry currents throughout read-and-write cycles.

Since only one NMOS transistor is on during a read cycle, it is possible to determine the transistor width in NMOS-gated V_{DD} by multiplying the width of that transistor by the total number of SRAM cells in the block. Because there are three NMOS transistors accessible between the bit lines and the ground wire, the stacking effect that results minimizes standby leakage current. Transistor width decreases in PMOS gated-V_{DD} due to minimal area overhead and the absence of any issues during the SRAM read cycle. Here, there is a lack of good isolation.

8.8.4 DRG (data retention gated ground) scheme

In this case, WL applies gated control. There is no need to switch off the entire block since before the read operation begins, a specific row of cells turns on and the rest rows are all dormant [22]. The "1" storage node stays at V_{DD} when the cell is not in use, and the "0" storage node is connected to the NMOS transistor's intermediate voltage. One of the bit lines discharges when the gated transistor is turned on, causing the "0" node to return to the ground. The gating transistor should be the proper size.

8.8.5 Self-controllable voltage level technique

In the DG FinFET SRAM cell with a shorted gate, self-controllable voltage level (SVL) technique is used to decrease leakage. Between V_{DD} and the 6TSRAM cell, a parallel circuit made up of two series NMOS transistors and a PMOS transistor is connected. A second circuit is made up of two series PMOS transistors and one NMOS transistor connected parallel between the SRAM cell and GND. The combined use of these two circuits, known as Upper SVL (USVL) and Lower SVL (LSVL) [23], reduces leakage current. The series combination of NMOS transistors in the USVL circuits reduces the value of the virtual supply compared to the actual supply. The series has an impact on the value of the virtual GND. Consequently, the SVL approach lowers the circuit's total voltage, reducing leakage power.

8.8.6 Adaptive voltage level technique

The adaptive voltage level (AVL) method is generally used on DG FinFET SRAM cell for reducing leakage current and power. This method combines AVLG (Adaptive Voltage Level Scaling at Ground), which raises the ground potential for the circuit in inactive mode, with AVLS (Adaptive Voltage Level Scaling at Supply), which decreases the circuit's power supply in static mode. Leakage power is significantly reduced by both a reduced power supply and a higher ground potential. When SRAM is used, it receives a full supply voltage of V_{DD}, but when it is not, it receives a decreased supply voltage of V_{DD}. This lowers the leakage current at access transistors. During standby mode in AVLG, a greater voltage is delivered, whereas, during active mode, a $0\,V$ is given at the ground node [24]. This strategy matches the diode-footed cache design strategy, which aims to prevent SRAM leakage. By combining a diode with a high threshold transistor, this design raises the ground level when in standby mode.

8.8.7 Hybrid technique

The hybrid SRAM cell technology uses a high threshold transistor to lessen the SRAM cell's subthreshold leakage. The high V_{th} transistor must be ON for a 6T SRAM cell to function properly in active mode. The high V_{th} transistor needs to be disabled in standby mode for improved leakage control [25].

8.8.8 Drowsy-cache technique

In this method, when only hold operations are needed, the SRAM cell is supplied with Low Supply voltage ($V_{DD}/2$). Multiple supply voltages are used for memory. Operations carried out in active mode are supplied with a high supply voltage. Leakage current decreases with decreasing V_{DD}, hence this technique uses high V_{DD} in active mode and low V_{DD} in hold mode to minimize leakage power dissipation. The simultaneous lowering of supply voltage and leakage current enhances the performance of the circuit and results in a quadratic reduction in leakage power.

8.9 CONCLUSION

This chapter helps to clarify how SRAM cells behave in situations where low power consumption and high performance are priorities. The devices' power consumption has greatly decreased, and this memory cell might be incorporated with any such memory-based devices that require less power. To reduce leakage, various techniques are discussed. The main aspects that are taken into consideration before designing include leakage, performance, transition overhead, and stability. In the design of SRAM, there is still a trade-off between parameters. Therefore, it is crucial to analyze all factors before selecting any strategy, after which the ideal option that best suits our application is chosen.

REFERENCES

[1] Piguet, C. (2018). *Low-Power Electronics Design*. Boca Raton: CRC Press.
[2] Dhanumjaya, K., Sudha, M., Prasad, M. G., & Padmaraju, K. (2012). Cell stability analysis of conventional 6T dynamic 8T SRAM cell in 45nm technology. *International Journal of VLSI Design & Communication Systems*, 3(2), 41.
[3] Butzen, P. F., & Ribas, R. P. (2006). *Leakage Current in Sub-Micrometer CMOS Gates*. Brazil: Universidade Federal do Rio Grande do Sul, pp. 1–28.
[4] Bai, P., Auth, C., Balakrishnan, S., Bost, M., Brain, R., Chikarmane, V., ... & Bohr, M. (2004). A 65nm logic technology featuring 35nm gate lengths, enhanced channel strain, 8 Cu interconnect layers, low-k ILD and 0.57/spl mu/m/sup 2/SRAM cell. In *IEDM Technical Digest. IEEE International Electron Devices Meeting, 2004*, pp. 657–660. San Francisco CA, USA: IEEE.
[5] Luo, Z., Steegen, A., Eller, M., Mann, R., Baiocco, C., Nguyen, P., ... & Wann, C. (2004). High performance and low power transistors integrated in 65nm bulk CMOS technology. In IEDM Technical Digest. IEEE International Electron Devices Meeting, 2004, pp. 661–664. San Francisco CA, USA: IEEE.
[6] Seevinck, E., List, F. J., & Lohstroh, J. (1987). Static-noise margin analysis of MOS SRAM cells. *IEEE Journal of Solid-State Circuits*, 22(5), 748–754.
[7] Morifuji, E., Yoshida, T., Tsuno, H., Kikuchi, Y., Matsuda, S., Yamada, S., ... & Kakumu, M. (2004). New guideline of Vdd and Vth scaling for 65nm technology and beyond. In *Digest of Technical Papers. 2004 Symposium on VLSI Technology, 2004*, pp. 164–165. San Francisco CA, USA: IEEE.
[8] Amrutur, B. S., & Horowitz, M. A. (2000). Speed and power scaling of SRAM's. *IEEE Journal of Solid-State Circuits*, 35(2), 175–185.
[9] Guo, Z., Balasubramanian, S., Zlatanovici, R., King, T. J., & Nikolić, B. (2005). FinFET-based SRAM design. In *Proceedings of the 2005 International Symposium on Low Power Electronics and Design*, San Diego CA USA, pp. 2–7.
[10] Raj, B., Saxena, A. K., & Dasgupta, S. (2011). Nanoscale FinFET based SRAM cell design: Analysis of performance metric, process variation, underlapped FinFET, and temperature effect. *IEEE Circuits and Systems Magazine, 11*(3), 38–50.

[11] Moore, G. E. (1965). *Cramming More Components onto Integrated Circuits.* IEEE.

[12] Azam, T., Cheng, B., Roy, S., & Cumming, D. R. S. (2010). Robust asymmetric 6T-SRAM cell for low-power operation in nano-CMOS technologies. *Electronics Letters*, 46(4), 273–274.

[13] Fan, M. L., Wu, Y. S., Hu, V. P. H., Su, P., & Chuang, C. T. (2010). Investigation of cell stability and write ability of FinFET subthreshold SRAM using analytical SNM model. *IEEE Transactions on Electron Devices*, 57(6), 1375–1381.

[14] Chung, Y. (2014). Stability and leakage characteristics of novel conducting PMOS based 8T SRAM cell. *International Journal of Electronics*, 101(6), 831–848.

[15] Ansari, M., Afzali-Kusha, H., Ebrahimi, B., Navabi, Z., Afzali-Kusha, A., & Pedram, M. (2015). A near-threshold 7T SRAM cell with high write and read margins and low write time for sub-20 nm FinFET technologies. *Integration*, 50, 91–106.

[16] Al-Hashimi, B. M. (Ed.). (2006). *System-On-Chip: Next Generation Electronics* (vol. 18). London, UK: IET.

[17] Gupta, R., Gill, S. S., & Kaur, N. (2014). A novel low leakage and high density 5T CMOS SRAM Cell in 45nm technology. In 2014 Recent Advances in Engineering and Computational Sciences (RAECS), pp. 1–6. Chandigarh, India: IEEE.

[18] Joshi, R. V., Williams, R. Q., Nowak, E., Kim, K., Beintner, J., Ludwig, T., ... & Chuang, C. (2004). FinFET SRAM for high-performance low-power applications. In *Proceedings of the 30th European Solid-State Circuits Conference (IEEE Cat. No. 04EX850)*, pp. 69–72. Leuven, Belgium: IEEE.

[19] Balasubramanium, J. Y. S. (2008). Design of sub-50 nm FinFET based low power SRAMs. *Semiconductor Science and Technology*, 23, 13.

[20] Sharma, V. K., Pattanaik, M., & Raj, B. (2014). PVT variations aware low leakage INDEP approach for nanoscale CMOS circuits. *Microelectronics Reliability*, 54(1), 90–99.

[21] Anis, M., Mahmoud, M., Elmasry, M., & Areibi, S. (2002). Dynamic and leakage power reduction in MTCMOS circuits using an automated efficient gate clustering technique. In *Proceedings of the 39th Annual Design Automation Conference*, pp. 480–485.

[22] Patel, S. R., Bhatt, K. R., & Jani, R. (2013). Leakage current reduction techniques in SRAM. *International Journal of Engineering Research & Technology (IJERT)*, 2(1).

[23] Akashe, S., & Sharma, S. (2013). Read write stability of dual-Vt 7T SRAM cell at 45 nm technology. *Journal of Computational and Theoretical Nanoscience*, 10(1), 69–72.

[24] Yadav, M., Akashe, S., & Goswami, Y. (2011). Analysis of leakage reduction technique on different SRAM cells. *International Journal of Engineering Trends and Technology*, 2(3), 78–83.

[25] Kushwah, R. S., & Sikarwar, V. (2015). Analysis of leakage current and power reduction techniques in FinFET based SRAM cell. *Radioelectronics and Communications Systems*, 58(7), 312–321.

Chapter 9

A comparative review on leakage power minimization techniques in SRAM

Appikatla Phani Kumar and Rohit Lorenzo
VIT-AP University

9.1 INTRODUCTION

Memory design has become an important part of modern VLSI systems. Semiconductor memories are an integral part of complex VLSI networks, as they occupy 90% of the total chip area. Power dissipation directly affects the building cost of the chip system. These points lead VLSI system design-ers to view power consumption as a critical problem [1,2]. Supply voltage scaling is one of the techniques that minimizes leakage and dynamic power. However, there are two problems associated with the scaled technique. First, subthreshold leakage increases exponentially in a low-V_{th} device. Every 0.1 V drop in threshold voltage causes a ten-fold increase in subthreshold leakage [3,4]. The second issue is the worst-case performance being reduced at lower supply because of threshold variation [5]. As the technology is scaling down, the leakage current in the deep nanometer region becomes increasingly substantial and is similar to the dynamic power dissipation [6,7]. To minimize leakage current for low-power applications, various leakage minimization techniques have been reported.

Random access memory (RAM) is categorized into two types. One is SRAM (Static RAM) and the other is DRAM (Dynamic RAM). A single transistor memory element is used to implement DRAM, and the state of a cell is stored in the form of charge on a capacitor [8]. The term dynamic means that the charge on an ideal storage capacitor needs to be refreshed periodically. In contrast, SRAM uses a bistable component like an inverter loop to store the cell state as a potential difference [9,10]. These compo-nents can maintain their state without any requirement to be refreshed as long as supply is given [11]. A basic SRAM cell is surprisingly complex and covers a much larger area than a DRAM cell. The use of SRAM or DRAM in your system design has various benefits. Some performance parameters to consider for choosing memory are power consumption, speed, density, cost, volatility, and reliability [12,13].

SRAM is used as a cache memory in information processing and in the latest portable hand-held devices like mobile phones and PDAs. Fast SRAMs have recently received a lot of attention [14]. With scaled submicron

DOI: 10.1201/9781003459231-9

technology, the area of each device on a chip decreases while the thickness of the chip raises [15]. Using this scaling method presents new difficulties like power consumption and reliability [16,17]. Many methods have been suggested to diminish the power dissipation of the SRAM cell. Various sources of leakage power dissipation and different techniques to reduce leakage power are discussed in this chapter.

9.2 HISTORY OF SRAM

Memory cells are the basic building blocks of the VLSI chip. A memory cell is an electronic circuit that stores a single bit of data in binary form [18]. To store logic-1, it should be set and to store logic-0, it should be reset. Its value is kept same until the set (or) reset process changes it. Reading and writing the value in the memory cell allows access to it [19]. We know that SRAM and DRAM are the two most common types of modern RAM.

SRAM stores a single bit of data with the help of the 6T memory cell architecture. This type of RAM is more expensive to manufacture, but its speed of operation is higher than DRAM. Further, it requires less dynamic power. SRAM is frequently employed in modern computers as the CPU cache memory. The SRAM memory cell is a flip-flop circuit, often built with FETs [20]. SRAM has a lower storage density and is more expensive, but it consumes less power while in hold mode. Both SRAM and DRAM are considered volatile because they lose their state when the system is turned OFF [21]. A 1-bit "SRAM" cell is comprised of two cross-connected inverters with two access transistors (NMOS), whose output depends on the cross-coupled inverter's storage node data [22]. It consists of 2-PMOS and 4-NMOS transistors as depicted in Figure 9.1. The data which is written in the memory can be read as logic-0 (or) logic-1, and the bit lines are used to access this data.

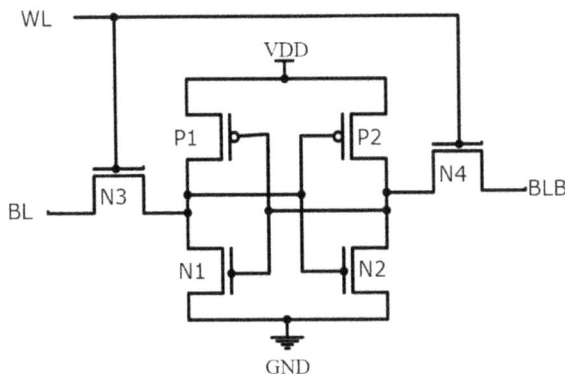

Figure 9.1 Schematic of conventional-6T SRAM.

9.2.1 Write operation

In write mode, BLB is made low, and BL is made high. Then the word line (WL) is enabled to turn ON the access transistors, and data is stored in the SRAM cell. Data is kept for read operation once the writing operation is completed and the access transistors will turn OFF.

9.2.2 Read operation

Both bit lines are precharged to VDD in read mode, and the WL will be activated. It is well known that the driving strength of the driver transistor is greater than the access transistor's strength. As a result, a memory cell pulls down the one-bit line. Then, the sense amplifier will sense the data at the bit lines, and the stored bits will finally be available at the sense amplifier's output.

9.2.3 Hold operation

The WL is disabled in hold mode. Hence, both access transistors are OFF, and there is no path between both bit lines and storage nodes. Therefore, both storage nodes remain at the previous value resulting in holding the data.

The operation makes it clear that the cell dissipates power in two situations: first, while reading and writing bits due to transistor switching activity. The second dissipates power due to leakage when the cell is in an idle state because one transistor from each inverter is continuously in the ON state [23]. In addition, peripheral circuits like sensing amplifiers, row decoders, and column decoders dissipate a lot of power.

9.3 SOURCES OF POWER DISSIPATION IN SRAM

Two different types of power dissipation occur in SRAM. They are static (or) leakage power dissipation and dynamic (or) active power dissipation. Power dissipation that is classified as static (or) leakage takes place while the device is in hold mode [24]. Power dissipation that occurs in write and read operations is referred to as dynamic power dissipation. Total power consumption is measured as the voltage and current product that is consumed from the source [25]. According to ITRS, Figure 9.2 shows that the total chip leakage dissipation is more significant when compared to dynamic power dissipation in the deep nanometer region.

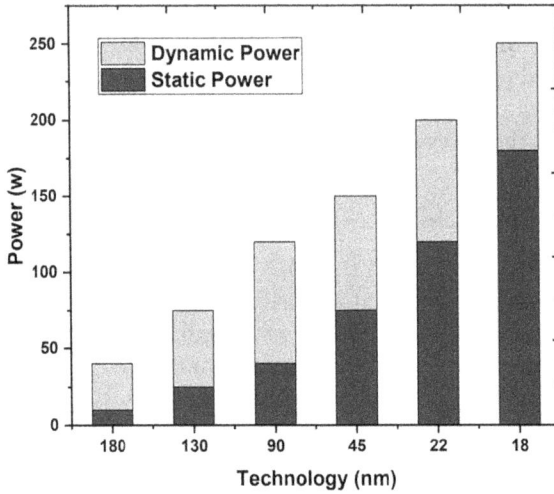

Figure 9.2 Importance of static power dissipation in total power dissipation.

9.4 CLASSIFICATION OF LEAKAGE COMPONENTS

Different types of leakage power dissipation are depicted in Figure 9.3. That includes subthreshold leakage, junction leakage, and gate leakage [26]. Hence the total leakage in a device can be expressed mathematically as,

$$I_{overall} = I_{sub} + I_{gate} + I_{junction} \tag{4.1}$$

9.4.1 Sub-threshold leakage

The current flowing between the source and drain of the device, when the gate to source voltage (V_{GS}) is lower than V_{th} is known as subthreshold leakage. It occurs when the device is working in a weak inversion region. Since the subthreshold current depends on V_{th} exponentially, small channel devices in particular experience significant subthreshold currents [27]. The concentration of minority carriers in the weak inversion zone is low and varies with the channel length. In strong and weak inversion regions, the drift current and diffusion currents are dominant, respectively. The sub-threshold current equation is represented in equation (4.2).

$$I_{sub} = I_o e^{\left(\frac{V_{gs} - V_{th0} - \eta V_{ds} - \gamma \frac{V_{sb}}{nV_\theta}}\right)} \left(1 - e^{\frac{-V_{ds}}{V_t}}\right) \tag{4.2}$$

here $I_o = \mu C_{ox} \dfrac{W}{L} V^2 e^{1.8}$, $V_e = \dfrac{KT}{q} = 0.026$ v and $\eta = 1 + \dfrac{C_{dm}}{C_{ox}}$.

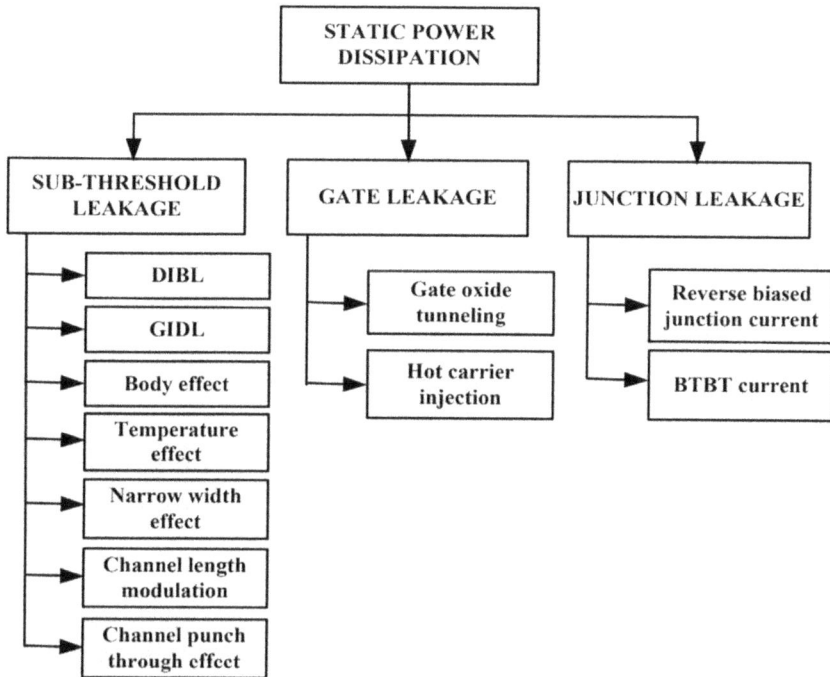

Figure 9.3 Classification of leakage power dissipation.

μ is the mobility of the carrier, K is the Boltzmann constant, W/L indicates the channel width/length ratio, C_{ox} denotes the gate oxide capacitance per unit area, T denotes the absolute temperature, q indicates the conduction electrical charge, η indicates the subthreshold factor, n indicates the coefficient of subthreshold swing, and C_{dm} denotes the depletion layer capacitance.

9.4.2 Gate leakage

Tunneling is another name for the current entering the transistor's gate. It happens when holes or electrons from the bulk of the silicon tunnel into the gate of an NMOS or PMOS, respectively. The electric field across the oxide increases as gate oxide thickness decreases [28]. As a result, the chance of electrons tunneling through the gate oxide increases exponentially. Further, it also increases the gate oxide tunneling current exponentially. The main components of gate leakage current are:

- Gate to channel currents
- Gate to source and gate to drain overlap current.
- Gate to substrate current.

The gate leakage current density can be expressed as in equation (4.3).

$$J_{gl} = A_g \left(\frac{V_{ox}}{T_{ox}}\right) \exp\left[\frac{-B_g\left(1-\left(1-\frac{V_{ox}}{\varphi_{ox}}\right)^{\frac{3}{2}}\right)}{V_{ox}\Big/T_{ox}}\right] \qquad (4.3)$$

Where, $A_g = \dfrac{Q^3}{16\pi^2\varphi_{ox}}$ and $B_g = \dfrac{4\sqrt{2m^*}\,\varphi_{ox}^{\frac{3}{2}}}{3\left(\dfrac{h}{2\pi}\right)\varphi_{ox}}$

J_{gl} denotes the gate leakage current density, V_{ox} indicates the potential drop across oxide, m^* is the electron's effective mass, φ_{ox} is the tunneling electron's barrier height and T_{ox} denotes the thickness of oxide.

9.4.3 Junction leakage

A PN junction reverse bias leakage current is created when the substrate-to-drain and substrate-to-source junctions are biased in the opposite direction. It contains two primary components:

- Minority carrier diffusion close to the depletion region's edge.
- The creation of electron hole pairs in the reverse-biased junction's depletion region.

The depletion junction area and doping concentration have the greatest impact on the reverse-biased PN junction leakage current [29]. The contribution of this element to the overall leakage current is insignificant.

9.5 LEAKAGE REDUCTION TECHNIQUES IN SRAM CELL

In this section, we will discuss and review different circuit-level standby leakage reduction techniques. The basic goal of all these methods is to lower subthreshold leakage currents. Each method has its own benefits and drawbacks. These methods fall under the following categories.

9.5.1 Power-gating techniques

Power gating is a reliable leakage reduction technology that is frequently utilized in industry. By including external transistors (header and footer) in the power-gating approach, the leakage currents are almost completely reduced [30]. When the devices are in idle mode, these transistors cut off the path from VDD to ground.

9.5.1.1 Sleep transistor

To minimize the static power dissipation in SRAM, the authors in [31] suggested a sleep transistor approach depicted in Figure 9.4. It is a power-gating method. This method connects a PMOS between the power supply and the SRAM cell, an NMOS between the SRAM and its ground. The sleep transistors will be turned ON in active mode, which has no effect on how the device operates because a path still connects VDD and GND. The voltage supply to the SRAM cell is cut off when the sleep transistors are turned OFF, generating a virtual VDD and GND link.

9.5.1.2 Forced stack

In addition to the original 6T SRAM, two more pull-down transistors and two additional pull-up transistors are placed in the forced stack SRAM cell [32], as depicted in Figure 9.5. During idle mode, one of each pair is activated depending on the stored bit value in the SRAM. This maintains supply to the ON transistors while cutting off power to the OFF transistors. When two or more transistors are turned off simultaneously, the subthreshold leakage current is reduced due to the impact of transistor stacking.

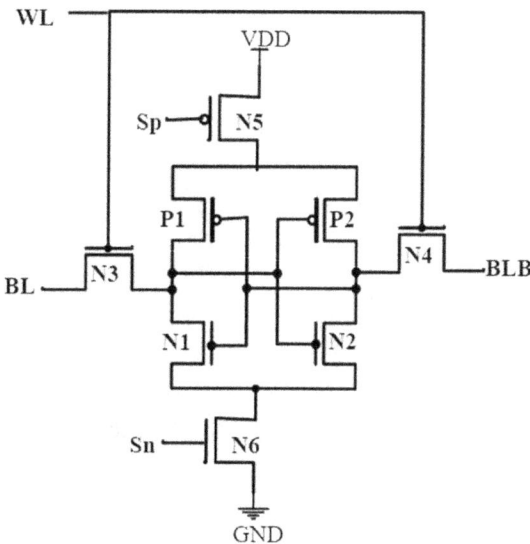

Figure 9.4 Sleep transistor SRAM.

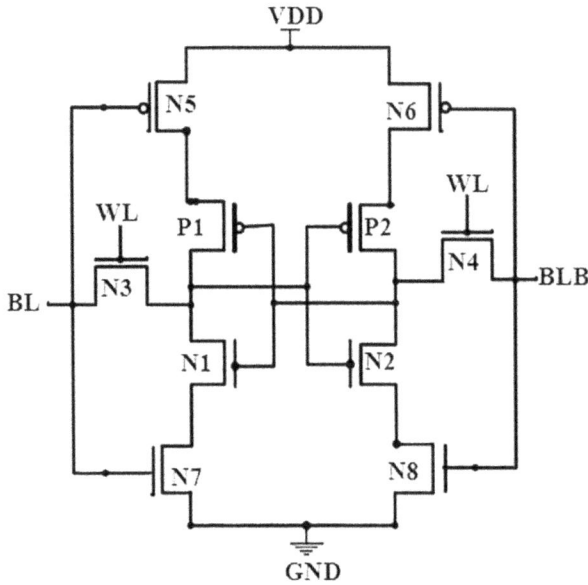

Figure 9.5 Forced stack SRAM.

9.5.1.3 Sleepy stack

The sleepy stack techniques, depicted in Figure 9.6, combine the sleep transistor approach and the forced stack approach [33]. The sleepy stack's sleep transistors function similarly to the transistors utilized in the sleep transistor approach, which turns them ON during active mode operation and OFF during standby mode. When active, we will provide $S=0$ and $S'=1$. Thus, each and every sleep transistor is turned ON. This structure can minimize the delay. This sleepy stack approach obtained lower switching time than the forced stack approach because the sleep transistors are constantly ON in active mode. We shall give $S=1$ and $S'=0$ during standby mode to turn OFF both of the sleep transistors. The sleepy stack method preserves the precise logic state even while the sleep transistors are OFF.

9.5.1.4 Data retention gated ground

The determination of the stored data and the increased pull-down path resistance caused by ground-gated transistors are the disadvantages of power-gating SRAM. To solve this problem and achieve minimal leakage power, the authors in [34] presented the DRG technique in SRAM cell

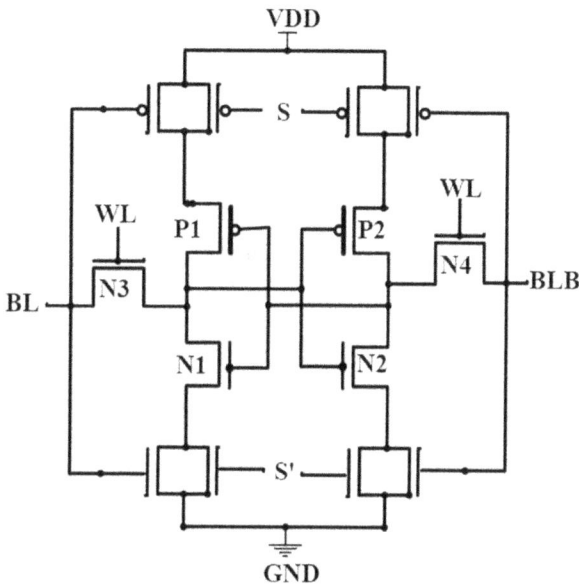

Figure 9.6 Sleepy stack SRAM.

design. The external signal attached to the WL controls the ground-gated transistor, depicted in Figure 9.7. By fundamentally "gating" the supply voltage of the cell and turning it OFF in standby mode and ON in active mode, the leakage current was dramatically reduced.

9.5.1.5 N-control SRAM with gated VDD

To further reduce leakage power dissipation, [35] presented an NC SRAM with a gated VDD method and a dual V_{th}, as shown in Figure 9.8. To achieve fast speed, the SRAM is built using lower threshold transistors. To reduce leakage power, more sleep transistors are constructed with high-threshold transistors. These transistors are indicated by dot circles.

9.5.2 Biasing techniques

To control the threshold voltage, the transistor's body terminal is often linked to ground in the case of NMOS, and in the case of PMOS, it is connected to power supply [36]. Based on the way the body terminal is connected to either VDD or GND, there are different types of biasing techniques, which are explained below.

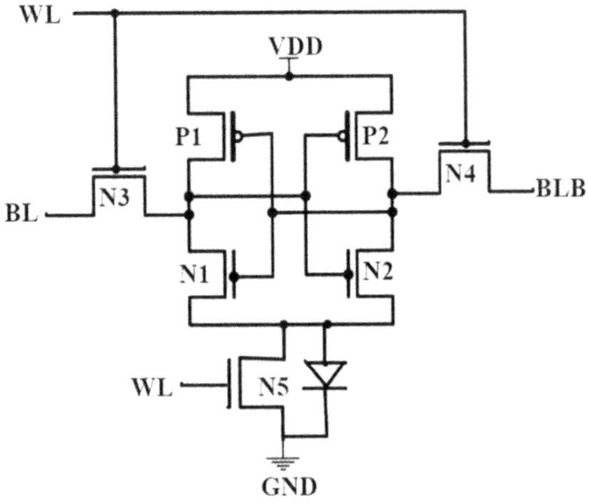

Figure 9.7 Data retention gated ground SRAM.

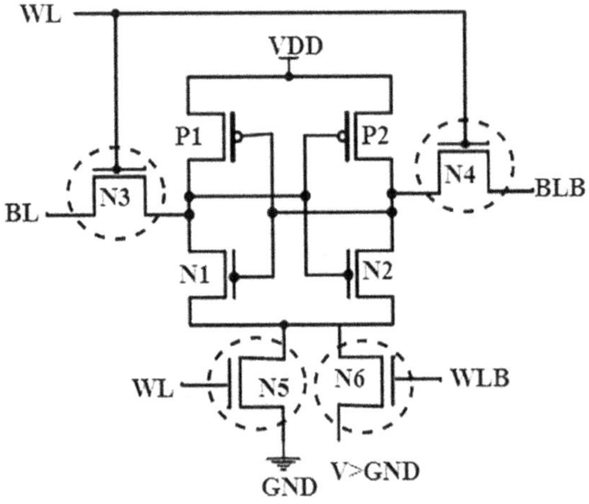

Figure 9.8 N-control SRAM cell with gated VDD.

9.5.2.1 Reverse body biasing

To lessen the latch-up issue and memory data degradation, reverse body biasing (RBB) has been frequently utilized in memory cells. RBB raises body voltage while decreasing subthreshold leakage. A better RBB thereby minimizes the overall leakage current [37]. However, the substrate BTBT, also known as GIDL, has a greater impact on junction leakage than the surface BTBT. To take advantage of RBB, new methods must be developed to limit substrate BTBT leakage. In addition, RBB is useless for low-V_{TH} device functionalities at high/room temperatures, as well as for controlling leakage in short channel devices.

9.5.2.2 Forward body biasing

Larger drain-substrate depletion layers are present in reverse body bias, which reduces the effects of short channels. However, a thicker depletion layer causes threshold variation across the chip. In addition, because of the lower DIBL and body coefficient (γ), the drain influences the channel potential more than the body in short channel devices. The low-threshold transistors are more susceptible to the SCEs and DIBL. As a result, RBB to FBB is the range that is motivated. Due to decreased switching capacitance, FBB lowers the V_{th} of high-threshold devices and enhances the performance of the circuit. Due to a smaller depletion width between the drain and source regions, the FBB devices create increased junction capacitance [38]. However, FBB causes a source-body junction to be in forward bias, which increases leakage current. The increased short circuit current at lower V_{TH} is caused by greater junction and gate capacitance. Therefore, the target forward body voltage is 0.45 V with ±50 mV tolerance for optimized design functioning at a maximum of 110°C temperature.

9.5.2.3 Variable body biasing

In the variable body biasing (VBB) technique, the PMOS body terminal is connected to a positive voltage (+VBB), higher than VDD. However, the substrate terminal of the NMOS is connected to a negative voltage (−VBB), lower than VSS to raise the V_{th} due to the body bias effect in standby mode [39]. Since raising the threshold voltage will have an impact on performance and the method used to control the body's input voltage supply, the reverse bias is smaller during the active mode of operation and stronger during standby. Therefore, it can lower the static power dissipation and subthreshold leakage current.

9.5.2.4 Transistor stacking

- As the drain-source voltage falls, the subthreshold leakage current and the DIBL decrease.
- When the gate-source voltage is negative, the subthreshold current falls off exponentially.
- When the substrate-source voltage is negative, the body effect raises the threshold voltage, which lowers the subthreshold current. Figure 9.9 (a) and (b) illustrate how the NMOS stacking transistors are implemented, including node voltage, and leakage current.

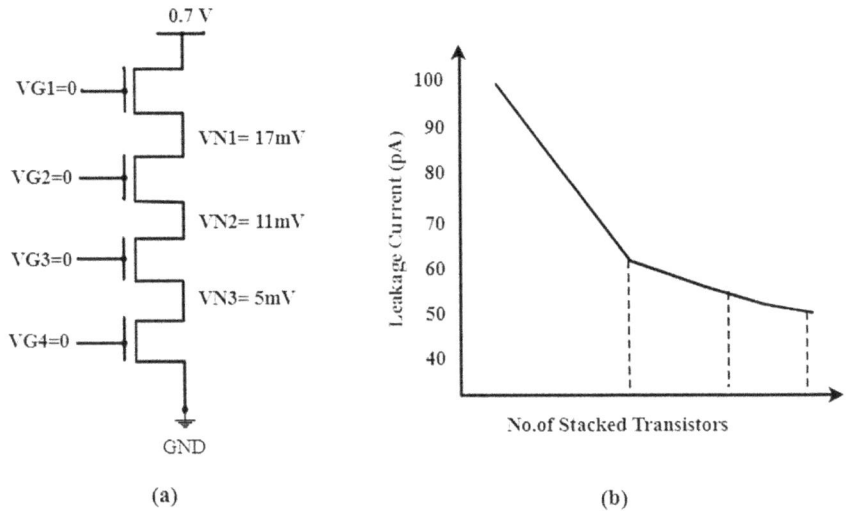

Figure 9.9 (a) Transistor stacking and (b) variation of leakage power with the number of stacked transistors.

Figure 9.10 illustrates the SRAM cell proposed in [40], using stacked transistors. The cell performs as a typical 6T SRAM when the WL made '0'. As a result of transistors P3 and P4 turning off when the WL is high, the self-reverse biasing (SRB) of transistors coupled in series lowers the subthreshold and gate leakage currents.

9.5.2.5 Dual V_{th}

Dual-V_{th} technology can be used to further enhance short channels and DIBL [41]. An apparatus with dual V_{th} alternates between high-V_{th} and low-V_{th}. By using FBB and a high-V_{th} device together, you can create a low-V_{th} device with no movement of the body. Because a dual-V_{th} method has a crucial manufacturing masking process, its complexity is minimized. There is no longer a need to fabricate separate devices with different threshold voltages.

9.5.2.6 VTCMOS

The body biasing approach is used in variable threshold CMOS (VTCMOS) to regulate the threshold voltage [42]. Using a self-body biasing transistor allows for the achievement of various threshold voltages (SBT). To achieve fast speed during active mode, hardly any body bias is used. RBB is utilized in idle mode to reduce leakage current and to increase threshold voltage.

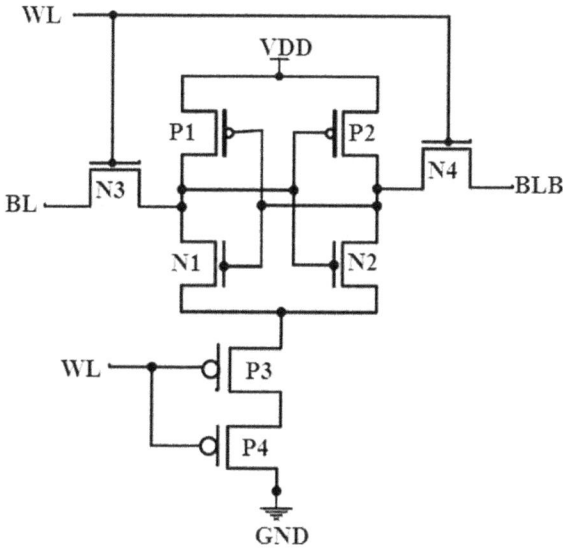

Figure 9.10 SRAM using transistor stacking.

In addition, a little amount of FBB is used in active mode to speed up performance while reducing the SCEs. For body biasing in VTCMOS, more circuitry is needed, increasing chip area.

9.5.2.7 MTCMOS

To minimize leakage power, one of the common topologies is MTCMOS (multi-threshold) technique. Transistors are added between SRAM and ground or VDD in this. Therefore, producing a virtual supply or ground [43,44]. All low V_{th} transistors are present in the logic block (SRAM) for the fastest switching rates. To reduce leakage, the header and footer transistors are constructed utilizing high-V_{th} transistors.

9.6 COMPARISON OF VARIOUS LEAKAGE POWER MINIMIZATION TECHNIQUES

In this section, we will see the comparison of different leakage minimization techniques which are discussed in the previous section. Table 9.1 describes this comparison which includes the advantages and limitations of each technique.

Table 9.1 Comparison of various leakage minimization Techniques

Technique	Advantages	Limitations
Power-gating techniques		
Sleep transistor [31]	Subthreshold leakage decreases	Data retention problem
Forced stack [32]	Substantial leakage saving, easy implementation	Circuit delay increases, need extra circuitry to retain the state.
Sleepy stack [33]	Delay is less with respect to the forced stacking	Leakage minimization is not significant compared to forced stacking, transistor count increases.
Data retention gated ground [34]	Data retention, low leakage	External control signal is required, delay penalty
N-control SRAM cell with gated VDD [35]	Low leakage and high speed	Virtual ground needs to be controlled
Biasing techniques		
Reverse body biasing [37]	Subthreshold leakage minimized	Sensitive to V_{th} variation & GIDL
Forward body biasing [38]	Small switching capacitance and speed is high	Sensitive to SCEs and DIBL
Variable body biasing [39]	Static power consumption is reduced	External voltage sources are required to provide body bias
Transistor stacking [40]	Subthreshold leakage and Gate leakage minimized	Area penalty
Dual-V_{th} [41]	Improved short channel and DIBL	Need of FBB
VTCMOS [42]	High speed, low leakage	Body bias is to be controlled by additional control circuitry
MTCMOS [43,44]	More reduction in leakage power	Need of high V_{th}

9.7 CONCLUSION

In this chapter, the origin of short channel device's leakage currents has been discussed. We show the various leakage minimization strategies put out for low-power SRAM cell construction. Analysis of various leakage minimization techniques is classified according to their basic structure and workings, such as biasing techniques and power-gating techniques. To manage subthreshold leakages, biasing strategies concentrate on adjusting the threshold voltage. Although the RBB design minimizes leakage, it has an impact on performance. A FBB design that has been optimized can aid in achieving great performance with little power consumption. The circuit's leakage path is where the power-gating technique places the most emphasis. To manage the leakage currents, additional circuitry is appended to establish a virtual GND in the leakage path. The one and only goal of all these methods is to reduce leakage power dissipation in the nanoscale region. The majority of methods focused in minimizing subthreshold leakage. Some solutions focus strongly on maintaining the data during standby mode. Table 9.1 provides a quick overview of the various leakage control techniques, their benefits and drawbacks, as well as the restrictions associated with their use. Researchers will be helped by the leakage minimization techniques provided in this chapter as they attempt to design new, low-power memory designs for ultralow power applications.

REFERENCES

[1] Lorenzo, R., & Chaudhury, S. (2017). Review of circuit level leakage minimization techniques in CMOS VLSI circuits. *IETE Technical Review, 34*(2), 165–187.

[2] Mead, C. A. (1994). Scaling of MOS technology to submicrometer feature sizes. *Analog Integrated Circuits and Signal Processing, 6*(1), 9–25.

[3] Amelifard, B., Fallah, F., & Pedram, M. (2008). Leakage minimization of SRAM cells in a Dual-$ V_t $ and Dual-$ T_{\rm ox} $ Technology. *IEEE Transactions on Very Large Scale Integration (VLSI) Systems, 16*(7), 851–860.

[4] Lorenzo, R., Pradeep, D. L., & Kumar, A. P. (2022). Low power 8T SRAM with high stability and bit interleaving capability. In *2022 2nd International Conference on Emerging Frontiers in Electrical and Electronic Technologies (ICEFEET)* (pp. 1–6). IEEE, NIT-Patna, India.

[5] Samandari-Rad, J., & Hughey, R. (2016). Power/energy minimization techniques for variability-aware high-performance 16-nm 6T-SRAM. *IEEE Access, 4*, 594–613.

[6] Rao, R., Srivastava, A., Blaauw, D., & Sylvester, D. (2004). Statistical analysis of subthreshold leakage current for VLSI circuits. *IEEE Transactions on Very Large Scale Integration (VLSI) Systems, 12*(2), 131–139.

[7] Xue, J., Li, T., Deng, Y., & Yu, Z. (2010). Full-chip leakage analysis for 65 nm CMOS technology and beyond. *Integration, 43*(4), 353–364.

[8] Lorenzo, R., & Chaudhury, S. (2017). Dynamic threshold sleep transistor technique for high speed and low leakage in CMOS circuits. *Circuits, Systems, and Signal Processing, 36*(7), 2654–2671.

[9] Prasad, G., & Anand, A. (2015). Statistical analysis of low-power SRAM cell structure. *Analog Integrated Circuits and Signal Processing, 82*(1), 349–358.

[10] Lorenzo, R., & Chaudhary, S. (2013). A novel all NMOS leakage feedback with data retention technique. In *2013 International Conference on Control, Automation, Robotics and Embedded Systems (CARE)* (pp. 1–5). IEEE, Jabalpur, India.

[11] Calimera, A., Macii, A., Macii, E., & Poncino, M. (2012). Design techniques and architectures for low-leakage SRAMs. *IEEE Transactions on Circuits and Systems I: Regular Papers, 59*(9), 1992–2007.

[12] Kumar, T. S., & Tripathi, S. L. (2021). Leakage reduction in 18 nm FinFET based 7T SRAM cell using self controllable voltage level technique. *Wireless Personal Communications, 116*(3), 1837–1847.

[13] Wang, J., An, H., Zhang, Q., Kim, H. S., Blaauw, D., & Sylvester, D. (2020). A 40-nm ultra-low leakage voltage-stacked SRAM for intelligent IoT sensors. *IEEE Solid-State Circuits Letters, 4*, 14–17.

[14] Lorenzo, R., & Chaudhury, S. (2017). A novel 9T SRAM architecture for low leakage and high performance. *Analog Integrated Circuits and Signal Processing, 92*(2), 315–325.

[15] Kumar, A. P., & Lorenzo, R. (2022). Performance analysis of DMG-GOS junctionless FinFET with high-k spacer. In *2022 IEEE Silchar Subsection Conference (SILCON)* (pp. 1–5), NIT-Silchar, India.

[16] Abbasian, E., Izadinasab, F., & Gholipour, M. (2022). A reliable low standby power 10T SRAM cell with expanded static noise margins. *IEEE Transactions on Circuits and Systems I: Regular Papers, 69*(4), 1606–1616.

[17] Kim, K. K., Kim, Y. B., & Choi, K. (2011). Hybrid CMOS and CNFET power gating in ultralow voltage design. *IEEE Transactions on Nanotechnology, 10*(6), 1439–1448.

[18] Vanhoof, B., & Dehaene, W. (2022). SRAM with stability monitoring and body bias tuning for biomedical applications. *IEEE Solid-State Circuits Letters, 5*, 29–32.

[19] Narendra, S., Keshavarzi, A., Bloechel, B. A., Borkar, S., & De, V. (2003). Forward body bias for microprocessors in 130-nm technology generation and beyond. *IEEE Journal of Solid-State Circuits, 38*(5), 696–701.

[20] Lorenzo, R., & Chaudhury, S. (2016). Optimal body bias to control stability, leakage and speed in SRAM cell. *Journal of Circuits, Systems and Computers, 25*(08), 1650096.

[21] Lorenzo, R., & Pailly, R. (2020). Single bit-line 11T SRAM cell for low power and improved stability. *IET Computers & Digital Techniques, 14*(3), 114–121.

[22] Turi, M. A., & Delgado-Frias, J. G. (2019). Effective low leakage 6T and 8T FinFET SRAMs: Using cells with reverse-biased FinFETs, near-threshold operation, and power gating. *IEEE Transactions on Circuits and Systems II: Express Briefs, 67*(4), 765–769.

[23] Yoshida, H., Shiotsu, Y., Kitagata, D., & Sugahara, S. (2021). Ultralow-voltage retention SRAM with a power gating cell architecture using header and footer power-switches. *IEEE Open Journal of Circuits and Systems, 2*, 520–533.

[24] Deshmukh, J., & Khare, K. (2012). Dynamic SVL and body bias for low leakage power and high performance in CMOS digital circuits. *International Journal of Electronics, 99*(12), 1717–1728.

[25] Yadav, S., Malik, N., Gupta, A., & Rajput, S. (2013). Low power SRAM design with reduced read/write time. *International Journal of Information and Computation Technology, 3*(3), 195–200.

[26] Roy, K., Mukhopadhyay, S., & Mahmoodi-Meimand, H. (2003). Leakage current mechanisms and leakage reduction techniques in deep-submicrometer CMOS circuits. *Proceedings of the IEEE, 91*(2), 305–327.

[27] Narendra, S., De, V., Borkar, S., Antoniadis, D. A., & Chandrakasan, A. P. (2004). Full-chip subthreshold leakage power prediction and reduction techniques for sub-0.18-/spl mu/m CMOS. *IEEE Journal of Solid-State Circuits, 39*(3), 501–510.

[28] Xu, Y., Chi, B., & Wang, Z. (2013). Gate-leakage compensation scheme for programmable SI-DAC of $\Sigma\Delta$ modulator in deep sub-micron. *Analog Integrated Circuits and Signal Processing, 76*(1), 155–160.

[29] Lorenzo, R., & Chaudhury, S. (2014). Body biasing scheme to control leakage, speed and stability in SRAM cell design. *International Journal of Computer Applications, 975*, 8887.

[30] Goel, A., Sharma, R. K., & Gupta, A. (2014). Area efficient diode and on transistor inter-changeable power gating scheme with trim options for SRAM design in nano-complementary metal oxide semiconductor technology. *IET Circuits, Devices & Systems, 8*(2), 100–106.

[31] Rajani, H. P., & Kulkarni, S. (2012). Novel sleep transistor techniques for low leakage power peripheral circuits. *International Journal of VLSI Design & Communication Systems, 3*(4), 81.

[32] Prasad, S. R., Madhavi, B. K., & Kishore, K. L. (2012). Design of 32nm forced stack CNTFET SRAM cell for leakage power reduction. In *2012 International Conference on Computing, Electronics and Electrical Technologies (ICCEET)* (pp. 629–633). IEEE, Nagercoil, India.

[33] Park, J. C., & Mooney III, V. J. (2006). Sleepy stack leakage reduction. *IEEE Transactions on Very Large Scale Integration (VLSI) Systems, 14*(11), 1250–1263.

[34] Bikki, P., & Karuppanan, P. (2017). SRAM cell leakage control techniques for ultra low power application: A survey. *Circuits and Systems, 8*(02), 23.

[35] Elakkumanan, P., Narasimhan, A., & Sridhar, R. (2003). NC-SRAM-A low-leakage memory circuit for ultra deep submicron designs. In *IEEE International [Systems-on-Chip] SOC Conference, 2003. Proceedings.* (pp. 3–6). IEEE, Portland, OR, USA.

[36] Shukla, N. K., Singh, R. K., & Pattanaik, M. (2011). Design and analysis of a novel low-power SRAM bit-cell structure at deep-sub-micron CMOS technology for mobile multimedia applications. *International Journal of Advanced, 2*(5), 43–49.

[37] Lorenzo, R., & Chaudhury, S. (2017). A novel SRAM cell design with a body-bias controller circuit for low leakage, high speed and improved stability. *Wireless Personal Communications, 94*(4), 3513–3529.

[38] Kim, C. H., Kim, J. J., Mukhopadhyay, S., & Roy, K. (2005). A forward body-biased low-leakage SRAM cache: Device, circuit and architecture considerations. *IEEE Transactions on Very Large Scale Integration (VLSI) Systems, 13*(3), 349–357.

[39] Kai, W. W., Ahmad, N. B., & Jabbar, M. H. B. (2017). Variable body biasing (VBB) based VLSI design approach to reduce static power. *International Journal of Electrical & Computer Engineering (2088-8708), 7*(6), 3010–3019.

[40] Neema, V., Chouhan, S. S., & Tokekar, S. (2010). Novel circuit technique for reduction of leakage current in series/parallel PMOS/NMOS transistor stack. *IETE Journal of Research*, *56*(6), 362–366.

[41] Ni, H., Hu, J., Yang, H., & Zhu, H. (2018). Comprehensive optimization of dual threshold independent-gate FinFET and SRAM cells. *Active and Passive Electronic Components*, *2018*, pp. 1–10.

[42] Deepaksubramanyan, B. S., & Nunez, A. (2007). Analysis of subthreshold leakage reduction in CMOS digital circuits. In *2007 50th Midwest Symposium on Circuits and Systems* (pp. 1400–1404). IEEE.

[43] Tripathi, T., Chauhan, D. S., Singh, S. K., & Singh, S. V. (2017). Implementation of low-power 6T SRAM cell using MTCMOS technique. In Sanjiv K. Bhatia, Krishn K. Mishra, Shailesh Tiwari, & Vivek Kumar Singh (Eds.), *Advances in Computer and Computational Sciences* (pp. 475–482). Springer, Singapore.

[44] Shibata, N., Kiya, H., Kurita, S., Okamoto, H., Tanno, M., & Douseki, T. (2006). A 0.5-V 25-MHz 1-mW 256-kb MTCMOS/SOI SRAM for solar-power-operated portable personal digital equipment-sure write operation by using step-down negatively overdriven bitline scheme. *IEEE Journal of Solid-State Circuits*, *41*(3), 728–742.

Chapter 10

Nonvolatile configurable logic block for FPGAs

Chee Hock Leong, T. Nandha Kumar,
and Haider A.F. Almurib
University of Nottingham Malaysia

10.1 INTRODUCTION

10.1.1 Field-programmable gate arrays

Field-programmable gate arrays (FPGAs) were first designed in the 1980s and are very-large-scale integration (VLSI) integrated circuits (ICs) that are highly configurable and can implement any digital hardware configuration. This configurability is achieved in the hardware level through look-up tables (LUTs) that can implement any Boolean logic and customizable connections between components and gates in the FPGA. This configurability on a dedicated hardware level gives the FPGA a speed advantage over microprocessors and is preferable over application-specific integration circuits (ASICs) as FPGAs allow reprogramming in case of a bug in the design, thus reducing non-recurring engineering costs and having shorter time-to-market [1,2].

The most basic architecture of FPGAs are programmable blocks of logic called Logic Blocks (LBs) which are interconnected through programmable routings called configurable interconnects. The LBs are the main digital processing resources of the FPGA and inside the LB, LUTs enable combinatorial logic operations while D Flip-Flops (DFFs) enable sequential logic (clock input-based) operations. New generation LUTs are capable of additional functions such as local storage (distributed RAM), shift register (SR), multiplexer, and adder/subtractor operations [3,4].

Since its conception, FPGAs have been based on three technologies namely, flash, anti-fuse, and static random-access memories (SRAMs) [1]. SRAM-based FPGAs are the most common commercial devices due to the cutting-edge process and maturity of transistor technology, leading to advantages in cost per function and silicon space compared to the previous technologies [5]. Moore's Law which is the ability for transistor technology nodes to shrink by half every 2 years has been the driving factor for the adoption of semiconductor transistors as the number of transistors in the same area of silicon wafer has doubled every generation for over three decades. However, the semiconductor transistor industry has been struggling to continue with the node downscaling (single-digit nanometer gate lengths) in recent times due to the physical limitations of the metal-oxide-silicon field-effect transistor (MOSFET) [6].

Numerous research studies have been undertaken on emerging new memory technologies that are capable of replacing the SRAM. Stand-out candidates for emerging memories are nonvolatile (NV) and include the phase change memory (PCM), spin-torque-transfer magnetic random-access-memory (STT-MRAM), ferroelectric field-effect transistor (FeFET), and resistive random-access memory (ReRAM). PCMs, STT-MRAMs, and ReRAMs are resistive memories, whereby they undergo a change in resistance to store or lose data [7]. Where they differ is the underlying physical mechanisms to achieve resistive switching (RS); PCM undergoes a phase transition of the active material [8], STT-MRAMs control the parallel/anti-parallel orientations of its ferromagnetic layers [9], and ReRAMs form/break a conductive filament (CF) in the active layer. FeFETs replace the transistor gate dielectric with a ferroelectric material that modulates the channel conductance [10]. Among these emerging devices, the ReRAM is very promising with good scalability, speed, energy efficiency, ease of fabrication, simple structure, and compatibility with complementary metal-oxide semiconductor (CMOS) technology [9,11,12]. ReRAMs have also demonstrated intrinsic multi-bit capability where more than two bits can be stored in a single cell.

This chapter will delve into NV circuit designs for the LB in the FPGA including single-bit (SB) and multi-bit (MB) designs for the NV LUT together with the NV LUT controller and an NV design of a DFF. Circuit simulation tests are carried out for the NV circuits to understand the circuit behaviour and to analyse the circuit delay, energy dissipation, and energy delay product (EDP).

10.1.2 The logic block

LBs are complex fundamental components in the FPGA where combinatorial and sequential logic operations are carried out. In modern FPGAs, the LBs contain clusters of basic logic elements (BLEs) that contain LUTs and DFFs that are responsible for the combinatorial and sequential operations, respectively; an example of a LB (called CLB) in a Xilinx Virtex5 FPGA is shown in Figure 10.1a and its BLEs (called Slice) in Figure 10.1b [13]. Any K-input Boolean function can be implemented in a K-input LUT which stores the function's truth table in the SRAM-LUT. An example of an SRAM-LUT is given in Figure 10.2 where the K-inputs select the appropriate output from the 2^K values in the truth table [3]. The Virtex5 BLE in Figure 10.1b has the 6-input LUT coupled with an adder component and a COUT output, a register that has DFF and latch functionalities, and bypassing multiplexers.

Thus, the BLE outputs can be adder outputs, COUT and DMUX, unregistered LUT output, D, or the registered output, DQ. The register contains a DFF which ties the LUT output to the FPGA clock signal for edge-triggered designs. As the LUT is made up of SRAMs, the LBs are fully configurable and allow implementation of any kind of circuit design in the FPGA by storing all Boolean logic function outputs in a truth table.

(a)

COUT   COUT

CLB

Switch
Matrix

Slice(1)

Slice(0)

CIN   CIN

(b)

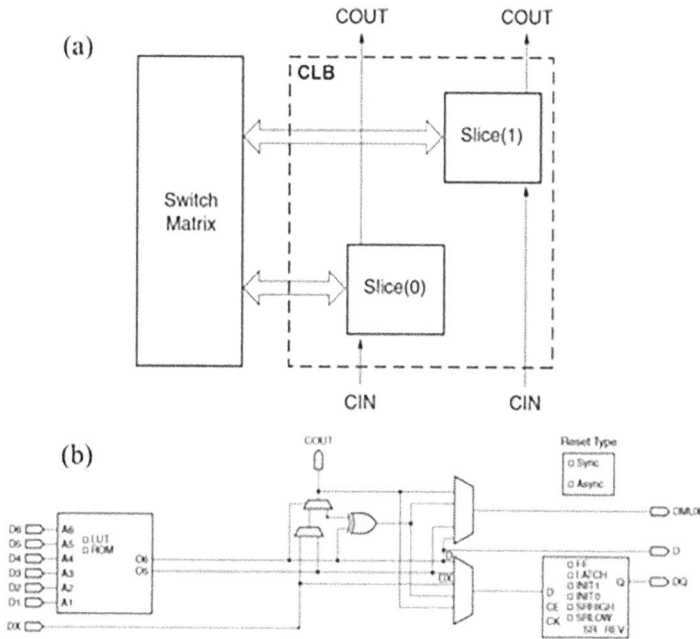

Figure 10.1 Xilinx Virtex5 FPGA (a) LB and (b) BLE [13].

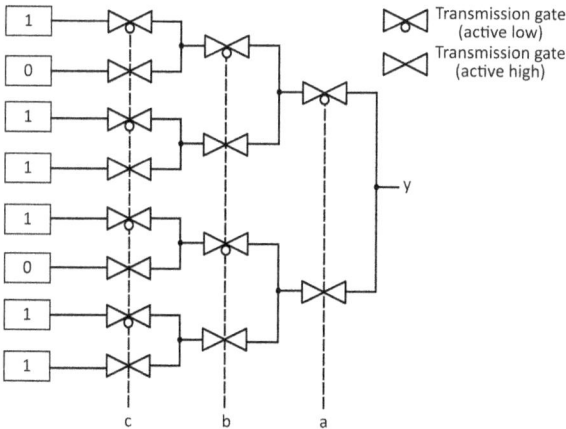

Figure 10.2 3-input SRAM-LUT. The a, b, and c inputs select the appropriate value for output y [1].

10.1.3 Resistive random-access memory

ReRAMs are memristor devices with two terminals which are made up of top and bottom metal electrodes. In between these electrodes is a transition-metal oxide (TMO) insulator which gives a metal-insulator-metal (MIM) structure (Figure 10.3) [14]. The TMO layer can switch between low and high resistance stable oxidation states, allowing the storage of binary data with a high '1' and a low '0'. ReRAMs typically require a forming process where a voltage supply is connected across the ReRAM to create a CF in the TMO layer that connects the two electrodes. A fully formed CF lowers the resistance of the ReRAM and rupturing the formed CF by supplying a voltage of reverse polarity increases the device resistance. Both CF rupture and formation processes are repeatable and allow the ReRAM to continuously switch between a low resistance state (LRS) and high resistance state (HRS).

As RS in ReRAMs operates behind the formation/rupture of CFs, it is possible to halt the process before the CF is fully formed/ruptured. Figure 10.3 shows the four possible CF states; CF1 is the fully ruptured HRS CF, CF1+CF2, and CF1+CF2+CF3 are intermediate ruptured CF states of different lengths, and CF1+CF2+CF3+CF4 is the fully formed LRS CF. This incomplete switching provides intermediate resistive states (IRSs) and opens up the possibility for multiple-bits-per-cell ReRAMs, also known as multibit-ReRAMs (MB-ReRAMs), thereby increasing the number of bits per cell and leading to higher density ReRAM arrays [15].

Another benefit of MB ReRAMs is that no additional components are required to achieve MB storage in ReRAMs since the only requirement is to halt the RS process before the CF reaches the fully formed/ruptured thresholds. The only requirement for MB ReRAMs is precise voltage control so that there is a wide margin between every resistive level in the cell to avoid data write errors and a modified READ scheme or sense amplifying circuit which is a general requirement for all MB devices.

Circuit models of the ReRAM have been created to facilitate the design and testing of ReRAM-based circuits. Generally, the formation and rupture of the CF represent the resistive state of the ReRAM and are captured

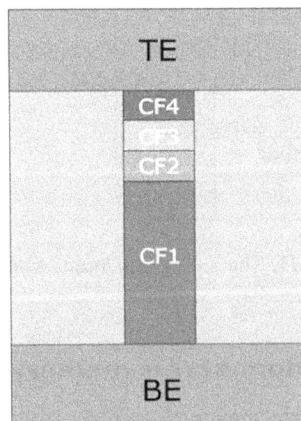

Figure 10.3 The MIM ReRAM cell. The forming/rupture of the CF is formed in the TMO layer.

by an electric-field-induced state variable in the models. This is then taken into account by the model's current equation to describe the current through the ReRAM. The differences between models come down to the physical accuracy of the model by using equations that describe the RS physical phenomena or generic mathematical models that model the device with mathematically fitting equations, sacrificing physical accuracy for simulation speed.

10.2 SINGLE-BIT NONVOLATILE LUT

The schematic for a single-bit ReRAM (SB ReRAM) 2-input NV LUT is given in Figure 10.4 switching provides IRSs and opens up the possibility for multiple-bits-per-cell ReRAMs, also known as multibit-ReRAMs (MB-ReRAMs), thereby increasing the x shared between two ReRAM cells (M11 and M21 are connected to BL1, M12 and M22 are connected to BL2). The separate wordlines prevent sneak current flowing from a selected cell to an unselected ReRAM cell in the array. The WRITE and READ operations are handled by the controller and its schematic is given in Figure 10.5.

The eight input pins, D0, D1, A, B, W_{EN}, R_{EN}, C, and Reset in the controller. The Reset line is always high during controller operation. The R_{EN} and W_{EN} inputs function as the respective READ and WRITE enable signals in the controller. The input pins, A and B are responsible for selecting the appropriate address of the ReRAM array as well as the output fed to the OUT pin in READ mode.

The C input selects the WRITE destination column in the LUT array during WRITE. Column-parallel WRITE operations are performed for all ReRAMs in a row in this design. The controller's logic is given in Table 10.1.

10.2.1 Circuit simulation

Simulation tests are carried out to check for any disturbance on the unselected cells when WRITE and READ operations are carried out for each ReRAM in the 2×2 array (M12, M21, and M22 states are measured when writing or reading from M11, etc.). The writing scheme will encompass writing '0' and '1' to each ReRAM. The initial state variable of the ReRAM model is 0.2 and a –2 V RESET input voltage is supplied to switch the ReRAM into HRS with a state variable of 0.

For WRITE operations, a pulsewidth of ± 2 V is supplied for 0.5 ms while a pulsewidth input of +2 V for 0.1 ms is supplied for READ. Although READ operations can be carried out with voltages of lower amplitudes, this work uses a higher amplitude to induce a higher READ current for presentation. Figure 10.6 shows each state variable for the ReRAMs in the array during consecutive WRITE '1' and '0'.

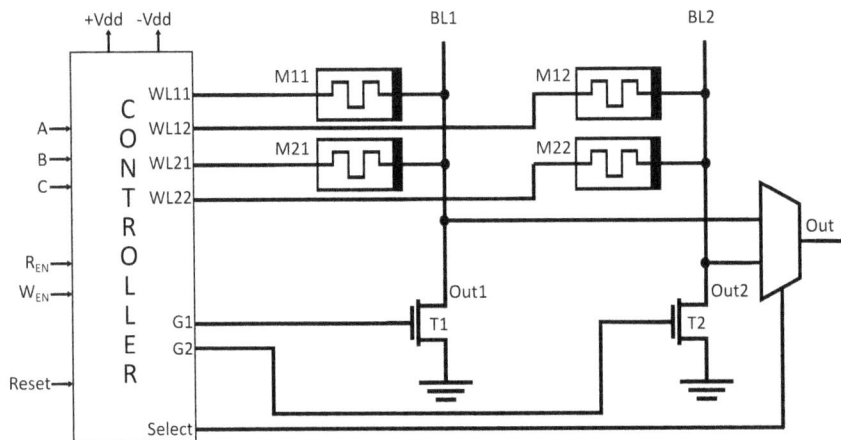

Figure 10.4 Circuit diagram of the SB NV LUT array taken from [17].

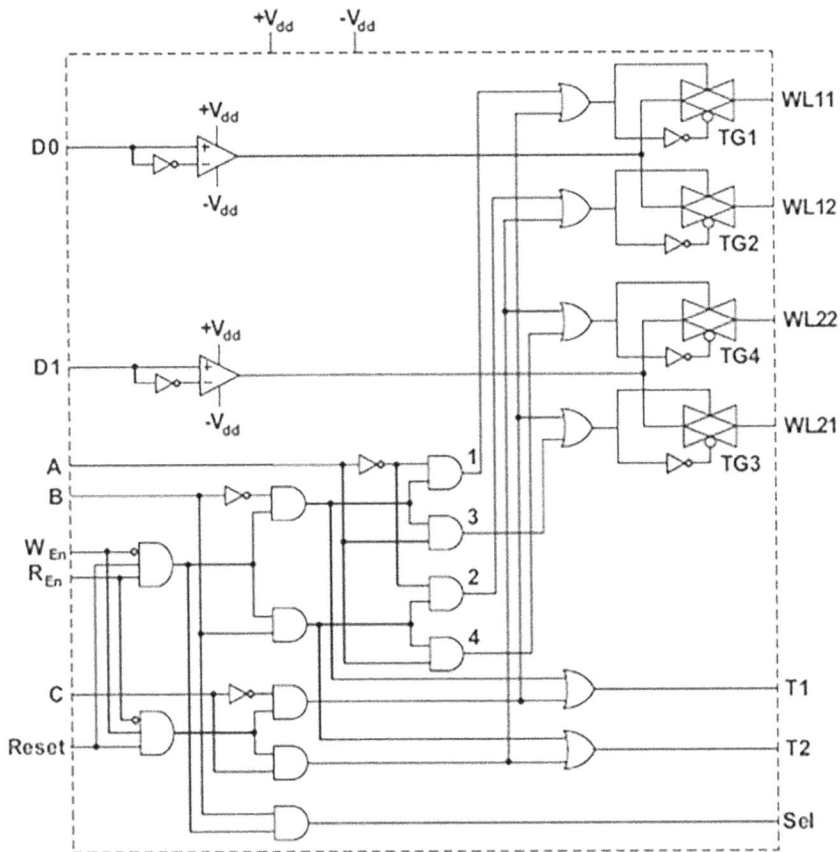

Figure 10.5 Circuit diagram of the controller taken from [17].

Table 10.1 MB NV LUT Controller Logic

Operation	Signals							ReRAM Input			
	W_{EN}	R_{EN}	T1	T2	A	B	C	M11	M12	M21	M22
READ	Low	High	High	Low	0	0	$-^a$	D0	X^b	X^b	X^b
	Low	High	Low	High	0	1	$-^a$	X^b	D0	X^b	X^b
	Low	High	High	Low	1	0	$-^a$	X^b	X^b	D1	X^b
	Low	High	Low	High	1	1	$-^a$	X^b	X^b	X^b	D1
WRITE	High	Low	High	Low	$-^a$	$-^a$	0		X^b	D1	X^b
	High	Low	Low	High	$-^a$	$-^a$	1	X^b	D0	X^b	D1

^{a} — is ignored.
^{b} x is floating

The selected ReRAM achieves full SET with the state variable reaching 1 and RESET with a state variable of 0. Additionally, the resistive states of the unselected ReRAMs are not disturbed (the highest recorded variation was 0.003), demonstrating that the design successfully eliminates the sneak current problem and half-select effect during WRITE.

In the READ test, the READ operation was carried out on the selected ReRAM for both '0' and '1' states. The selected ReRAM is initially switched to '0' before writing '1' or '1' before writing '0'. The measured output current then indicates the state of the selected ReRAM. Figure 10.7a and b

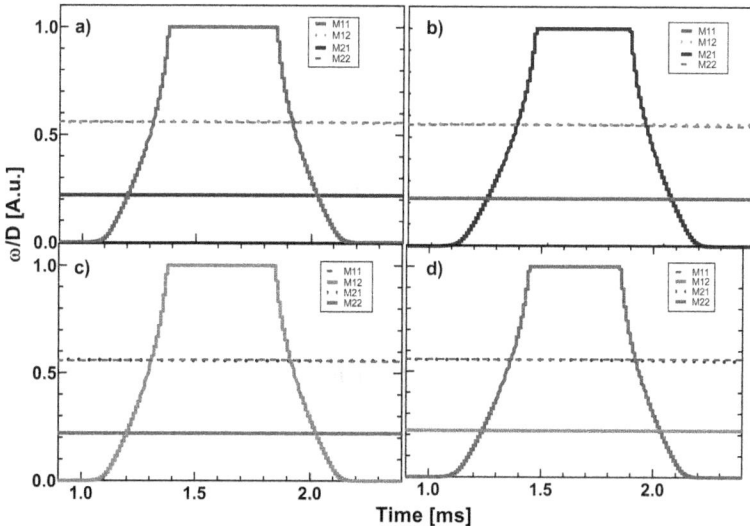

Figure 10.6 Effect of WRITE '1' followed by WRITE '0' on the states of M11, M21, M12, and M22 (a, b, c, d). The states of the unselected ReRAMs are also plotted.

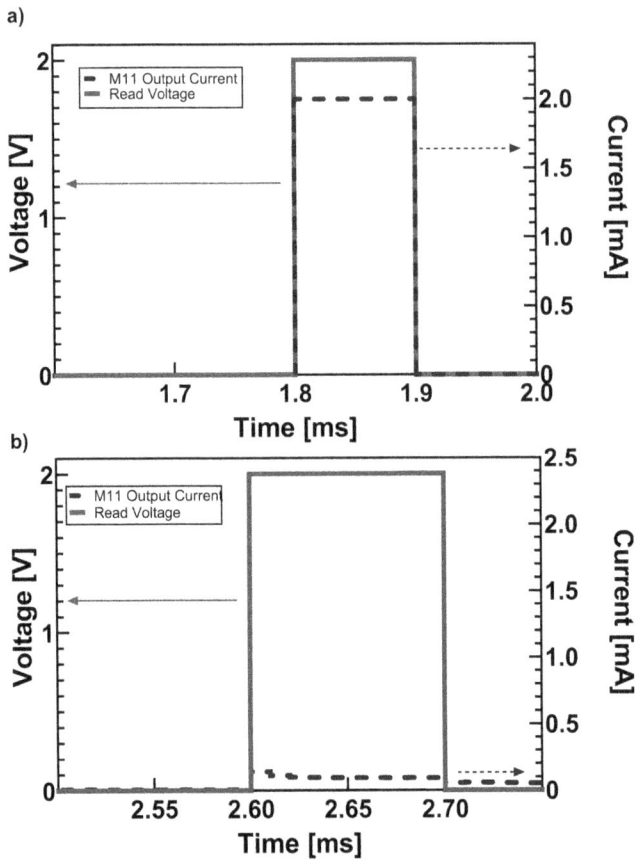

Figure 10.7 Current output for READ for a ReRAM cell in state (a) 1 and (b) 0.

show the results for READ '1' and '0' respectively. The correct output current values demonstrate the controller's ability to achieve successful writing and reading.

NV is implemented into the LUT through the use of SB ReRAMs that retain their states even with the removal of power supply. The LUT array contains the 2^K number of cells for K inputs, similar to the SRAM LUT. An opportunity exists to utilize the intrinsic MB capability of ReRAMs to increase the density of the LUT array.

10.3 MULTI-BIT NONVOLATILE LUT

In this section, the MB NV LUT design is presented with the analysis of the NV LUT's electrical behaviour obtained and compared with the SB

NV LUT. The implementation of MB ReRAMs in (c and d respectively). A MB NV LUT first requires the selection of an appropriate MB switching scheme.

Figure 10.8 depicts the pule-width input voltage used in this scheme with these paramaters: 0.4 ns of +1 V for LRS, 0.46 ns of −0.5V for HRS to implement SB switching and 0.4 ns of +1 V for LRS, 0.06 ns of −0.5 V for IRS-1, 0.1 ns of −0.5 V for IRS-2, and 0.3 ns of −0.5 V for HRS switching. HRS switching utilizes lower voltages compared to LRS switching to enable accurate control of IRS switching for the MB cell.

Figure 10.9 shows a comparison between the SB and MB current and power consumption. The highest current and power consumption for LRS switching in SB and MB devices are both ~440fA (Figure 10.9a and b) and ~500fJ (Figure 10.9c and d) respectively as they undergo the same LRS switching input voltages.

The highest device current for SB and MB cells in HRS switching is ~200fA because the cells are in LRS just before HRS switching is initiated (Figure 10.9a and b). The similarity in overall HRS power consumption for both ReRAMs at 4.7pJ despite the additional IRS states that exist in the MB switching scheme can be attributed to the identical durations of the total HRS voltage supply time for both ReRAMs.

However, the MB ReRAM switches two bits per switching operation ('11'→'00') compared to the SB ReRAM which only switches a single bit per operation ('1'→'0').

10.3.1 Multi-bit ReRAM LUT array

The MB NV LUT array schematic is given in Figure 10.10 [16]. Each wordline contains a ReRAM cell (Mn) which has an output transistor (TGn) controlled bitline connected to the opposite side. WRITE and READ voltages

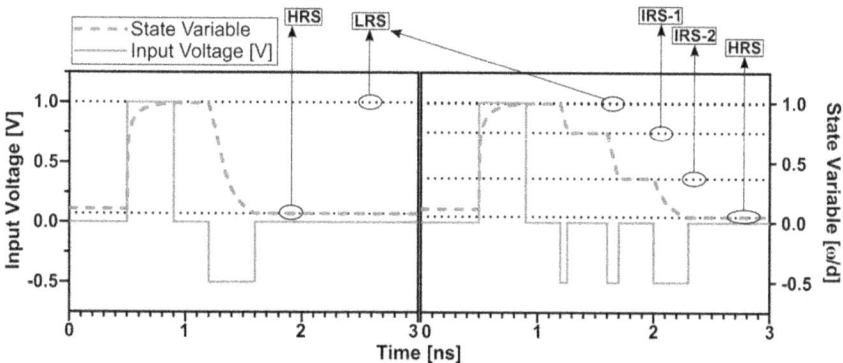

Figure 10.8 Switching behaviour comparison between SB-switching (left) and MB-switching (right).

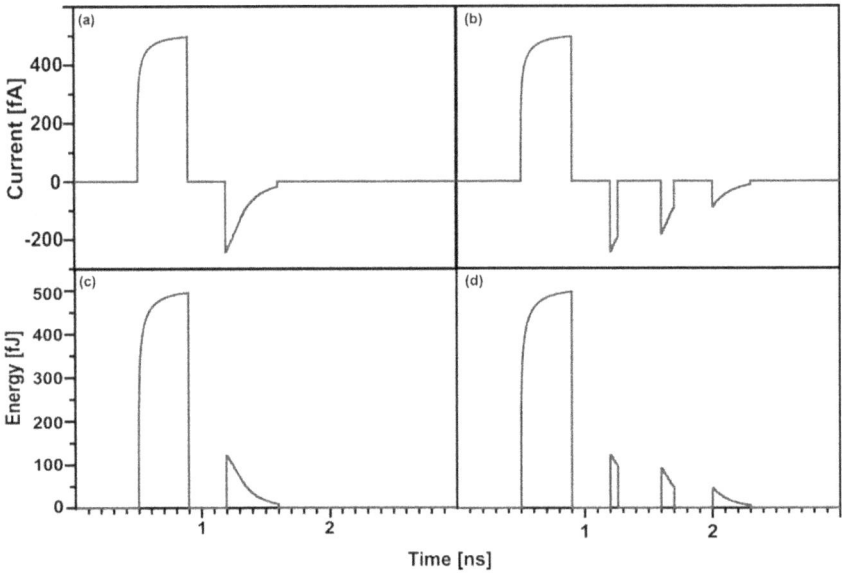

Figure 10.9 Plot of SB and MB device currents (a and b respectively) and power consumptions.

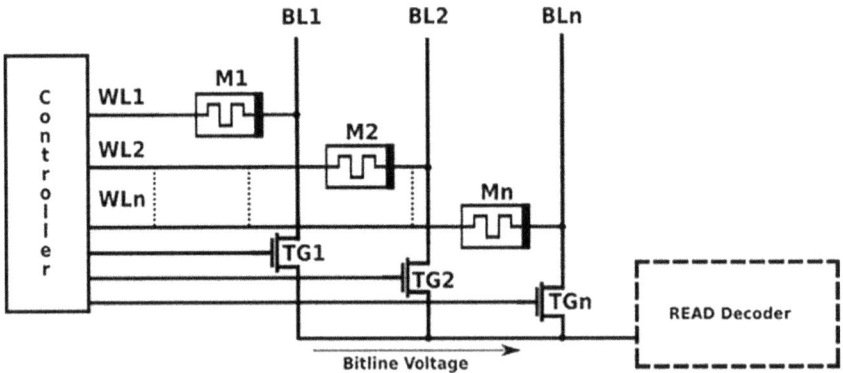

Figure 10.10 The MB ReRAM array.

are fed into the ReRAM cells from the wordline and the ReRAM output voltages passes through the bitlines to the READ decoder. The controller is responsible for activating the wordlines and the bitlines.

The MB ReRAM holds two address locations, and the four possible output-bit combinations for the two addresses can be represented by the four resistive states of the MB ReRAM. For example, M1 holds the addresses AB='00' and AB='01' while AB='10' and AB='11' are located in M2 in

Figure 10.10. For typical LUT operations, a single output bit, b of either '0' or '1' is stored in a single address such as AB='00'. Therefore a combination of two addresses like

AB='00' and AB='01' in M1 can only contain four output-bit combinations for $b_{00}b_{01}$ which are '00', '01', '10', and '11'. This is shown in the logic table in Table 10.2.

The A and B columns for the memory address are the typical A and B inputs in conventional LUTs. However, the addresses for AB='00' and AB='01' are now contained in M1 while the addresses for AB='10' and AB='11' are now contained in M2 which accounts for all the necessary addresses for a 2-input LUT. Each resistive state now contains the output bits of two addresses and covers all the possible output bit combinations like described above.

In Table 10.2, the HRS ReRAM has an xy output of '00' where AB='00's output bit is '0' and AB='01's output bit is '0'.

In another example, the IRS-1 ReRAM has an xy output-bit of '10', meaning that AB='00's output bit is '1' and AB='01's output bit is '0'. For the sake of comparison, the SB NV LUT logic table is provided in Table 10.3 where a single memory address and output bit are stored in a solitary ReRAM cell. The number of MB NV LUT array cells is now 2^{K-1} using this storage scheme and is reduced compared to the 2^K of the SB NV LUT.

10.3.2 The MB LUT controller

The different WRITE and READ requirements of the MB NV LUT array necessitate the design of a novel switching controller circuit which is presented in Figure 10.11a. The controller consists of sub-blocks where each block controls two wordlines and enables larger arrays through the addition of further sub-blocks.

In Figure 10.11a, the signals DATA1, DATA2, DATA3, DATA4, W_{EN}, and R_{EN} and an input from an external decoder make up the input lines for Controller Block 1 while the output lines contain the wordlines, WL1 and WL2 and outputs, G1 and G2. As it supplies the outputs for two wordlines, it can be seen that Controller Block 1 functions as a 2-input LUT on its own. The input decoder signal follows the logic in Table 10.2, selecting control lines WL1 and TG1 for M1 when A='0' and WL2 and TG2 when A='1'. This way, only the required wordline and bitline are activated, thus ensuring that the unselected output lines remain in high impedance.

The x and y output bit values are supplied by DATA1 and DATA2 lines for M1 and by DATA3 and DATA4 for M2 which activates the corresponding WRITE transistors, T11-T14 and T21-T24. This operation is coupled with the enable input for WRITE, W_{EN}. The enable input for READ, R_{EN} activates the T15 or T25 READ transistor depending on address selection.

Figure 10.11b shows an example of how multiple controller blocks can be cascaded to accommodate a larger-sized LUT, in this case, four controller blocks are used to implement a 4-input MB NV LUT array.

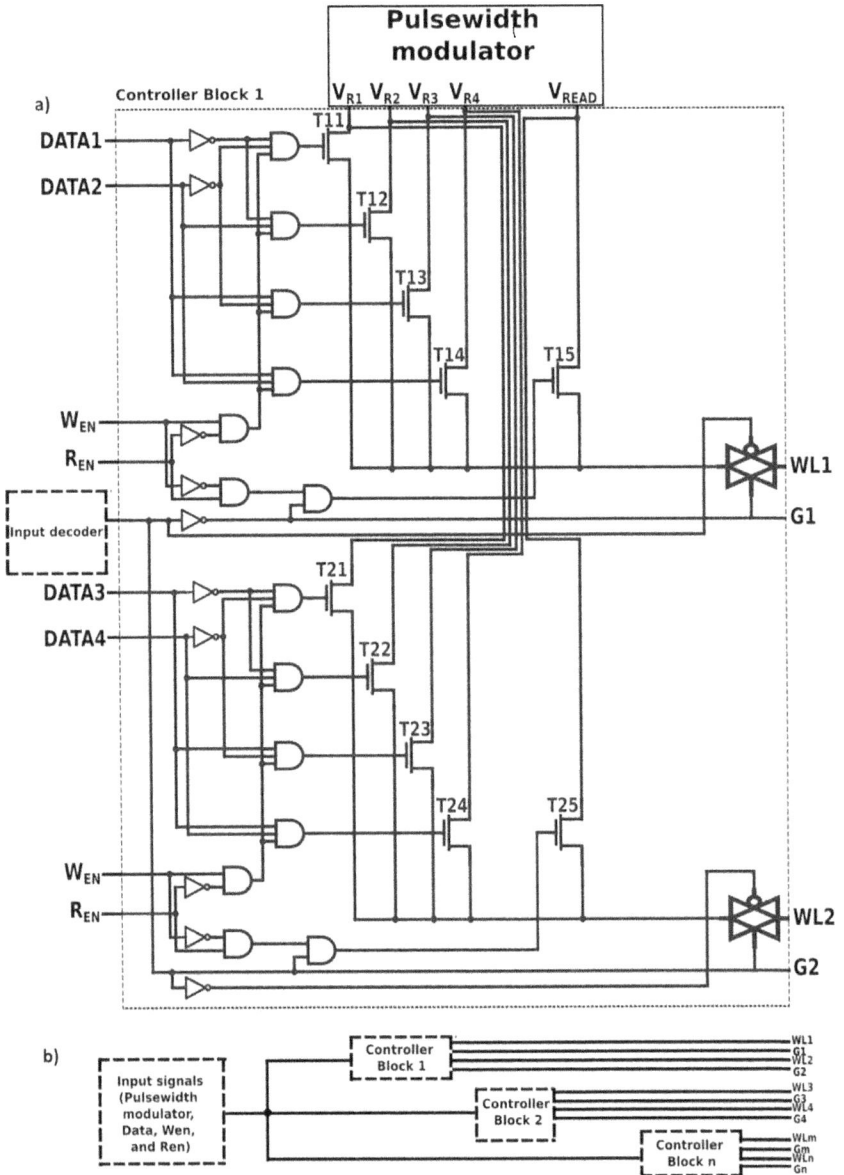

Figure 10.11 (a) MB NV LUT controller block schematic. This block is capable of controlling a 2-input LUT. (b) Higher-input-LUTs can be implemented through additional controller blocks.

10.3.3 WRITE operation

To initiate the WRITE operation, the W_{EN} signal is raised and either WL1 or WL2 is selected through the input at A. Depending on this selection, DATA1 and DATA2 select the appropriate resistance state for M1 or DATA3 and DATA4 select the appropriate resistance state for M2. DATA1 and DATA3 are the x bits while DATA2 and DATA4 are the y bits in Table 10.2, respectively. These xy values then switch on the correct transistor gates (T11, T12, T13, or T14 for M1; T21, T22, T23, or T24 for M2) to direct the voltages for WRITE operation, V_{R1}, V_{R2}, V_{R3}, or V_{R4} to the array cell. The voltage values follow the scheme in Figure 10.8 in which V_{R1} writes the cell to HRS, V_{R2} writes the cell to IRS-2, V_{R3} writes the cell to IRS-1, and V_{R4} writes the cell to LRS.

10.3.4 READ operation

Similar to the WRITE operation, the R_{EN} signal operates as the READ enable and is raised to initiate the READ operation and either WL1 or WL2 are selected based on the input at A. The pass transistors TG1 or TG2 (Figure 10.10) which are connected at the bitline are also activated this way. Depending on which WL is selected, the READ transistors, TG14 or TG15 are activated to pass through READ voltage to the array cell. The output voltage in the bitline is then directed into the READ decoder (Figure 10.12) where a comparator compares with four reference voltages for the four resistive states (four combined output bits). Referring to Table 10.2, the reference voltage values correspond to V_{ref1} for HRS when $xy = $ '00', V_{ref2} for IRS-2 when $xy = $ '01', V_{ref3} for

IRS-1 when $xy = $ '10', and V_{ref4} for LRS when $xy = $ '11'. The comparator's analog output is then converted into digital through an analog-digital

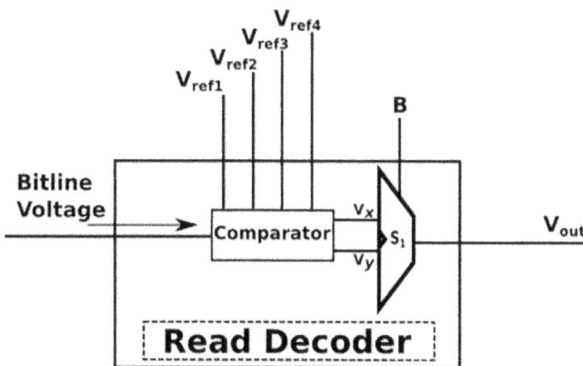

Figure 10.12 Block diagram of the READ controller circuit designed for two input MB LUT.

converter (ADC). The x and y bits can then be extracted from the digital voltage by the selector, S_1.

10.3.5 Simulation results

Simulations for 2-, 4-, and 6-, and 8-input LUTs are carried out to analyse the design of the MB NV LUT. The average delays, energies, and EDPs are measured for WRITE and READ operations to comprehend and illustrate the abilities of the MB NV LUTs and comparisons with the SB NV LUTs' results are performed. The SB NV LUT design is taken from [17] with SB ReRAM cells behaviour in Figure 10.8. The SB NV LUT has 2^K number of array cells where K is the number of LUT inputs whereas the MB NV LUT has 2^{K-1} number of array cells. Circuit simulations were carried out in LTSpice EDA using 100 μm transistor technology node.

The comparison between SB and MB NV LUTs was performed for five condition tests which are *Condition 1*: WRITE '0'→'1' to all array cells, *Condition 2*: WRITE '1'→'0' to all array cells, *Condition 3*: flip two bits at a time (WRITE '01'→'10'), *Condition 4*: flip two bits at a time (WRITE '10'→'01'), and *Condition 5*: READ '00', '01', '10', '11' from the array. Half of the SB-array cells are written from '0'→'1' and the other half are written from '1'→'0' for *Condition 3* and *4* tests.

10.3.5.1 1-bit WRITE operation results

All LUT array cells are written from '1'→'0' for *Condition 1* test. This way each MB ReRAM cell in the array is switched from '11'→'00' and each SB ReRAM cell in the array is switched from '1'→'0'. *Condition 2* is the reverse of *Condition 1*. *Condition 1* and *Condition 2* results are plotted in Figures 10.13 and 10.14 and listed in Tables 10.4 and 10.5, respectively. The WRITE delays of the MB NV LUT in both conditions are half that of the SB NV LUT, showing the advantage of storing two bits per ReRAM.

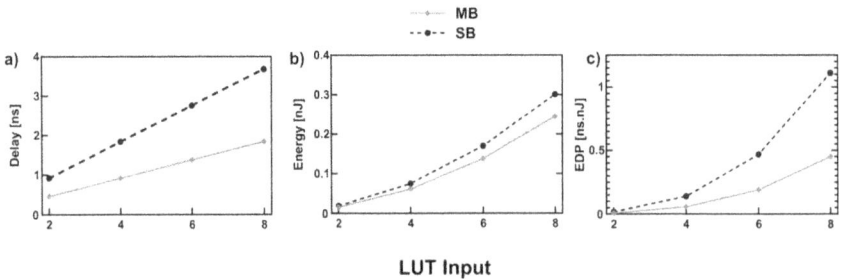

Figure 10.13 2-, 4-, 6-, and 8-input array delay, energy, and EDP measurements for *Condition 1*: SB vs MB WRITE '0'.

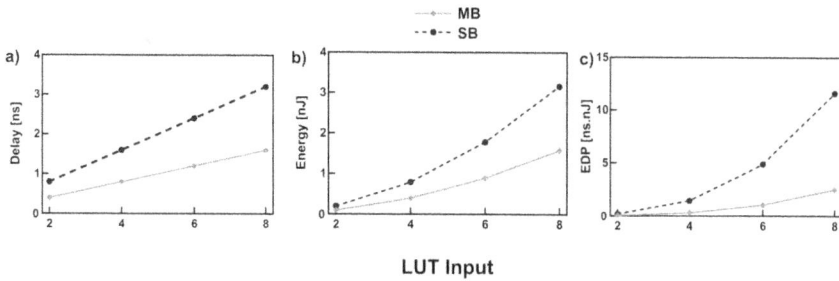

Figure 10.14 2-, 4-, 6-, and 8-input array delay, energy, and EDP measurements for Condition 2: SB vs MB WRITE '1'.

For *Condition 1* and *Condition 2*, the MB NV LUT's energy consumption is respectively 1.22× and 2× lower on average compared to the SB NV LUT. This can be attributed to the combination of the output bits (xy) in the MB ReRAM array cell and the lower number of ReRAM cells in the MB NV LUT array. The differences in energy consumption between the MB and SB NV LUTs increase exponentially with the array size. As for the EDP, the MB NV LUT is lower by an average of 2.46× and 4.6× for *Condition 1* and *Condition 2* respectively. Switching into a higher cell resistivity for *Condition 1* results in an overall lower energy consumption compared to *Condition 2*.

10.3.5.2 2-bit WRITE operation results

The NV LUTs performance during simultaneous 2-bit WRITE is tested in *Condition 3* and *Condition 4*. Since SB ReRAMs only contain a single bit, two SB ReRAMs are simultaneously switched ('0'→'1' for one SB ReRAM cell and '1'→'0' for another SB ReRAM cell) for both conditions.

In *Condition 3*, the MB ReRAM is required to switch from IRS-2 ('01') to IRS-1('10'). Since IRS-2 succeeds IRS-1 in the switching scheme in Figure 10.8, the MB ReRAM has to be switched back to LRS first following the switching scheme sequence of '01'→'11'→'10'. However, as shown in Figure 10.15 and Table 10.6 there is no significant penalty on the MB NV LUT delays which averages half the delay of the SB NV LUT. The MB NV LUT's energy consumption is on par with the SB NV LUT, averaging slightly lower by 1.03×. The EDP of the MB NV LUTs average 2× lower than the SB NV LUTs as a result of the lower WRITE delays.

The MB NV LUT has an advantage over the SB NV LUT in *Condition 4* tests as shown in Figure 10.16 and Table 10.7. A 2-input MB NV LUT has 0.1 ns delay while the 2-input SB NV LUT has a delay of 0.92 ns.

Condition 4 has an average of 9.2× for the WRITE delay. As the MB ReRAM only switches from IRS-1 to IRS-2, the energy consumption of the MB NV LUT is 128× lower on average compared to the SB NV LUT. This

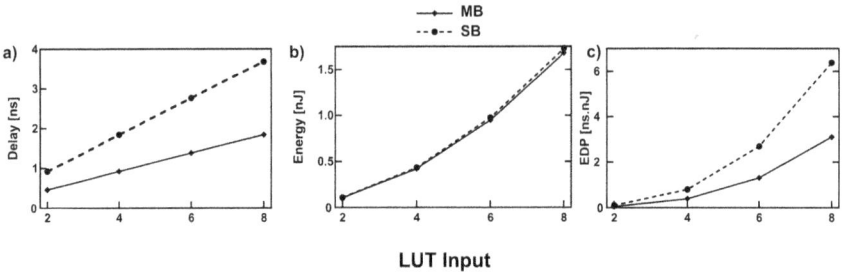

Figure 10.15 2-, 4-, 6-, and 8-input array delay, energy, and EDP measurements for *Condition 3*: SB vs MB 2-bit WRITE performance (01 to 10).

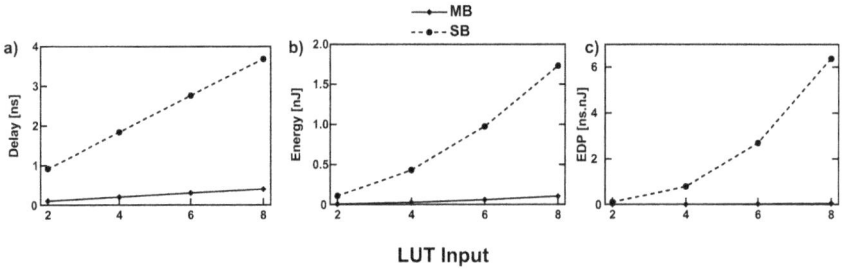

Figure 10.16 2-, 4-, 6-, and 8-input array delay, energy, and EDP measurements for *Condition 4*: SB vs MB 2-bit write performance (10 to 01).

advantage is carried forward to the EDP where the MB NV LUT is 153× lower compared to the SB NV LUT on average.

The MB NV LUT demonstrates consistent performance improvements over the SB NV LUT in all WRITE tests.

10.3.5.3 RED operation results

For the READ analysis, an MB ReRAM cell containing 2-bits is compared with two SB cells which make up 2-bits as well. Two cells from different columns are selected to READ the 2-bits from the SB NV LUT.

The average READ delay is 5× higher for the SB NV LUT than the MB NV LUT. The reading of two cells in the SB NV LUT compared to just a single cell in the MB NV LUT also results in an increase in energy consumption. The average READ energy and average READ EDP for the SB NV LUT are 3.2× and 14.6× higher than the MB NV LUT respectively. The ability of the MB NV LUT to store two bits per cell makes it more efficient than the one bit per cell SB NV LUT.

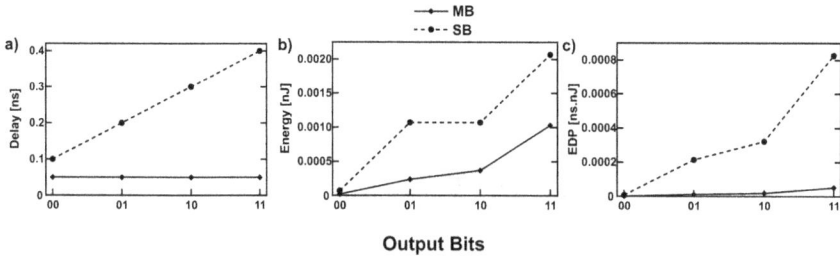

Figure 10.17 Delay, energy, and EDP measurements for *Condition 5*: SB vs MB READ 00, 01, 10, 11; (a) Delay, (b) Energy, and (c) EDP.

Further improvements of efficiency can be achieved by using ReRAMs with >2 bits per cell. The READ delay, energy, and EDP values for *Condition 5* are plotted in Figure 10.17 and given in Table 10.8.

10.4 NONVOLATILE D FLIP-FLOP

The design of the NV DFF in this section is capable of passive STORE/ RESTORE operations and negates the need for additional sequences that come with delay and energy penalties as well as external control circuitry. Figure 10.18 shows the schematic of the NV DFF. It is comprised of a master and slave latch and an NV segment composed of ReRAMs.

Figure 10.19 shows the switching scheme of the ReRAMs in this design; application of a positive polarity 1.2 V pulsewidth for 3ns switches the ReRAM into LRS while application of a negative polarity −1.2 V pulsewidth for the same duration switches the ReRAM into HRS. The ReRAM's state variable in Figure 10.19a reaches 1 when the ReRAM is in LRS with an LRS resistance of ~100 Ω and maximum device current of 10 mA Figure 10.19b. Figure 10.20 shows the normal operation process waveform of the NV DFF. The master and slave components function similarly to a regular DFF, passing input data, V_D to the Master latch when Clk is high. The transmission gate, TG2 is activated when Clk is low, passing through the data into the Slave latch so that Q takes the same value as V_D and Q_b takes the inverted V_D.

When Q is high ('1') and Q_b is low ('0'), the current flow direction in the NV segment is from BE to the TE for M1 and reversed for M2. This switches M1 into HRS and M2 into LRS as can be seen in Figures 10.18 and 10.19. HRS M1 holds the Q line high while LRS M2 pulls Q_b to ground, pulling Q_b line low. The reverse happens when Q is low and Q_b is high as M1 will be in LRS and M2 in HRS.

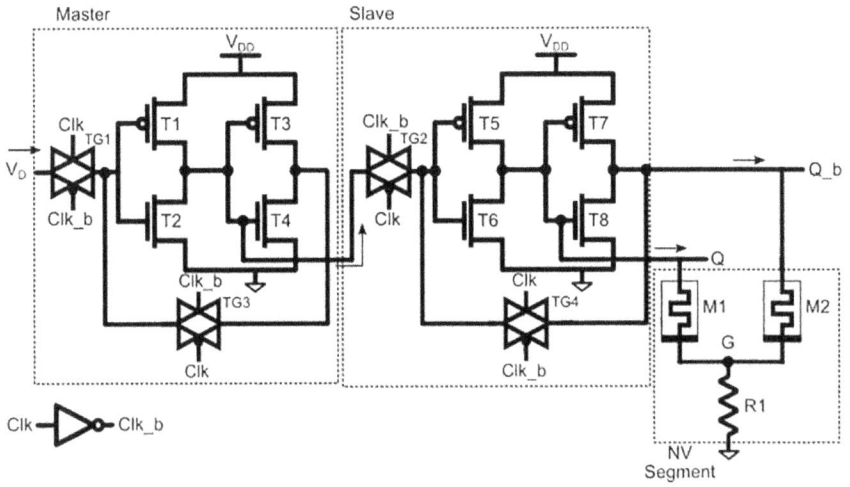

Figure 10.18 NV DFF schematic.

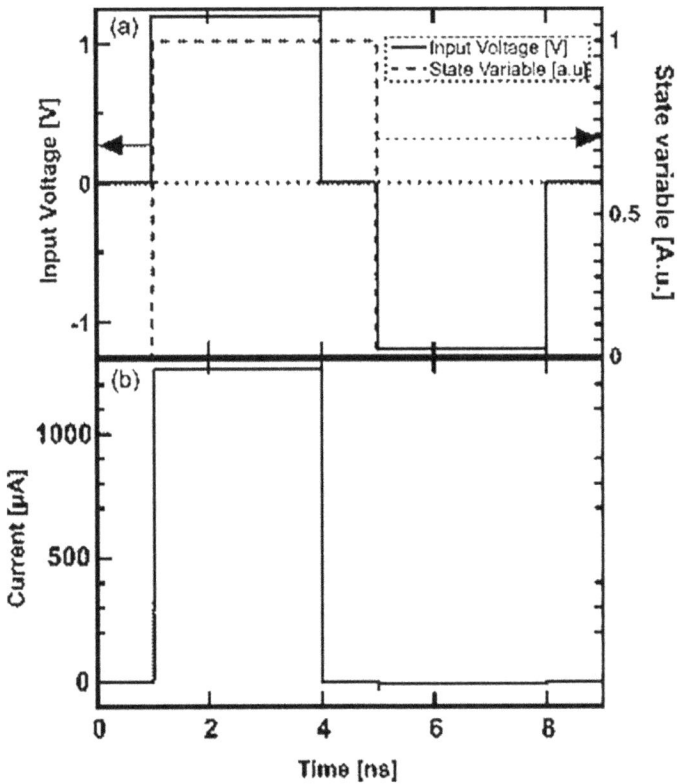

Figure 10.19 Electrical characteristics of the ReRAM model showing (a) The response of the state variable to the input voltage and (b) the device current.

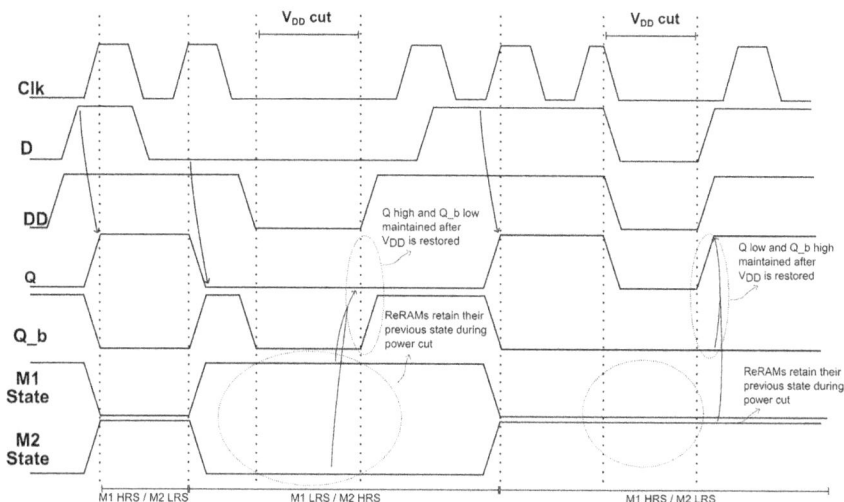

Figure 10.20 NV DFF process waveforms with two VDD cut scenarios

NV is introduced in this design as the ReRAMs M1 and M2 will retain their resistive states when V_{DD} is disrupted. Upon restoration of V_{DD}, M1 and M2 will pull the Q and Q_b lines to their pre-power cut values depending on the ReRAM's resistance. Figure 10.20 shows two power cut scenarios, the first occurs when Q (Q_b) is '1'('0') and the second when Q (Q_b) is '0'('1').

10.4.1 Simulation results

Analysis of the NV DFF is performed for three transistor technology node sizes; 32, 45, and 65nm using Predictive Technology Models in EDA software, LTSpice [18].

Figures 10.21 and 10.22 depict the simulation results of the NV DFF storing and restoring after power disruption for Q (Q_b) is '0'('1') and Q (Q_b) is '1'('0') respectively. Both the storage into the NV segment during normal operation and the recovery of Q and Q_b lines after power restoration processes are done passively, and the design does not require additional STORE/RESTORE sequences or control circuits.

Table 10.9 lists the measurements for Q/Q_b rising/falling times, clock-to-q delay (tCQ), maximum frequency (F_{max}), and average and worst-case RESTORE times of the NV DFF for 32, 45, and 65 nm transistor technology nodes. Measurements for the V DFF are also included for comparison. It should be noted that the NV DFF's minimum required STORE times which ensures both ReRAMs are fully written into their required states are equivalent to the NV DFF's switching times as the STORE process is passive.

The RESTORE times for the NV DFF is an average 31.11 ps for RESTORE0 and RESTORE1 operations owing to the passive mechanism.

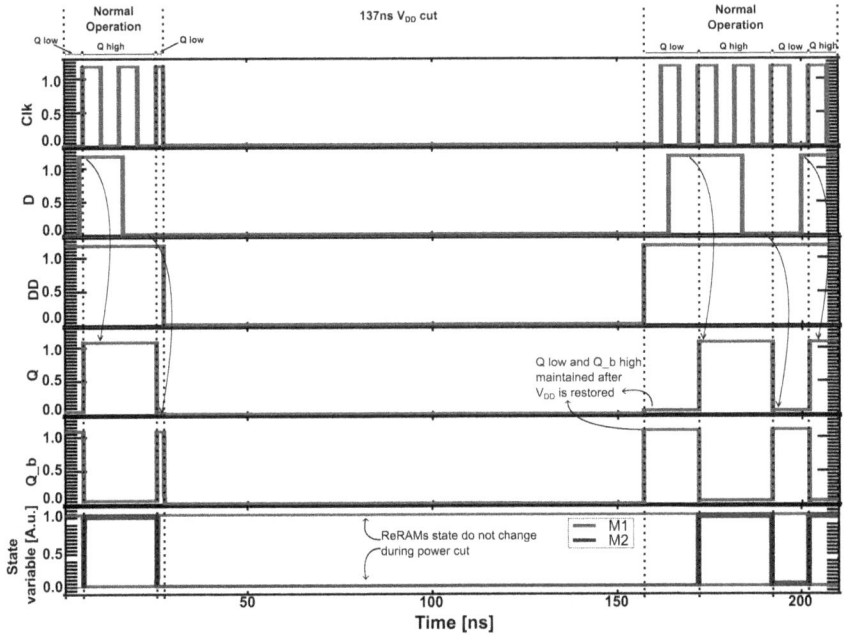

Figure 10.21 Successful restore of Q='0' and Q_b='1' after 137ns V_{DD} cut. Q switches from '1' to '0' at 25ns before V_{DD} interrupt

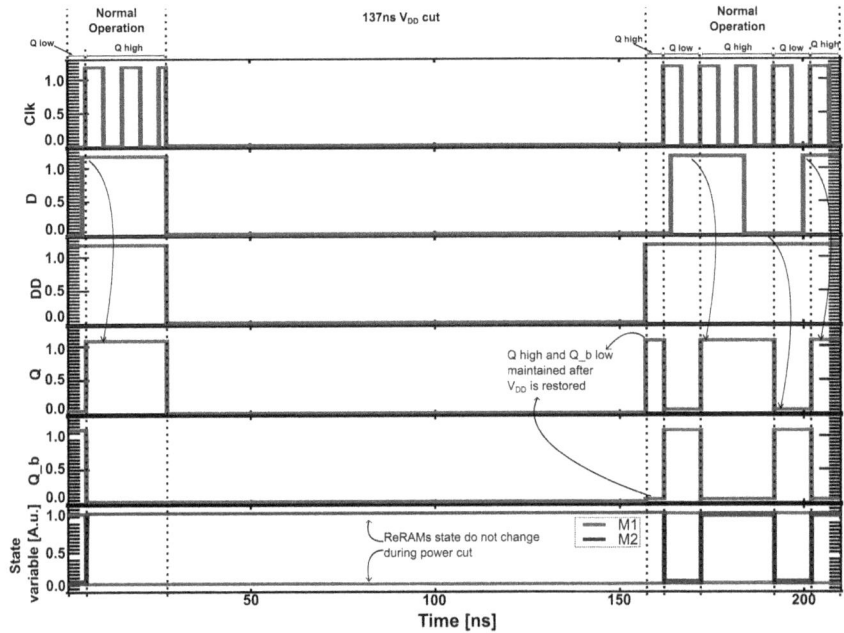

Figure 10.22 Successful restore of Q='1' and Q_b='0' after 137ns V_{DD} cut.

Worst-case RESTORE values are taken from RESTORE operations with the largest delay and average at 38.79 ps. Clk-to-Q timings are strongly dominated by transistor characteristics, and the results of both NV DFFs and V DFFs are thus close to each other. However, the NV DFFs have higher timings for low-to-high and high-to-low operations compared to the V DFFs due to the additional time required to switch the ReRAMs in the NV segments. At average 1.1× and 1.01× slower timings for low-to-high and high-to-low switching compared to the V DFFs, the NV DFFs' performance is close to the V DFFs and can be viable with the implementation of fast-switching ReRAMs.

10.5 THE NONVOLATILE LOGIC BLOCK

As discussed in the beginning of this chapter, the basic LB consists of clusters of BLEs which comprises the LUT and the DFF (Figure 10.1). To implement NV into this structure, the 6-input CMOS-based LUTs and DFFs are replaced with the MB NV LUTs and NV DFFs designs presented in this chapter. The block diagram of this NV CLB is shown in Figure 10.23 [19].

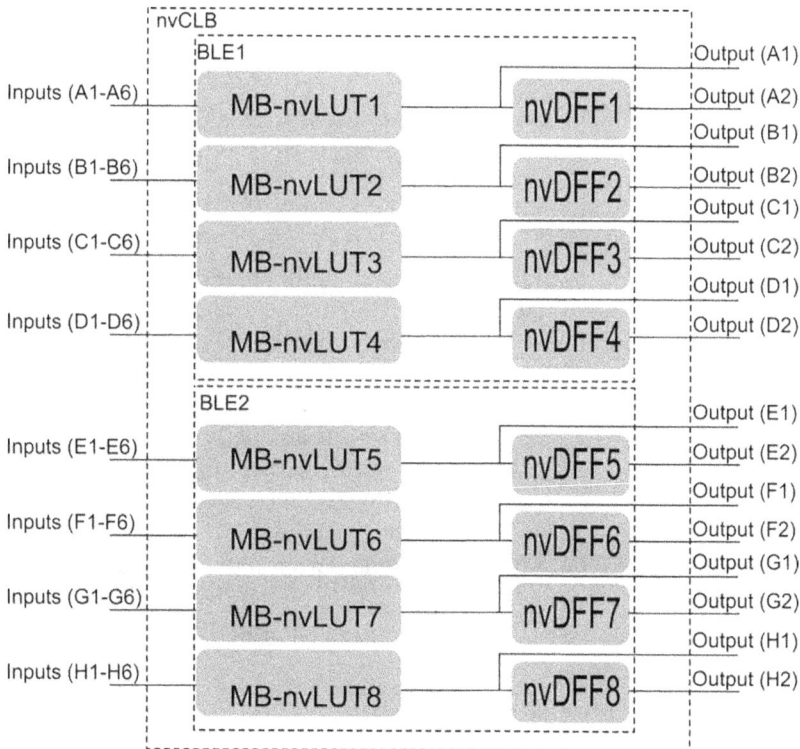

Figure 10.23 Block diagram of NV CLB.

The NV CLB's passive storage ability is tested out with an application with the following Boolean expression:

$$\text{Output} = A\left[\overline{BC}\left[\overline{D}\left(\overline{E}F+E\overline{F}\right)+D\overline{EF}\right]+\overline{DEF}\left(\overline{B}C+B\overline{C}\right)\right] \qquad (10.1)$$

In this expression, the A input is checked first where the LUT output will be '0' if A is '0' and the remaining five bits determine the LUT output if A is '1'. If the sum of the remaining five bits is '1' and only '1' then the output will be '1' otherwise the LUT's output is '0'. The A1 input comes from an external signal for the first LUT, LUT1 in the NV CLB and the output of LUT1 will be fed into the first NVDFF, LUT1_NVDFF. The output of LUT1_NVDFF is fed into the second LUT, LUT2 as the A2 input. Expression (1) is stored in both LUTs.

In this simulation, the input '100010' is fed into LUT1 which gives an output of '1' which is then fed into LUT1_NVDFF and into the A2 input of LUT2. The selected address for LUT2 is '111111' which gives the output, '0' which is fed into LUT2_NVDFF.

Figure 10.24 shows the written states for three cells, LUT1_M17, LUT1_M18, and LUT2_M32;

LUT1_M17 and LUT2_M32 are the locations of the selected addresses and LUT1_M18 is plotted to show the state condition of an unselected cell in the array. As a 6-input LUT, the MB NV LUT has 32 cells in the array, half the number of array cells in a SB LUT array.

Cell LUT1_M17 stores the data bits for addresses '100001' and '100010' which are '0' and '1' respectively based on (1) while cell LUT1_M18 stores '1' and '0' for addresses '100010' and '100011', and cell LUT2_M32 stores '0' and '0' for addresses '111110' and '111111'.

Therefore, cells LUT1_M17, LUT1_M18, and LUT2_M32 are written into IRS_2, IRS_1, and HRS, respectively, following the WRITE logic in Table 10.2.

This is shown in Figure 10.24 where both LUT1_M17 and LUT1_M18 are undergo SET switching into LRS and are then written into their respective required states. As LUT2_M32 is initially in HRS, no operations are performed on the cell. V_{DD} is then cut and resupplied at 285 µs. The resistance states of the ReRAM cells are not disturbed and the cells successfully retain their states during power disruption and restoration. The same power disruption tests are performed for the NV DFFs, LUT1_NVDFF and LUT2_NVDFF in the NV CLB. The output data from the LUT1 and LUT2 are respectively supplied to the LUT1_NVDFF and LUT2_NVDFF. Similar V_{DD} conditions to Figure 10.24 are used and the results are plotted in Figure 10.25.

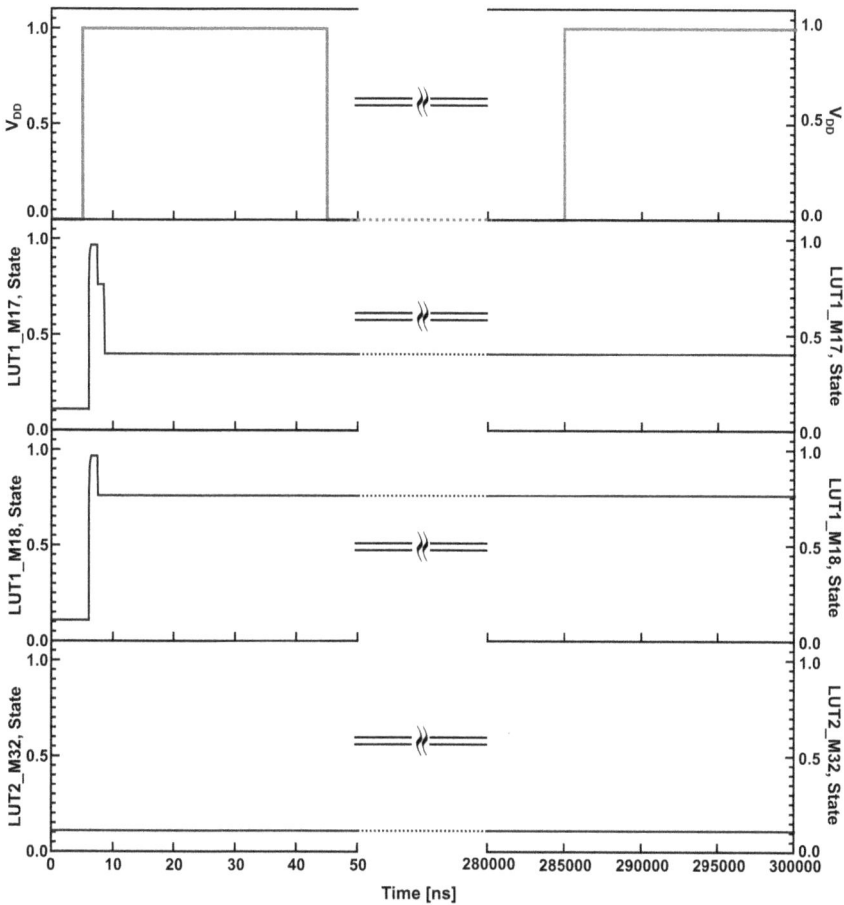

Figure 10.24 ReRAM states written with data '100111' retain their data during power cut and after power resumption.

Data '1' is fed to NV DFF1 from the MB NV LUT and Q rises while Q_b falls. On the other hand, data "0" is fed to NV DFF2 and Q falls while Q_b rises. During power cuts and restoration, the ReRAMs successfully preserve their respective states, and the $Q/Q–b$ values of the corresponding NV DFFs are recovered.

An interesting outcome from NV implementation in the BLE is the opportunity for the reduction of multiplexers; the bottom multiplexer in Figure 10.1 can provide a direct output from the LUT or feed the LUT output into the DFF for sequential operations. This multiplexer is removed in the NV CLB so that the NV LUT output is split into the direct output and the input into the NV DFF. If the NV DFF is not in use, power supply to the NV DFF can be halted and the NV DFF retains the previous Q/Q_b values.

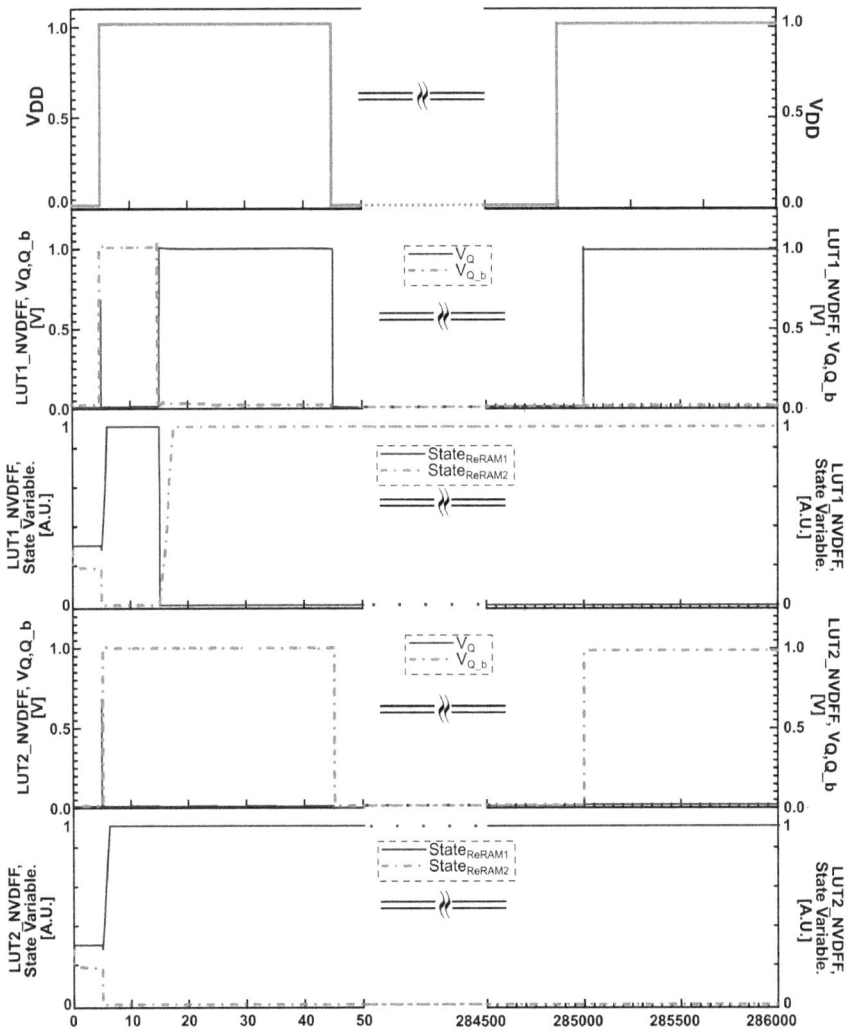

Figure 10.25 NV DFFs' Q and Q_b outputs and their respective ReRAM states. ReRAMs retain their states during power disruption and both Q and Q_b values are restored after power resumption.

REFERENCES

[1] C. Maxfield, *FPGAs: Instant Access*, 1st ed. Newnes, London, 2008.
[2] V. Betz, J. Rose, and A. Marquardt, *Architecture and CAD for Deep-Submicron FPGAS*. Springer New York, NY, 1999.

[3] A. Boutros and V. Betz, "FPGA architecture: Principles and progression," *IEEE Circuits Syst. Mag.*, vol. 21, no. 2, pp. 4–29, 2021, doi: 10.1109/MCAS.2021.3071607.

[4] E. Monmasson, L. Idkhajine, M. N. Cirstea, I. Bahri, A. Tisan, and M. W. Naouar, "FPGAs in industrial control applications," *IEEE Trans. Ind. Informatics*, vol. 7, no. 2, pp. 224–243, 2011, doi: 10.1109/TII.2011.2123908.

[5] S. M. Trimberger, "Three ages of FPGAs: A retrospective on the first thirty years of FPGA technology," *Proc. IEEE*, vol. 103, no. 3, pp. 318–331, 2015, doi: 10.1109/JPROC.2015.2392104.

[6] P. Ye, T. Ernst, and V. M. Khare, "The last silicon transistor," *IEEE Spectr.*, vol. 58, no. 8, pp. 31–35, 2019.

[7] A. Lekshmi Jagath, C. Hock Leong, T. N. Kumar, and H. F. Almurib, "Insight into physics-based RRAM models - review," *J. Eng.*, vol. 2019, no. 7, pp. 4644–4652, 2019, doi: 10.1049/joe.2018.5234.

[8] N. H. El-Hassan, T. N. Kumar, and H. A. F. Almurib, "Phase change memory cell emulator circuit design," *Microelectronics J.*, vol. 62, no. February, pp. 65–71, 2017, doi: 10.1016/j.mejo.2017.02.006.

[9] A. Chen, "A review of emerging non-volatile memory (NVM) technologies and applications," *Solid. State. Electron.*, vol. 125, pp. 25–38, 2016, doi: 10.1016/j.sse.2016.07.006.

[10] H. Amrouch et al., "ICCAD tutorial session paper ferroelectric FET technology and applications: From devices to systems," *IEEE/ACM Int. Conf. Comput. Des. Dig. Tech. Pap. ICCAD*, vol. 2021, no. CiM, 2021, doi: 10.1109/ICCAD51958.2021.9643578.

[11] Y. Lu, J. H. Lee, and I.-W. Chen, "Scalability of voltage-controlled filamentary and nanometallic resistance memory devices," *Nanoscale*, vol. 9, no. 34, pp. 12690–12697, 2017, doi: 10.1039/C7NR02915B.

[12] F. Zahoor, T. Z. Azni Zulkifli, and F. A. Khanday, "Resistive random access memory (RRAM): An overview of materials, switching mechanism, performance, multilevel cell (MLC) storage, modeling, and applications," *Nanoscale Res. Lett.*, vol. 15, no. 1, 2020, doi: 10.1186/s11671-020-03299-9.

[13] Xilinx, "Xilinx UG190 Virtex-5 FPGA user guide," vol. 190. pp. 1–385, 2012, [Online]. Available: https://www.xilinx.com/support/documentation/user_guides/ug190.pdf.

[14] P. W. C. Ho, F. O. Hatem, H. A. F. Almurib, and T. N. Kumar, "Comparison between Pt/TiO$_2$/Pt and Pt/TaO$_X$/TaO$_Y$ /Pt based bipolar resistive switching devices," *J. Semicond.*, vol. 37, no. 6, 2016, doi: 10.1088/1674-4926/37/6/064001.

[15] H. L. Chee, T. N. Kumar, and H. A. F. Almurib, "Electrical model of multilevel bipolar Ta2O5/TaOx Bi-layered ReRAM," *Microelectronics J.*, vol. 93, no. March, p. 104616, 2019, doi: 10.1016/j.mejo.2019.104616.

[16] H. L. Chee, T. N. Kumar, and H. A. F. Almurib, "Low energy non-volatile look-up table using 2 bit ReRAM for field programmable gate array," *Semicond. Sci. Technol.*, vol. 37, no. 6, 2022, doi: 10.1088/1361-6641/ac6903.

[17] H. A. F. Almurib, F. Lombardi, and T. N. Kumar, "Design and evaluation of a memristor-based look-up table for non-volatile field programmable gate arrays," *IET Circuits, Devices Syst.*, vol. 10, no. 4, pp. 292–300, 2016, doi: 10.1049/iet-cds.2015.0217.

[18] Predictive Technology Model, "Nanoscale Integration and Modeling (NIMO) Group, ASU, 2007. https://ptm.asu.edu.

[19] H. L. Chee, T. Nandha Kumar, and H. A. F. Almurib, "A ReRAM-based nonvolatile FPGA," In *2022 IEEE 20th Student Conference on Research and Development (SCOReD*, 2022, pp. 62–67, doi: 10.1109/SCOReD57082.2022.9973999.

Chapter 11

Efficient layout techniques to design physically realizable quantum-dot cellular automata circuits

Marshal R
Indian Computer Emergency Response Team

Raja Sekar K
Centre for Development of Advanced Computing

Lakshminarayanan Gopalakrishnan
National Institute of Technology-Tiruchirappalli

Seok-Bum Ko
University of Saskatchewan

Anantharaj Thalaimalai Vanaraj
Senior Technologist, R& D, Western Digital

11.1 INTRODUCTION

The ever-rising demand for smaller, compact and high-performance devices is increasing design complexity using CMOS [1,2]. This has shifted the focus on non-transistor-based design technologies that can achieve high demands of circuit complexity and performance. Quantum-dot cellular automata (QCA)-based nanotechnology is a methodology to fabricate circuits with superior densities and ultra-high speed. As transfer of inputs and processing is performed through electronic repulsion, the power consumption demands are minimal [3].

To have efficient and controlled information flow, clocks are used in QCA. The clocks are used with different delays to facilitate the information flow and processing. Design rules are proposed in the literature to design QCA circuits [4,5]. The design rules suggest having a minimum of two cells connected with a clock signal. However, clocking is a major issue in such small-sized circuits. With the crossovers, the clocking also increases the complexity of QCA circuit fabrication.

DOI: 10.1201/9781003459231-11

Several technologies are proposed to realize QCA circuits to operate them at room temperatures. Metal, molecular and magnetic-based fabrication techniques are proposed in the literature. However, molecular and magnetic realizations are showing signs of possible realization at room temperature. To overcome the boundaries levied by clocking, novel cocking mechanisms are proposed in the literature to design physically realizable circuits. The large-sized clock zones reduce the clocking complexity and increase the robustness of clock connections. The clocking schemes have different clock zone patterns, hence, the layout complexity is increased. The lack of proper automated layout tools in QCA also increases the complexity of designing QCA circuits in novel clocking schemes [6–10]. In this work, layout techniques are proposed to facilitate the easy layout of designs using novel clocking schemes. The proposed techniques are validated by designing a multiplexer. The designs are simulated and verified using QCADesigner.

The remainder of this chapter is arranged as follows. An insight into different parameters that influence QCA circuit design is provided in Section 11.2. Section 11.3 briefs about various clocking schemes in the literature to design physically realizable QCA circuits. Section 11.4 presents the layout techniques that can be followed to design physically realizable QCA circuits in an effective manner. Section 11.5 presents the designs of multiplexer developed using the layout techniques and discusses the results. Section 11.6 delivers the conclusion.

11.2 QCA FUNDAMENTALS

11.2.1 Cell

QCA cell entails four quantum dots with two electrons in it. The dots hold the electrons, and the electrons travel between them based on neighbouring cells. The angle between the dots classifies and determines the cells as normal and rotated cells. The electron-occupied positions are encoded as logic '0' and logic '1'. The encoding and cells are presented in Figure 11.1. Each cell has different states of operation such as Release, Relax, Hold and Switch [11].

11.2.2 Wire

Information transfer is carried out by wires constructed using cells. Normal cell-based wires carry values without any modification. However, the cell count in rotated cell-based wires determine the value as presented in Figure 11.2.

11.2.3 Majority gate

Majority logic is deployed for QCA circuits by using gates with odd number of inputs and single output. These gates are utilized to develop Boolean logic AND and OR gates by deploying fixed polarization cells as presented

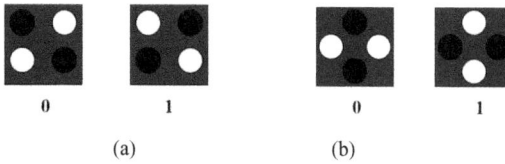

Figure 11.1 Cell. (a) Rotated and (b) normal.

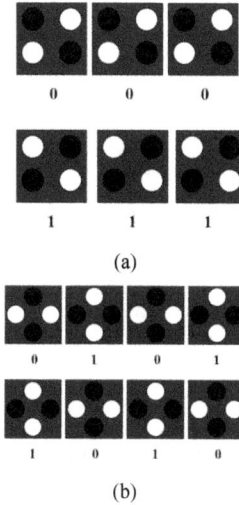

Figure 11.2 Wire. (a) Rotated and (b) normal.

in Figure 11.3. To complement and facilitate the majority gates, inverters are developed by varying cell combinations and orientations as presented in Figure 11.4.

11.2.4 Information flow and delay

To ensure synchronized operation, circuits are separated into four zones with each coupled to a clock having phase difference with its neighbouring zones. The clocks control the operational state of the cells and ensure that no zones have identical state of operation to facilitate controlled flow of information. The zones travelled by an input towards reaching the output determines the delay. In general, delay is represented by clock zones travelled (clock phases) or clock cycles (1 clock cycle=4 clock phases). The differences in cell connection of a wire to different clocks and the associated delay are presented in Figure 11.5. The wire carries '0' from one wire to another with a delay of four clock phases or one clock cycle.

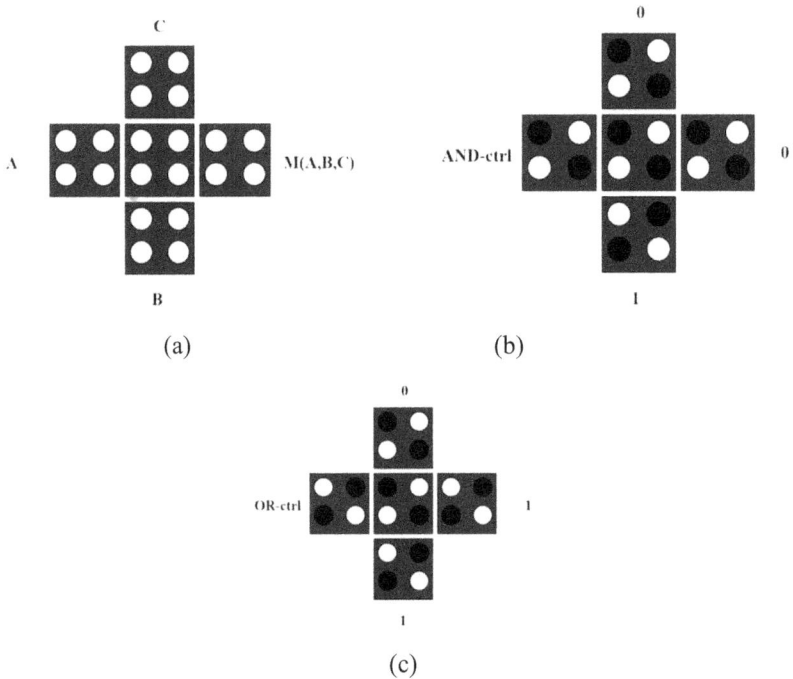

Figure 11.3 Gates. (a) 3-input, (b) 2-bit, and (c) 2-bit OR.

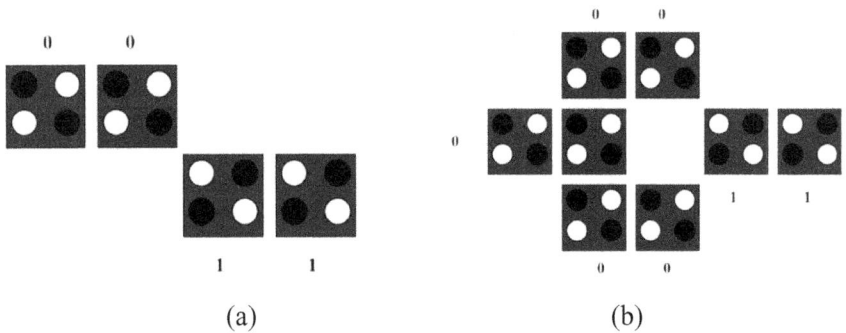

Figure 11.4 Inverters. (a) Type-1 and (b) Type-2.

11.2.5 Crossover

Information-carrying wires jump over other wires in the same layer by exploiting the non-interaction between rotated and normal cell or by using the non-interaction by non-neighbour clock zones. Additional layers can also be deployed to facilitate the crossings, as presented in Figure 11.6. However, each technique comes with its pros and cons.

Clock 0   Clock 1   Clock 2   Clock 3

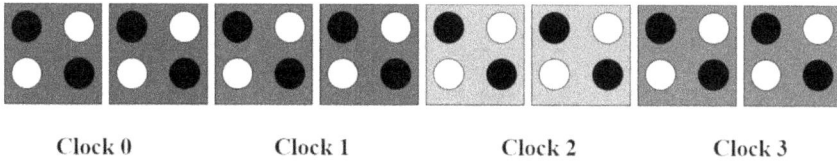

Figure 11.5 Information flow in a wire through different clock zones.

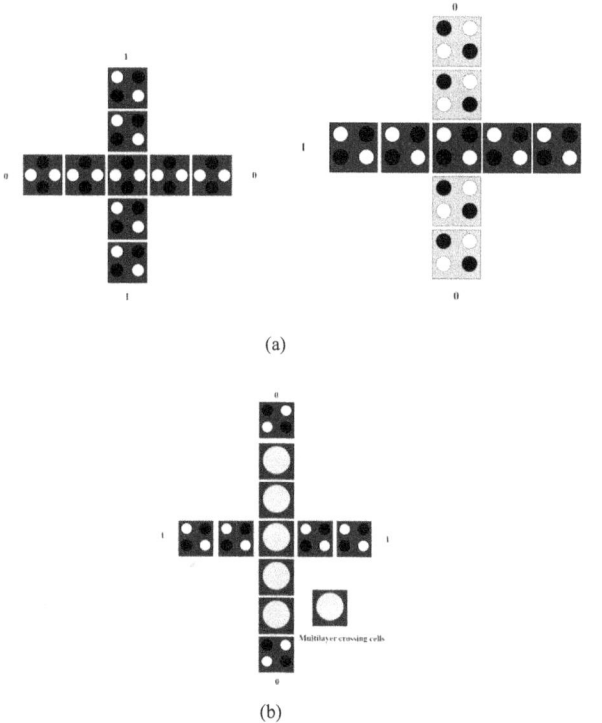

(a)

(b)

Figure 11.6 Crossover. (a) 1-layer and (b) 3-layer.

11.2.6 Fault tolerance

QCA fabrication comprises synthesis and deposition stages. Individual cells encompassing the dots with electrons are fabricated in the synthesis stage initially. Later, the cells developed in the synthesis stage are placed in their defined places in deposition stage. The nano-sized deposition is highly complex, and the defects are more likely to occur. They include missing cell, stuck-at-fault, additional cell and misalignment defects as presented in Figure 11.7. Missing cell defect due to the failure in deposition stage. Stuck-at fault defect happens due to cells getting fixed to a particular logic. Misalignment defect happens due to deposition in wrong location. These defects can lead to faulty outputs. More detailed description on the defects and computation of their impact on circuit performance is presented in [12–14].

(a)

(b)

(c)

(d)

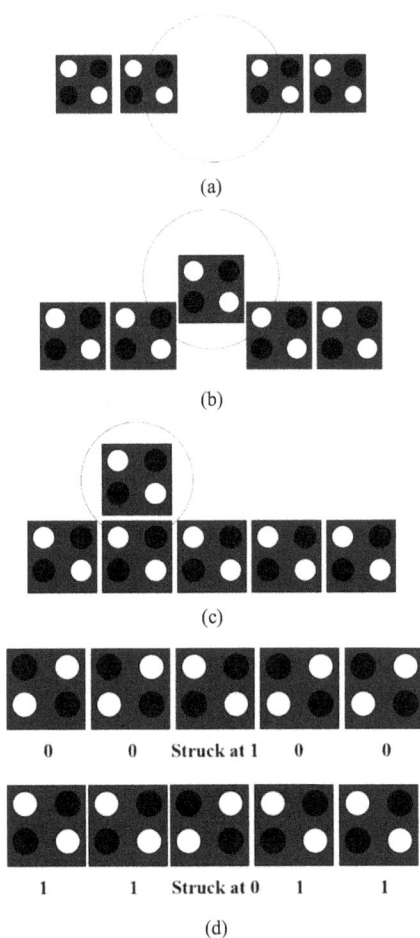

Figure 11.7 Defects. (a) Missing, (b) misalignment, (c) additional, and (d) stuck at fault.

11.3 CLOCKING SCHEMES IN QCA

The small size of circuits and the high complexity involved with the clocking for different clock zones increases the fabrication complexity. To overcome this complexity, clocking schemes are proposed in the literature. The clocking schemes are deployed by arranging cells in a grid pattern. It reduces the clocking complexity and the crossovers involved in clocking. Several such schemes are proposed in the literature [6–10]. Each scheme has its advantages and limitations. The schemes are presented in Figure 11.8.

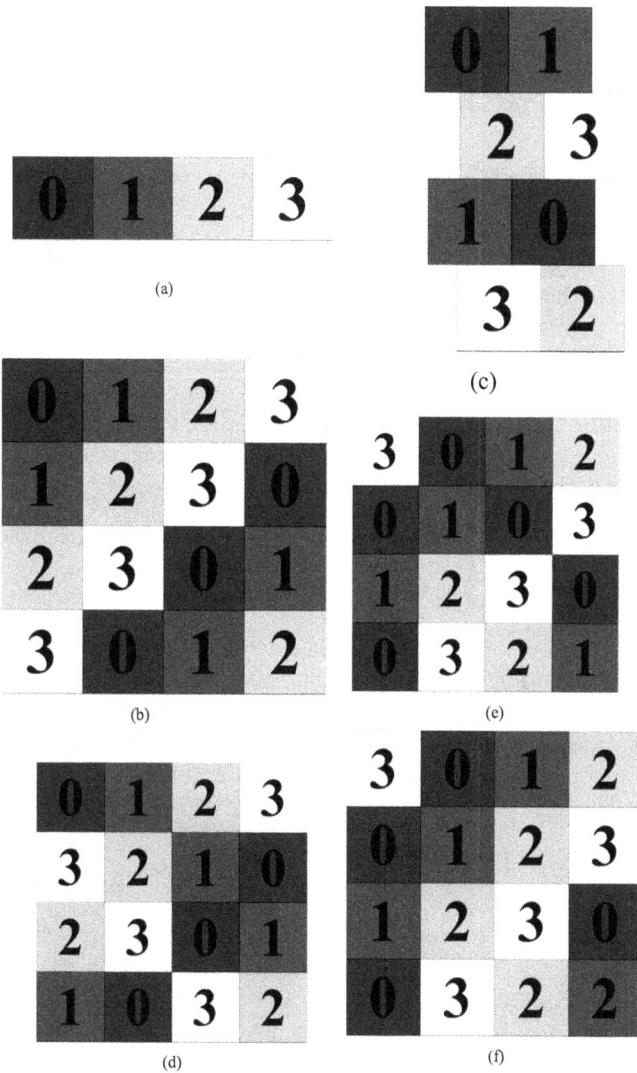

Figure 11.8 Clocking. (a) 1-D, (b) 2-DDW, (c) optimized 2-D, (d) USE, (e) RES, and (f) ESR.

11.4 LAYOUT TECHNIQUES

Design rules are available in the literature for designing QCA circuits. The rules focus on the spacings, cell count in clock zones [4]. Layout strategies are also proposed to design efficient circuits [5]. However, automated layout tools are not available for QCA design [15]. In addition to that, the proposed strategies and techniques focus on the design using conventional clocking techniques and not by using the modern clocking schemes proposed to increase the reliability and reduce the fabrication complexity. The major complexity involved in the clocking schemes is the layout of circuits as there are no automated layout tools available for QCA. In this work, basic guidelines and techniques are proposed to design efficient QCA circuits using different clocking schemes.

11.4.1 Circuit segmentation

The first and foremost key technique is to have a knowledge on the circuit to be realized. The circuit to be realized has to be studied from the input toward the output. The delay of the inputs reaching each of the gates needs to be studied, and the circuit must be divided into stages to ensure that the inputs reach each gate with uniform delay. This must also consider the feedback paths to ensure that the feedback paths also travel similar delay paths and that no delay mismatch is faced by the inputs reaching the gates. This understanding is very crucial in designs using modern clocking techniques, as the feedback paths need to be changed in accordance to the scheme.

11.4.2 Placement of I/O cells

The inputs must be placed in the same clock zone to ensure that the inputs travel uniform number of zones while reaching the output. Care has to be taken to place all the inputs at the corners of the circuit to facilitate the utilization of the circuit as a building block or as an element in other larger circuits. The output cells should also be placed at the corners of the circuit to facilitate the utilization of the output by other blocks.

11.4.3 Crossovers

The crossovers must be decided based on the clocking scheme chosen for implementation. Some clock schemes do not support all the clocking schemes. The crossovers increase the complexity as well as cost of the circuit [16,17]. Hence, the crossovers must be chosen in accordance with the design constraints such as cost, area, delay and fabrication complexity.

11.4.4 Layout approach

It is always preferable to have the layouts generated with inputs on one side of the circuit and outputs at the other end of the circuit. However, circuits having multiple outputs or circuits which need to be connected with other blocks may have the outputs in the corners of the circuit to facilitate efficient area utilization.

11.5 MULTIPLEXER DESIGN

Multiplexers are key elements used in designing combinational and sequential circuits. To show the effectiveness of the techniques suggested in the previous section, a multiplexer is realized using USE clocking scheme. USE scheme is chosen for implementation as it supports feedback path and as the initial stage of layout strategy, the multiplexer circuit is divided into segments as presented in Figure 11.9. The layout developed using the techniques is presented in Figure 11.10 with its results in Figure 11.11. It can be seen in Figure 11.10 that all the inputs are in the same clock zone and are at the corner of the circuit. The multiplexer developed using the techniques is compared with existing QCA multiplexers [5,18–22] in Table 11.1.

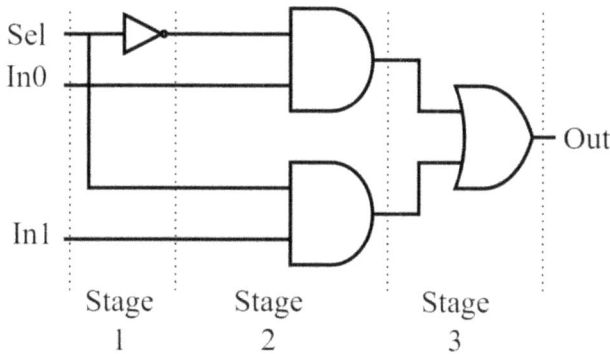

Figure 11.9 Segmentation of multiplexer circuit.

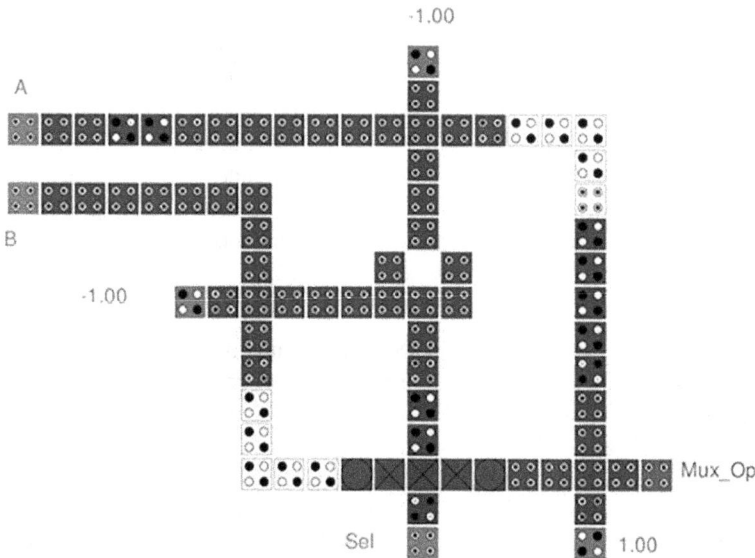

Figure 11.10 USE multiplexer.

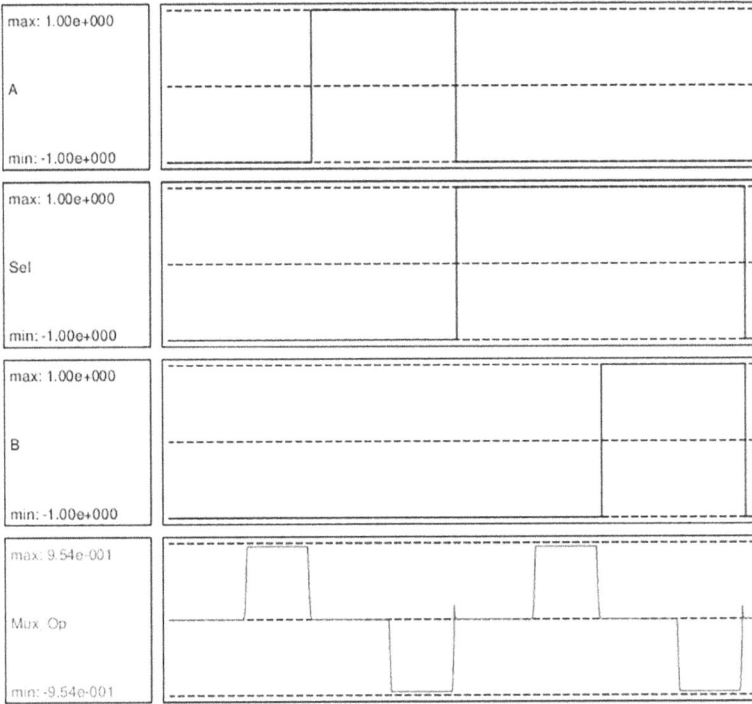

Figure 11.11 USE multiplexer results.

Table 11.1 Multiplexer comparison

Ref.	Cells	Area in µm²	Clock phases	Clocking	Realization complexity
[18]	224	0.5	6	Conventional	High
[19]	26	0.02	2	Conventional	High
[20]	23	0.02	2	Conventional	High
[21]	19	0.02	2	Conventional	High
[5]	16	0.01	2	Conventional	High
[22]	15	0.01	2	Conventional	High
Proposed	83	0.12	6	USE	Less

It is clear that the design using USE scheme has more area and cells. This is due to the increased grid size followed in modern clocking schemes compared to conventional scheme. However, these designs have better clocking robustness and reduced complexity.

The modern clocking schemes promise better reliability and physical realization compared to circuits designed using conventional schemes. The layout techniques proposed in this chapter will be helpful in efficient designing using modern schemes. In future, the techniques can be further refined and improved for automated layout generation.

11.6 CONCLUSION

In this chapter, the layout techniques are proposed to design efficient QCA circuits using modern clocking schemes. To validate the techniques, a multiplexer is developed using USE scheme. Designs developed using modern clocking schemes occupy more area and cells. However, they reduce the clocking crossings and increase the reliability of the designs. QCADesigner is utilized for designing the circuits. The techniques can be elaborated and improved to develop automated QCA layout tools that support different clocking techniques.

REFERENCES

[1] Das, S., Chen, A., and Marinella, M. (2021). Beyond CMOS. In *2021 IEEE International Roadmap for Devices and Systems Outbriefs*, Santa Clara, CA, 1–129. https://doi.org/10.1109/IRDS54852.2021.00011.

[2] Bohr, M. T., and Young, I. A. (2017). CMOS scaling trends and beyond. *IEEE Micro*, 37(6), 20–29. https://doi.org/10.1109/MM.2017.4241347.

[3] Lent, C. S., Tougaw, P. D., Porod, W., and Bernstein, G. H. (1993). Quantum cellular automata. *Nanotechnology*, 4(1), 49–57. https://doi.org/10.1088/0957-4484/4/1/004.

[4] Liu, W., Lu, L., O'Neill, M., and Swartzlander, E.E. (2011). Design rules for quantum-dot cellular automata. *2011 IEEE International Symposium of Circuits and Systems (ISCAS)*, Rio de Janeiro, Brazil, 2361–2364. https://doi.org/10.1109/ISCAS.2011.5938077.

[5] Raj, M., Gopalakrishnan, L., Ko, S. B., Naganathan, N., and Ramasubramanian, N. (2020). Configurable logic blocks and memory blocks for beyond-CMOS FPGA-based embedded systems. *IEEE Embedded Systems Letters*, 12(4), 113–116. https://doi.org/10.1109/LES.2020.2966791.

[6] Rani, S., and Sasamal, T.N., (2017). Design of QCA circuits using new 1D clocking scheme. *International Conference on Telecommunication and Networks*, 1–6. https://doi.org/10.1109/TEL-NET.2017.8343540.

[7] Vankamamidi, V., Ottavi, M., and Lombardi, F. (2007). Two-dimensional schemes for clocking/timing of QCAcircuits. *IEEE Transactions on Computer-Aided Design of Integrated Circuits and Systems*, 27(1), 34–44. https://doi.org/10.1109/TCAD.2007.907020.

[8] Campos, C. A. T., Marciano, A. L., Neto, O. P. V, and Torres, F. S. (2016). USE: A Universal, scalable and efficient clocking scheme for QCA. *IEEE Transactions on Computer-Aided Design of Integrated Circuits and Systems*, 35(3), 513–517. https://doi.org/10.1109/TCAD.2015.2471996.

[9] Goswami, M., Mondal, A., Mahalat, M. H., Sen, B., and Sikdar, B.K. (2019). An efficient clocking scheme for quantum-dot cellular automata. *International Journal of Electronics Letters*, 8(1), 1–14. https://doi.org/10.1080/21681724.2019.1570551.

[10] Pal, J., Pramanik, A.K., Sharma, J.S., Saha, A. K., Sen, B. (2021). An efficient, scalable, regular clocking scheme based on quantum dot cellular automata. *Analog Integrated Circuits Signal Processing*, 107, 659–670. https://doi.org/10.1007/s10470-020-01760-4.

[11] Lent, C. S., and Tougaw, P. D. (1997). A device architecture for computing with quantum dots. *Proceedings of the IEEE*, 85(4), 541–557. https://doi.org/10.1109/5.573740.

[12] Raj, M., Gopalakrishnan, L., Ko, S-B. (2021). Design and analysis of novel QCA full adder-subtractor, *International Journal of Electronics Letters*, 9(3), 287–300. https://doi.org/10.1080/21681724.2020.1726479.

[13] Dhare, V., and Mehta, U. (2015). Defect characterization and testing of QCA devices and circuits: A survey. *2015 19th International Symposium on VLSI Design and Test*, Ahmedabad, India, 1–2. https://doi.org/10.1109/ISVDAT.2015.7208060.

[14] Vanaraj, A. T., Raj, M., and Gopalkrishnan, L. (2020). Reliable coplanar full adder in quantun-dot cellular automata using five-input majority logic. *Journal of Nanophotonics*, 14(2), 026017. https://doi.org/10.1117/1.JNP.14.026017.

[15] Kumaresan, R. S., Raj, M., and Gopalakrishnan, L. (2022). Framework for QCA layout generation and rules for rotated cell design. *Journal of Circuits, Systems and Computers*. https://doi.org/10.1142/S0218126623501141.

[16] Liu, W., Lu, L., O'Neill, M., and Swartzlander, F.F. (2014). A first step toward cost functions for quantum-dot cellular automata designs. *IEEE Transactions on Nanotechnology*, 13(3), 476–487. https://doi.org/10.1109/TNANO.2014.2306754.

[17] Raj, M., Gopalakrishnan, L. and Ko, S.-B. (2021), Reliable SRAM using NAND-NOR gate in beyond-CMOS QCA technology. *IET Computers and Digital Techniques*, 15, 202–213. https://doi.org/10.1049/cdt2.12012.

[18] Sen, B., Dutta, M., Singh, D. K., Saran, D., and Sikdar, B. K. (2012). QCA multiplexer based design of reversible ALU. *IEEE International Conference on Circuits and Systems*, 168–173, https://doi.org/10.1109/ICCircuitsAndSystems.2012.6408309.

[19] Nadooshan, R.S., and Kianpour, M. (2014). A novel QCA implementation of MUX-based universal shift register. *Journal of Computational Electronics*, 13(1), 198–210. https://doi.org/10.1007/s10825-013-0500-9.

[20] Sen, B., Goswami, M., Mazumdar, S., and Sikdar, B. K. (2015). Towards modular design of reliable quantum-dot cellular automata logic circuit using multiplexers. *Computers and Electrical Engineering*, 45, 42–54. https://doi.org/10.1016/j.compeleceng.2015.05.001.

[21] Sen, B., Goswami, M., Mazumdar, S., and Sikdar, B. K. (2014) Modular design of testable reversible ALU by QCA multiplexer with increase in programmability, *Microelectronics Journal*, 45(11), 1522–1532. https://doi.org/10.1016/j.mejo.2014.08.012.

[22] Raj, M., Kumaresan, R. S., and Gopalakrishnan, L. (2019). Optimized Multiplexer and Exor gate in 4-dot 2-electron QCA using novel input technique. *2019 10th International Conference on Computing, Communication and Networking Technologies (ICCCNT)*, Kanpur, 1–4. https://doi.org/10.1109/ICCCNT45670.2019.8944782.

Chapter 12

Low power reduction techniques for a 6T-SRAM cell design using CNTFET technology

G. Karthy
Dhanalakshmi Srinivasan University

P. Sivakumar
Kalasalingam Academy of Research and Education

T. Jayasankar
University College of Engineering, Anna University

12.1 INTRODUCTION

The reduction of modern CMOS technologies' physical gate length (element size) to the nanoscale presents a number of crucial and reliable concerns, potentially limiting their ability to build applications with improved energy efficiency. To overcome physical and technological restrictions such as augmented effects of short channel, mitigated gate control, increasing leakage currents, large process deviation, and elevated power consumption, scientists and researchers are developing new emerging solutions to replace conventional CMOS technology [1]. As the density of transistors increases, so does the power dissipate by the microcircuit. Because high power consumption reduces battery life, modern battery-powered mobile devices such as mobile phones, gadgets, and personal digital assistants (PDAs) are particularly vulnerable. As a result, power dissipation has emerged as a critical design factor. In integrated circuits, dynamic or switching power components are employed to predominate the overall power dissipated.

In nanoscale mode, static power, also known as leakage power, accounts for a significant fraction of total power loss [2]. When there is no input contact, the leakage current flows through the transistors and enters a steady state, dissipating the leakage power. Scaling the voltage level is the most efficient technique to reduce dynamic power dissipation, and the transistor threshold voltage must also be scaled appropriately to preserve performance [3]. This reduces the leakage's power. The leakage power is not negligible in nanometer mode due to the lower supply voltage. Several methods for reducing static power losses by effectively minimizing lost leakage power have been proposed [4]. Carbon nanotubes (CNTs) have been proposed as

DOI: 10.1201/9781003459231-12

(a)   (b)

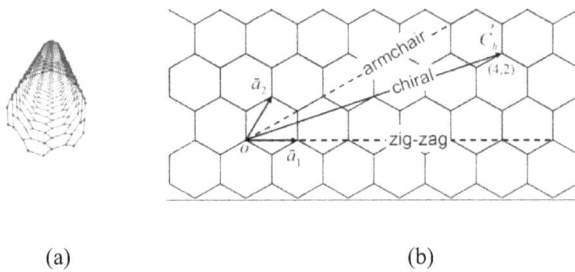

Figure 12.1 (a) SWCNT (b) Graphine sheet.

an alternative design material for next-generation SRAMs in recent years, with the goal of achieving higher density, performance, reliability, and lower power [5–11]. Techniques for decreasing leakage power dissipation in CNTFET-based SRAM cells are thus required. The hollow cylinders of carbon nanotube (CNT) comprised one or more concentrated layers of carbon atoms in a hierarchical structure. The structure of the CNT made a graphite sheet rolling in a tube, and a connection at the end of the sheet, as indicated in Figure 12.1, generates a tube clamp. CNT can have either semiconductor or metallic behavior according to its chirality (rolling direction).

12.2 LITERATURE WORK

A Cadence Virtuoso-based 6T SRAM 90 nm cell was proposed by Shruti H. Choudhary et al. [12]. This study analyzed the performance of five SRAM cell topologies, including the traditional use of SRAM cells such as 6T to 10T. Parameters such as leakage flow, leakage power, and read behavior of each SRAM cell are investigated separately. Read and write latency, average power, and leakage power are the primary criteria for recognition. With a 1 V power supply, SRAM has 40%–60% lower leakage power and 10%–20% lower read and link latency. A unidirectional CNFET based on a multi-stable 9T SRAM cell was demonstrated by Pramod Kumar Patel et al. [13]. Error-free operation at low supply voltages is ensured by the excellent design and bit-interleaving approach. The suggested SRAM cell implementation for 16 nm CNFET technology provides high integration and reduces power consumption by 3.2× when compared to conventional cells. Elangovan et al. [14] recommended comparing the proposed 10T CNTFET SRAM cell's power utilization and performance to that of a regular 10T CNTFET SRAM cell. The voltage ranges from 0.6 to 0.9 V, and the temperature ranges from 27°C to 125°C. The suggested SRAM CNTFET cell has written, hold, and read modes, according to simulation results. Shital Joshi [15] utilized SRAM cells, which, because of the exclusive capacity of data storage, are one of the most prevalent memory elements in most digital systems.

SRAM's design is being put to the test in terms of latency and dependability as technology improves. This paper compares the static noise margin (SNM), write margin (WM), read latency, and consumption of power in several CNTFET-based SRAM cell topologies, including 6T to 10T cells. Arusha Srivastava [16] examined the performance of 6T, 7T, 8T, and 10T SRAMs built in 32 nm CNTFETs and CMOS. As a result, the Stanford library is used to create SRAMs with varying chiral indices for HSPICE modeling. This chapter compares CMOS and CNTFETs manufactured in 32 nm technology. A CNTFET has a 98% lower power consumption than a CMOS transistor. As a result, power is measured at different voltages for different chiral numbers in order to design 6T to 10T SRAM cells under the best possible SRAM design conditions.

12.3 BASIC 6T SRAM CELL DESIGN

A 1-bit memory cell is a 6-transistor SRAM. It has six transistors, two of which are designated as access transistors based on what they do through or through the transistors, and the remaining four transistors are connected by two inverters to power the output of one inverter, as the name implies. As illustrated in Figure 12.2, this form of association is known as a cross-coupled inverter. Both active and standby modes are possible with SRAM cells. Reading data, manipulating data, and storing data are the three types of SRAM cell operations. Data is accessed or modified in active mode, while data is saved in standby mode on the two inverters. The output is provided by one of the inverter outputs, denoted Q, and the other as an inverted Q value, denoted Qbar, depending on the mode.

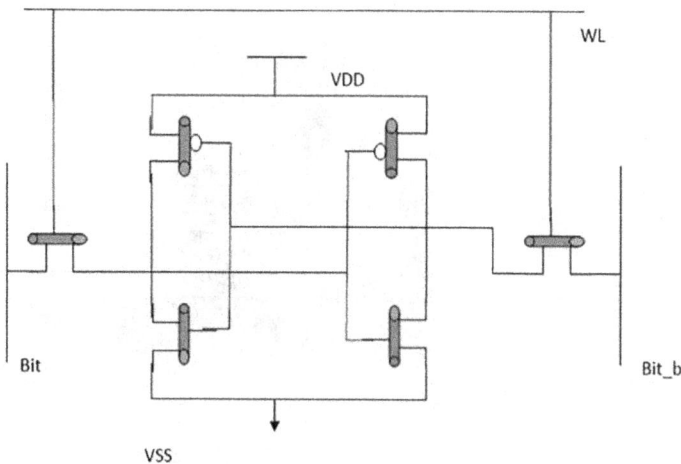

Figure 12.2 Conventional 6 transistor SRAM cell design.

12.3.1 Read operation

Bit strings (Bit) and word strings (WL) are the two types of strings used in SRAM to access data in cells (WL). Charging the bit line to logic "1" allows data to be read from SRAM. Because the Q values complement each other, the Qbar values will be "1" if the Q values are initially "0." Logic "1" must continue to be followed by Qbar and bit b. When the line bit is set to logical "0," the WL line is set to logical "1." As the Q value decreases, the Bit line's value increases.

12.3.2 Write operation

The use of the system should update SRAM cells. It sets the bit to a logic "1" after setting the memory cell to a logic "1" with logic "0" assigned to bit "b". Assume that Q begins with a value of "0". Because line bit is 1, line b bit is set to 0. The Qbar value is also "1" because the Q value is "0". Their values will then be "0" due to the voltage drop across the Qbar node, and the Q value will be 1. The Simulation waveforms of read and write operations are shown in Figure 12.3.

12.4 METHODS FOR REDUCING POWER LEAKAGE

Low power dissipation is one of the main issues with VLSI technology, which necessitates a large memory capacity. When the device is switched

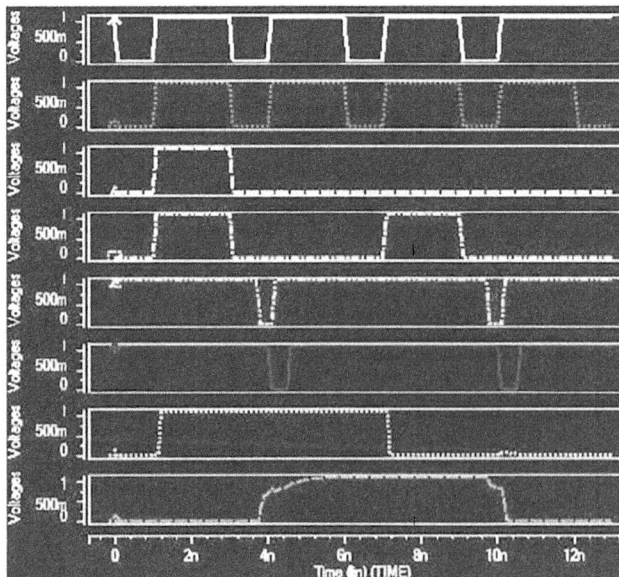

Figure 12.3 Simulation waveform of 6T SRAM write/read operation.

off, no current would be passing through the circuit and possess elevated barrier potential stimulating the attraction of electrons to the "+" terminal of the battery, resulting in reverse leakage current. Temperature, aspect ratio, and supply voltage all enhance leakage power. To reduce leakage power in SRAM cells, many approaches are being explored. This chapter discusses two major approaches for lowering leakage power.

a. Voltage divider technique
b. Power down transistor technique

12.4.1 Voltage divider technique

A voltage divider is a technology that is based on the premise that the power required by the cell in hold mode is lower, lowering the source voltage to reduce power dissipation in the circuit. Two transistors serve as power gateways in voltage divider technology. The function of the cell determines the input to this transistor. As shown in Figure 12.4, the input signal V is applied to the gate inputs of both transistors. The active region of this signal is low, indicating that the PMOS is turned on and the circuit is powered by VDD. SRAM is behaving like a regular cell. In standby mode, the input V signal is high because the power requirements are lower than in active mode. The voltage is halved as a result and fed to the NMOS drain. The NMOS transistor is turned on in standby mode, supplying VDD/2 to the SRAM cell. The simulation output of voltage divider circuit is shown in Figure 12.5.

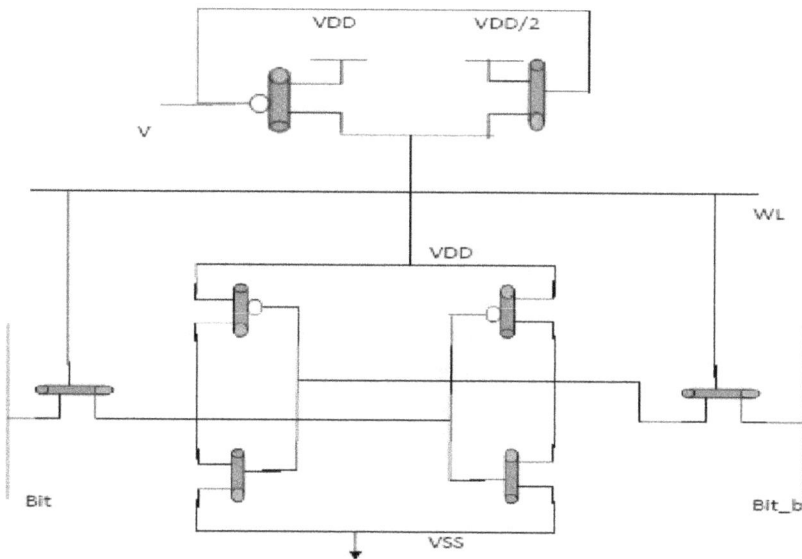

Figure 12.4 Voltage divider technique in low power reduction SRAM cell.

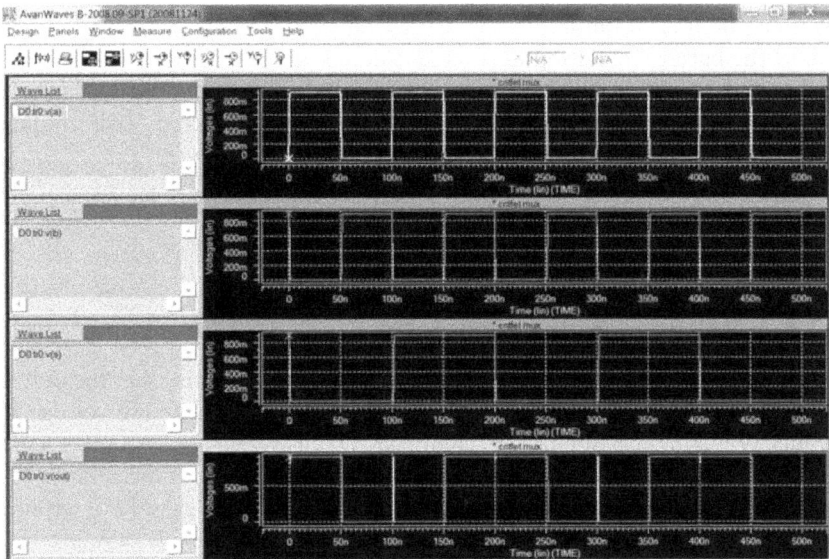

Figure 12.5 Simulation output waveform of the voltage divider technique.

12.4.2 Power down transistor technique

This power down technique is employed in SRAM cells using CNTFETs. The power-down transistor successfully avoids any leakage energy dissipation in the unused SRAM sector by "turning off" the supply voltage to the CNTFET SRAM cell. The fundamental scheme of the work is adding an additional transistor across Drain voltage (VDD) or ground circuit of a CNTFET SRAM cell (GND). In the used section, additional transistors turn on, while in the unused section, they turn off. The power-down strategy reduces leakage and leakage energy losses while retaining the performance benefits associated with lower supply and threshold voltages.

Figure 12.6 depicts a proposed CNTFET SRAM cell with a power down transistor. PPWD connects the CNTFET SRAM PMOS transistor to VDD, while NPWD connects the CNTFET SRAM NMOS transistor to GND. When the cell is in "active" mode, the PPWD and NPWD power-down transistors turn on, and when the cell is in "standby" mode, they turn off. These two Power Down transistors are driven by a signal called Power Down (PWD). The Power Down transistor is triggered in "ON" mode and "off" during the idle mode because this control signal for standby mode is "1" and "0" in active mode. This occurs during "W_L" is put to "1." When a "W_L" control signal is set to "0," a "Power Down" signal is set to "1," and a "Power Down Bar" is set to "0," power down transistors are turned off, lowering leakage currents and consequently leaking power.

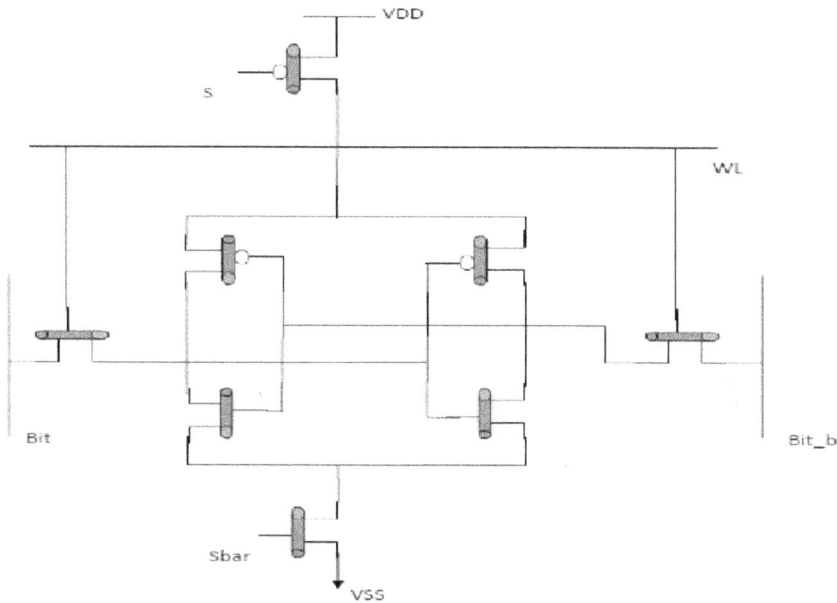

Figure 12.6 6T CNTFET SRAM with power down transistor.

12.5 RESULTS AND DISCUSSION

Synopsis HSPICE is a simulation tool for estimating latency and power usage. For simulations, a Stanford CNTFET model with a device size of 32 nm and a VDD supply voltage of 0.9 V was employed [17]. The proposed power reduction method mitigates the rate of leakage by 24.6% when contrasted to the traditional CNTFET 6T SRAM cell, and the simulation results of the CNTFET SRAM cell utilizing the previous approach are summarized in Table 12.1 for evaluation.

Power leakage, latency, and SNM parameters were analyzed and compared with conventional approaches. The proposed power-down transistor method improved leakage power by 38.4% and delay by 9%, while the voltage divider method improved leakage power by 19.3% and delay by 4%. Figures 12.7–12.9 compare the results of the proposed method with the parameters of the previous approach for leakage power, delay, and SNM.

SRAM cell dynamic power consumption measured during read/write operations. Power is consumed by the SRAM cell when it is in active mode. The findings are summarized in Table 12.2.

Table 12.1 Comparison results of simulated low power CNTFET SRAM cell designs

No.	Parameter	8T CNTFET SRAM cell [18]	6T CNTFET SRAM cell [19]	Proposed voltage divider technique	Proposed power down transistor technique
1	Power leakage (W)	9.56E-9	6.95E-9	5.67E-9	4.36E-9
2	Delay (S)	5.64E-10	2.401E-10	2.320E-10	2.199E-10
3	SNM (mV)	196	154	126	104
4	No of transistor	8	6	8	8

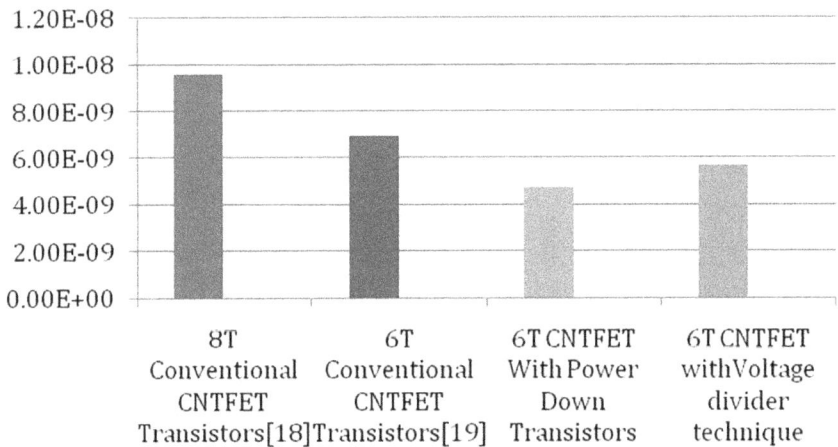

Figure 12.7 Leakage power comparison of CNTFET SRAM cell designs.

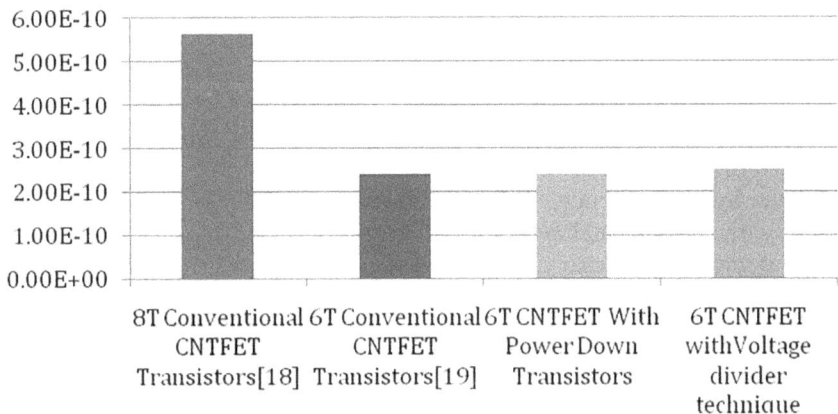

Figure 12.8 Delay comparison of CNTFET SRAM cell designs.

12.6 CONCLUSION

This CNTFET technology has progressed to the point where it is now necessary to reduce SRAM cell power consumption, which is a major concern. As databases grow in size, today's world wants to store large amounts of data. To meet these goals, a large number of SRAM cells must be wired in

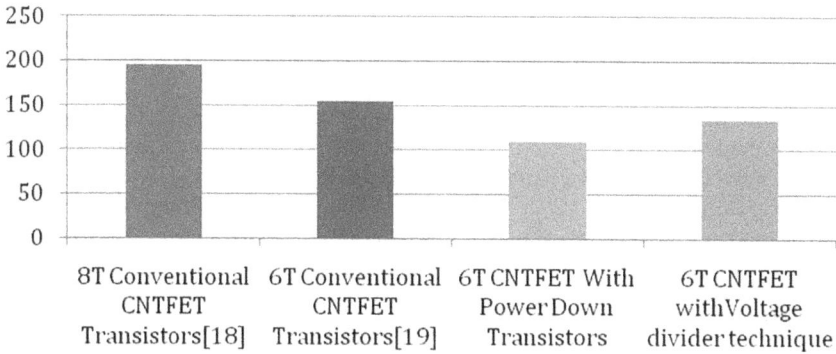

Figure 12.9 SNM comparison of CNTFET SRAM cell designs.

Table 12.2 Dynamic power consumption comparison

No.	Operation	Conventional 6T SRAM (W)	Proposed power down transistors(W)	Proposed voltage divider technique(W)
1	Write 0	6.456E-5	4.635e-07	4.987E-7
2	Write 1	6.388E-5	4.125e-07	4.564E-7
3	Read 0	6.154E-5	4.234e-07	4.656E-7
4	Read 1	6.256E-5	4.335e-07	4.545E-7

a low-power manner. Some power-saving technology can assist in meeting this need. To limit power leakage, the proposed 6T CNTFET-based SRAM cell employs a voltage divider and power down transistor approach. Leakage power is reduced by 38.4% and delays are reduced by 9% using the proposed step-down transistor method, while leakage power is reduced by 19.3% and delays are reduced by 4% using the voltage divider method. The results show that by keeping the delay, this design significantly reduces leakage power. The proposed cell can be used to create ultra-small storage device SRAM based on CNTFETs. Using CNTFET technology, these cells can be used to create low-leakage memories. This technology shall be employed in registers of high-speed capacity, and tiny storage locations to mitigate overall power dissipation and enhance electronics efficiency.

REFERENCES

[1] Roy, K., Mukhopadhyay, S., & Mahmoodi-Meimand, H. (2003). Leakage current mechanisms and leakage reduction techniques in deep-submicrometer CMOS circuits. *Proceedings of the IEEE, 91*(2), 305–327.
[2] Semiconductor Industry Association. (2009). International technology roadmap for semiconductors. https://www. itrs. net.
[3] Mutoh, S. I., Douseki, T., Matsuya, Y., Aoki, T., Shigematsu, S., & Yamada, J. (1995). 1-V power supply high-speed digital circuit technology with multithreshold-voltage CMOS. *IEEE Journal of Solid-state circuits, 30*(8), 847–854.

[4] Powell, M., Yang, S. H., Falsafi, B., Roy, K., & Vijaykumar, T. N. (2000). Gated-Vdd: A circuit technique to reduce leakage in deep-submicron cache memories. In *Proceedings of the 2000 International Symposium on Low Power Electronics and Design*, New York, United States (pp. 90–95).

[5] Kim, K. K., Nan, H., & Choi, K. (2010). Power gating for ultra-low voltage nanometer ICs. In *Proceedings of 2010 IEEE International Symposium on Circuits and Systems*, Paris, France (pp. 1472–1475). IEEE.

[6] Yu, Z., Chen, Y., Nan, H., Wang, W., & Choi, K. (2011). Design of a novel low power 6-T CNFET SRAM cell working in sub-threshold region. In *2011 IEEE International Conference on Electro/Information Technology*, Mankato, MN, USA (pp. 1–5). IEEE.

[7] Wang, W., & Choi, K. (2010). Novel curve fitting design methodology for carbon nanotube SRAM cell optimization. In *2010 IEEE International Conference on Electro/Information Technology*, Normal, IL, USA (pp. 1–6). IEEE.

[8] Kureshi, A. K., & Hasan, M. (2009). Performance comparison of CNFET-and CMOS-based 6T SRAM cell in deep submicron. *Microelectronics Journal*, 40(6), 979–982.

[9] Lin, S., Kim, Y. B., Lombardi, F., & Lee, Y. J. (2008). A new SRAM cell design using CNTFETs. In *2008 International SoC Design Conference*, Busan, Korea (South) (vol. 1, pp. 1–168). IEEE.

[10] Lin, S., Kim, Y. B., & Lombardi, F. (2009). Design of a CNTFET-based SRAM cell by dual-chirality selection. *IEEE Transactions on Nanotechnology*, 9(1), 30–37.

[11] Choudhari, S. H., & Jayakrishnan, P. (2019). Structural analysis of low power and leakage power reduction of different types of SRAM cell topologies. *In 2019 Innovations in Power and Advanced Computing Technologies (i-PACT)*, Vellore, India (Vol. 1, pp. 1–7). IEEE.

[12] Patel, P. K., Malik, M. M., & Gupta, T. K. (2018). Reliable high-yield CNTFET-based 9T SRAM operating near threshold voltage region. *Journal of Computational Electronics*, 17, 774–783.

[13] Patel, P. K., Malik, M. M., & Gupta, T. K. (2019). Performance evaluation of single-ended disturb-free CNTFET-based multi-Vt SRAM. *Microelectronics Journal*, 90, 19–28.

[14] Elangovan, M., & Gunavathi, K. (2020). High stable and low power 10T CNTFET SRAM cell. *Journal of Circuits, Systems and Computers*, 29(10), 2050158.

[15] Joshi, S., & Alabawi, U. (2017). Comparative analysis of 6T, 7T, 8T, 9T, and 10T realistic CNTFET based SRAM. *Journal of Nanotechnology*, 2017, Article ID 4575013, 9 pages.

[16] Shrivastava, A., Damahe, P., Kumbhare, V. R., & Majumder, M. K. (2019). Designing SRAM Using CMOS and CNTFET at 32 nm technology. In *2019 IEEE International Symposium on Smart Electronic Systems (iSES) (Formerly iNiS)*, Rourkela, India (pp. 284–287). IEEE.

[17] Stanford University CNFET Model website, https://nano.stanford.edu/model.php?id=23.

[18] Kim, Y., Patel, S., Kim, H., Yadav, N., & Choi, K. K. (2021). Ultra-low power and high-throughput SRAM design to enhance AI computing ability in autonomous vehicles. *Electronics*, 10, 256.

[19] Murotiya, S. L., Matta, A., & Gupta, A. (2014). Performance evaluation of CNTFET-based SRAM cell design. In *International Conference on Computer and Communication Technology (ICCCT)*, 02(04), 11–15.

Chapter 13

Design and development of integrated low power high performance MOSFET structure

Debasis Mukherjee
Brainware University

13.1 CHALLENGES DUE TO LEAKAGE CURRENT IN MOS TRANSISTOR

With time, the charming low leakage quality of MOSFET decayed. To achieve low cost, high speed, and low power-consuming CMOS circuits, transistors have been scaled down continuously following Moore's law [1,2] and targeting more than Moore (MtM) [3] and More-Moore [4] for about 40 years. In 1971, the half node of transistor was 10 μm, which became 22 nm in 2012, and approaching towards 7 and 5 nm for the years 2018 and 2020, respectively [5]. With such tiny dimensions, the ideal behavior of long channel transistors became no longer valid, and the gate terminal lost its active control over the channel area. Apart from gate voltage, threshold voltage became dependent on drain voltage as well, and this phenomenon is known as drain-induced barrier lowering or DIBL [6]. Due to this, when the applied gate voltage is below threshold voltage, ideally transistor should remain in OFF condition; however, the conduction current flows as an effect of drain voltage and is known as subthreshold leakage current [7]. Each time a new technology was introduced, due to short channel effect (SCE), the threshold voltage further reduced in sequence, resulting in more subthreshold leakage current. This phenomenon is known as rolling off of threshold voltage [6]. Due to all these reasons, subthreshold voltage has increased exponentially with technology scaling and occupied almost 50% of the total power consumption [8]. This leakage has the highest contribution to the transistor OFF current [6]. Consequently, analysis and modeling of subthreshold leakage current; and designing of novel structure with self-resisting capability from this leakage current became extremely important in deep nanometer technology. Modern electronic circuits primarily in the area of hand held low power devices such as smart phones, portable digital assistant (PDA), palmtops, laptops, tabs, etc., leakage current poses a challenge to the modern semiconductor industry in the form of battery drain and deteriorating noise margin.

DOI: 10.1201/9781003459231-13

13.2 DRAWBACKS OF AVAILABLE LEAKAGE REDUCTION TECHNIQUES

To diminish incessantly escalating leakage current, methodologies have been projected such as (1) transistor stacking, (2) multiple threshold voltages, (3) dynamic threshold voltages, (4) scaling of supply voltage, etc. [9]. But, these methods need extra circuitry and are applicable only in particular cases.

Transistor stacking method either needs an extra transistor to be inserted [10], increasing the area overhead, or this method is applicable to only particular situations where at least two series transistors are in off condition simultaneously by a particular combination of input vector, so it is not applicable for all input combinations [11,12].

Multiple threshold voltages can be obtained by varying the (1) channel doping, (2) gate oxide thickness, (3) channel length, and (4) body bias [9]. But all of these methods produce different transistors for critical and non-critical paths. The same transistor is not suitable for both the operations. **Multiple channel doping** technique requires two extra masks, increasing the process cost. Moreover, in nanometer technologies, it is very likely to have nonlinear distribution of doping, resulting in different threshold voltage than the requirement. In dual threshold voltage system, as both the threshold voltage are very nearby, this process often results faulty threshold voltage transistors [9,13]. **Multiple gate oxide thickness** technique uses lower gate oxide thickness in critical path to enhance the speed. But lower thickness of gate oxide increases the subthreshold leakage, gate oxide tunneling leakage and dynamic power consumption [13]. Moreover, as the CMOS technology has already reached nanometer range, thickness of gate oxide is already approaching towards the technology limit, and farther reduction is not an easy process. Farther reduction can introduce heavy tunneling current and normal operation of transistor may not be possible [14]. Not only that, the use of high gate oxide thickness in non-critical path also increases SCE [15]. So, both reduction and increase of gate oxide thickness have unfavorable effect. **Multiple channel length** technique is also not very effective process for 0.1 μm and less technology generations. Moreover, the increase of channel length increases gate capacitance, which affects the performance of the transistor [9,16]. **Multiple body bias** technique needs costly triple well mechanism, as different transistors cannot have the same well potential [9].

Dynamic threshold voltage design is done by (1) hopping and (2) scaling of threshold voltage. The threshold voltage hopping method cannot be executed without software interference [17,18]. The dynamic threshold voltage scaling method requires full setup of controller with feedback loop [19].

Supply voltage scaling method is of two types: (1) static and (2) dynamic. Static method costs multiple voltage sources [20]. Dynamic method requires operating system, regulation loop and microprocessor [18].

In short, all these circuit-level techniques for leakage reduction require additional mechanism, not fit for all conditions and have drawbacks. So a strong need was felt for a transistor structure, which is capable of leakage current reduction without additional circuitry and which can work in all conditions.

Not only circuit level but also **process-level techniques** have been proposed for reduction of leakage current. (1) Retrograde doping and (2) halo doping are popular process-level techniques [9,21–23]. As both techniques work on channel area, they are not capable to reduce p-n junction leakage current, which is one of the three most dominating leakage components [24–28]. Both techniques are not capable to reduce subthreshold leakage current at other parts than channel area. **So, a better technique is needed to reduce leakage from all parts of transistor.**

13.3 IMPORTANCE OF AREA MINIMIZATION OF TRANSISTOR

Area minimization of individual transistor is another vital work in semiconductor industry, apart from leakage minimization. Up to 1958, transistors were manufactured in discrete form only. In 1960, production of logic gate chips was possible. In 1962, chips were made containing 2–4 logic blocks. In 1964, the number reached 5–20. In 1967, single chip contained 20–200 logic blocks, and it was popularly known as medium scale integration (MSI). Large-scale integration (LSI) started in 1972 with 200–2,000 logic blocks were placed in a single chip. Since 1978, very large-scale integration (VLSI) era started with 2,000–20,000 logic blocks per chip. If the number is more than 20,000, then it is called ultra large-scale integration (ULSI), which started in 1989 [7]. This continuous increase in transistor density was possible due to scaling. With advancement of technology, fabrication of transistor with smaller features was possible. When all dimensions of transistor geometry shrink following a particular ratio formula, the process is called scaling [29,30]. **So, apart from scaling process, if it is possible to decrease the feature size of transistor farther, without compromising the performance, it will be a forward thrust for semiconductor industry.** In this work, additional area minimization was achieved.

13.4 PROCESS SIMPLIFICATION AND MINIMIZING THE COST OF FABRICATION

Process simplification and minimizing the cost of fabrication are another key area of semiconductor industry, apart from leakage minimization and area minimization. To cope up with the tremendous leakage, additional manufacturing complexity has been added to bulk planner CMOS technology, especially for 28 nm and smaller technology [31]. These complicated technology increased the production cost a lot. **In this work, some critical process steps, which are otherwise necessary for conventional MOSFET, are eliminated.**

13.5 DESIGN AND DEVELOPMENT OF INTEGRATED LOW POWER HIGH PERFORMANCE MOSFET STRUCTURE

13.5.1 Minimization of reverse bias p-n junction leakage current and band to band tunneling leakage current

Reverse bias p-n junction leakage current is denoted by

$$I_{\text{reverse}} = AJ_s\left(e^{\frac{qV_{\text{bias}}}{kT}} - 1\right) \tag{13.1}$$

Here A is the area of junction. So, to minimize this type of leakage current, junction area should be minimized. Insertion of insulator between p and n region eliminates the formation of p-n junction. As silicon dioxide is the most compatible with silicon technology, it is used as insulator. Recalling the graph of Chapter 3, between drain voltage and substrate current with gate voltage as zero volts, this structure is very useful to prevent reverse bias p-n junction leakage current. The structure is again presented in Figure 13.1.

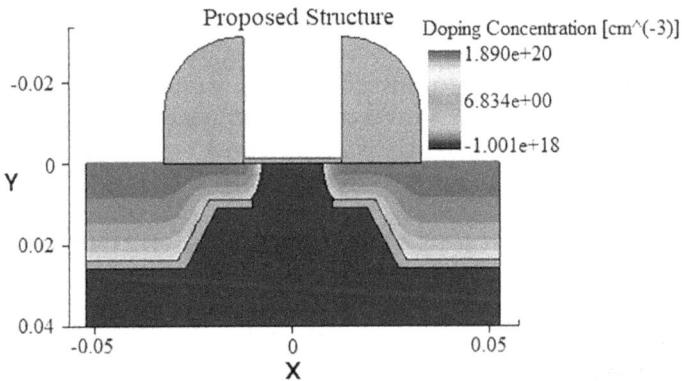

Figure 13.1 Proposed structure.

The band-to-band leakage current flows when electric field across the reverse biased pn junction exceeds 10^6 V/cm [9], which is the case of maximum modern devices. The equation for this kind of leakage current is given by [9]:

$$J_{\text{band-to-band}} = A \frac{E V_{\text{applied}}}{E_g^{\frac{1}{2}}} \exp\left(-B \frac{E_g^{\frac{3}{2}}}{E}\right) \tag{13.2}$$

$$\text{where } A = \frac{\sqrt{2m^*} q^3}{4\pi^3 h^2} \tag{13.3}$$

$$\text{and } B = \frac{4\sqrt{2m^*}}{3qh} \tag{13.4}$$

Form Figure 13.2, it is clear that this type of leakage current flows only in pn junction. As Figure 13.1 eliminates the p-n junction, so this type of leakage current also cannot flow through semiconductor. This can again be confirmed from the huge potential barrier due to inserted SiO_2 structure, given in Figure 13.3.

13.5.2 Minimization of subthreshold leakage current

Subthreshold leakage current is given by [7,15]:

$$I_{\text{sub}} = \mu_0 C_{\text{oX}} \frac{w}{L} (m-1)(v_T)^2 \times e^{\frac{V_g - V_{\text{th}}}{\text{mv}_T}} \times \left(1 - e^{-v_{\text{DS}}/v_T}\right) \tag{13.5}$$

So threshold voltage decreases exponentially with increase of threshold voltage.

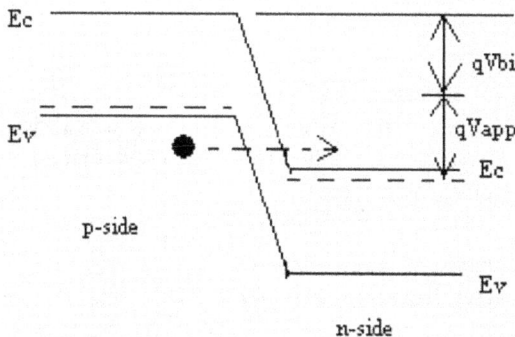

Figure 13.2 Band to band tunneling leakage current.

C1(Proposed)

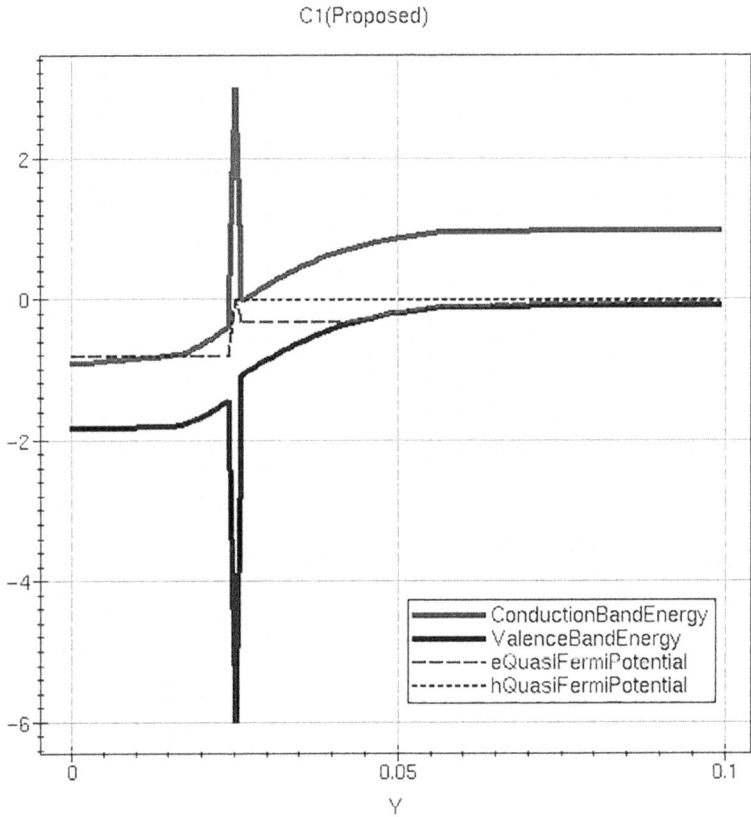

Figure 13.3 Energy band diagram of proposed structure. Very high potential barrier due to inserted SiO₂ structure is clearly visible. It is almost impossible for charge carriers to jump across such huge potential barrier.

$$V_{th} = V_{th\text{-long}} - \left(V_{ds} + 0.4 \text{ V}\right) \cdot \frac{c_d}{c_{oxe}} \qquad (13.6)$$

Below diagram shows position of C_d

So the proposed structure of Figure 13.1 is also valid for reduction of subthreshold leakage current. The same is also confirmed from the following simulation result (Figure 13.4).

13.5.3 Reduction of gate oxide tunneling leakage current

This kind of leakage current is generated due to tunneling through and into gate oxide. This phenomenon is increasing due to very low thickness of gate

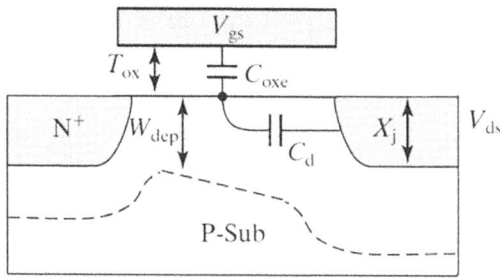

Figure 13.4 Oxide capacitors [6].

oxide. There are mainly two types of tunneling currents: Fowler Nordheim tunneling or FN tunneling and direct tunneling.

$$J_{FN} = \frac{q^3 E_0^2 x}{16\pi^2 h\phi_{ox}} \exp\left(-\frac{4\sqrt{2m^*}\,\varnothing_{ox}^{3/2}}{3hqE_{ox}}\right) \qquad (13.7)\,[9,21]$$

$$J_{DT} = AE_{ox}^2 \exp\left\{-\frac{B\left[1-\left(1-\frac{V_{ox}}{\varnothing_{ox}}\right)^{3/2}\right]}{E_{ox}}\right\} \qquad (13.8)\,[9,14]$$

This gate oxide tunneling current can be controlled by replacing the gate oxide with high-k dielectric material [14] (Figure 13.5). Different gate materials are given in Figure 13.6. HfO_2 may be considered as a good choice as gate material.

13.5.4 Reduction of leakage current due to hot carrier injection from bulk to SiO_2

This type of leakage current can also be minimized by replacing SiO_2 with HfO_2 since this dielectric has more thickness for equivalent functionality of SiO_2 [14].

13.5.5 Reduction of gate induced drain leakage or GIDL

Figure 13.7 shows the mechanism of gate-induced drain leakage. High V_{DD} and narrow oxide width enhance GIDL [9]. But in 7 nm technology, V_{DD} is very less, 0.8 V [5]. Moreover, if oxide is replaced by high-k dielectric, then thickness is increased. Also, high doping in drain region decreases GIDL. With every new technology, doping concentration is continuously

Comparison of Drain TotalCurrent

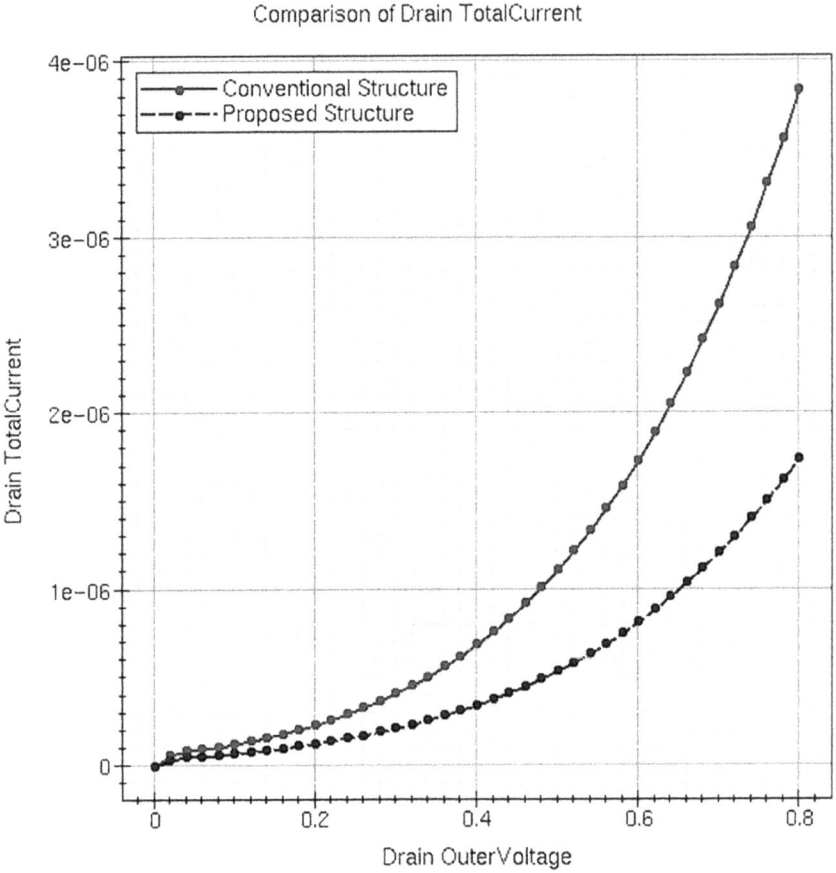

Figure 13.5 Drain current graph.

Gate Dielectric Material	Dielectric Constant	Energy Band Gap(Eg)
SiO$_2$	3.9	9
TiO$_2$	80	3.5
HfO$_2$	25	6
Ta$_2$O$_5$	25	4.4
Al$_2$O$_3$	8	8.8
ZrO$_2$	25	5.8
ZrSi$_x$O$_y$	8-12	6.5
Y$_2$O$_3$	13	6
Ya$_2$O$_3$	27	4.3

Figure 13.6 Different dielectric materials for gate [14].

$$V_d = V_{DD} \qquad V_g < 0$$

Tunnel created
minority carrier

n+ ploy gate

n+ drain

GIDL

Depletion edge

p-substrate

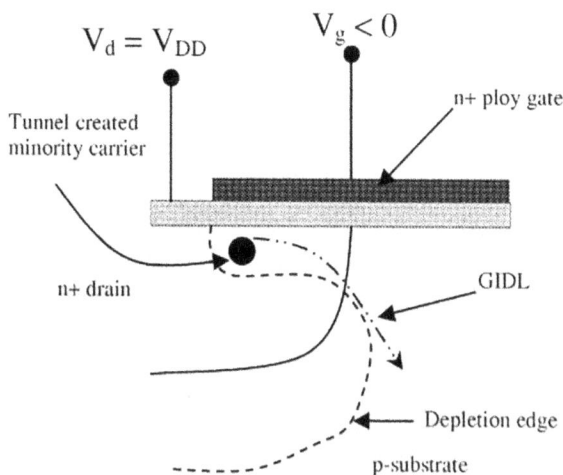

Figure 13.7 Gate induced drain leakage [9].

increasing. In 7 nm technology, top layer doping of source and drain is $1.0 \times 10^{20}/cm^3$ and bottom layer doping of source and drain is $8.4 \times 10^{18}/cm^3$. High doping of drain region lessen depletion width, as a result tunneling current is reduced and so is the GIDL.

13.5.6 Reduction of channel punchthrough leakage current

Punchthrough means when depletion regions of drain and source merge together, forming direct connection between source and drain, resulting uncontrolled current flow between source and drain. But the proposed transistor structure of has insulator layer at source and drain boundary. So flow of charge particle not possible between source and drain in proposed structure. So in any condition, uncontrolled current flow between source and drain region is not possible for proposed structure, eliminating any chance of punchthrough leakage current.

13.5.7 Improvement of proposed structure

Proposed structure of Figure 13.1 was improved and presented as Figure 13.8. About 55% leakage reduction was noted on the cost of about 4% reduction of ON current. However, about 61% leakage reduction was done with same amount of ON current loss. Not only that, different combination of leakage current reduction and ON current loss was formulated. According to requirement, any combination can be used. For very low power design, vertical gap between gate oxide and inserted SiO_2 strip along

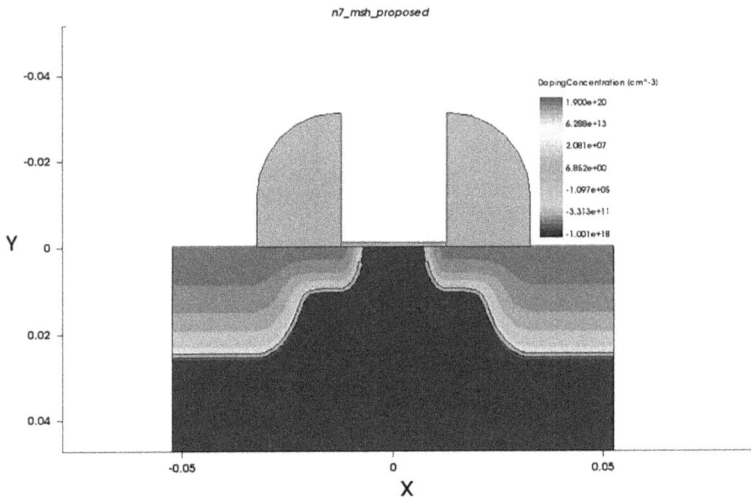

Figure 13.8 Improved proposed structure for leakage reduction.

p-n junction may be taken as two times of gate oxide width. It would reduce leakage current more than 96%. For a balance requirement, three times gap may be used. It reduces leakage current more than 92% while less than 10% decrease in ON current is noted. For high-performance transistor design, six times gap may be incorporated. It reduces leakage current nearly 75% for about 5% loss of ON current (Figure 13.8).

13.5.8 Reduction of area

Proposed structure is useful for a lot of area reduction with almost similar performance. The same is again given in Figure 13.10 for a ready reference. It consumes visibly less amount of area than Figure 13.9, conventional structure.

13.5.9 Similar performance

The proposed structure of Figure 13.10 has almost same performance to the conventional structure given in Figure 13.9. So the proposed structure of Figure 13.10 has benefit of less area and less cost with almost similar performance than conventional MOSFET structure.

13.6 PROTOTYPE

The proposed transistor structure of Figure 13.8 is capable of minimization of all six types of leakage current with replacement of gate oxide with high-k material. The proposed structure of Figure 13.10 is capable of area

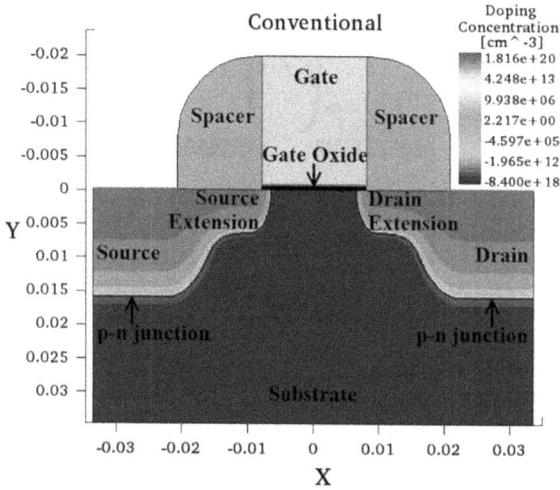

Figure 13.9 Conventional MOSFET.

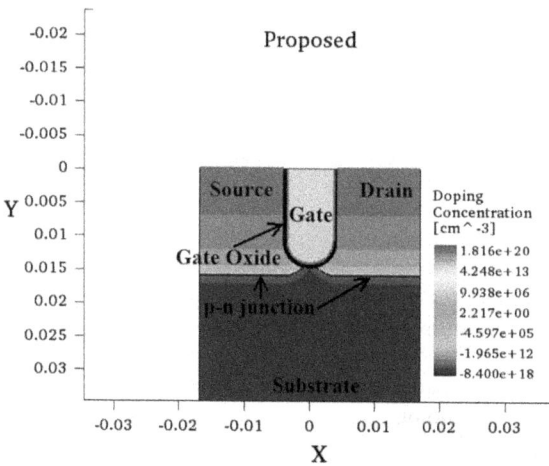

Figure 13.10 Vertical gate MOSFET.

reduction and cost reduction with almost the same performance than the conventional one. Combining both the structures, Figure 13.11 is the proposed prototype for all types of benefits. Here thin SiO_2 layer of Figure 13.8 is replaced by thick SiO_2 layer as this layer is used only for insulation purpose as demonstrated in Figure 13.3. Thickness equal to gate oxide requires sophistic fabrication process. Only for insulation purpose, such costly process is not required. Taking thick layer would reduce process cost as well. Second change incorporated is replacing gate oxide with high-k dielectric

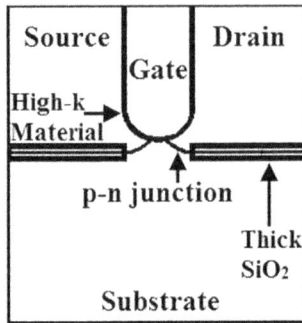

Figure 13.11 Prototype.

to reduce gate leakage current. So the structure of Figure 13.10 is changed in these two ways. First, thick SiO_2 layer is placed in the p-n junction except channel area just under gate region. Second, gate oxide is replaced with high-k material. Insulator layer in p-n junction would reduce maximum type of leakage currents as described. Replacing gate oxide with high-k material would reduce gate leakage current as described in [14]. Vertical gate structure reduces the size a lot with almost no compromise in performance. So, this prototype of Figure 13.11 is for low power high-performance CMOS circuits.

REFERENCES

[1] G. E. Moore, "Cramming more components onto integrated circuits, reprinted from electronics, volume 38, number 8, April 19, 1965, pp.114 ff.," *IEEE Solid-State Circuits Newsl.*, vol. 20, no. 3, pp. 33–35, 2006.
[2] C. Mac, "The multiple lives of Moore's law," *IEEE Spectr.*, vol. 52, no. 4. pp. 31–37, 2015.
[3] S. J. Wind et al., "'More-than-Moore' white paper," 2012.
[4] International Technology Roadmap for Semiconductors 2.0, "More Moore," *ITRS*, pp. 1–52, 2013. https://www.semiconductors.org/wp-content/uploads/2018/06/5_2015-ITRS-2.0_More-Moore.pdf
[5] "2013 ITRS - International Technology Roadmap for Semiconductors." [Online]. Available: https://www.itrs2.net/2013-itrs.html [Accessed: 04-Oct-2018].
[6] C. C. Hu, "MOSFETs in ICs-scaling, leakage, and other topics," In *Modern Semiconductor Devices for Integrated Circuits*, pp. 259–289, 2010. https://www.chu.berkeley.edu/wp-content/uploads/2020/01/Chenming-Hu_ch7.pdf
[7] S.-M. Kang and Y. Leblebici, *CMOS Digital Integrated Circuits Analysis and Design*, 3rd ed. New Delhi: McGraw-Hill, 2003.
[8] S. Borkar, "Getting gigascale chips," *Queue*, vol. 1, no. 7, p. 26, 2003.
[9] K. Roy, S. Mukhopadhyay, and H. Mahmoodi-Meimand, "Leakage current mechanisms and leakage reduction techniques in deep-submicrometer CMOS circuits," *Proc. IEEE*, vol. 91, no. 2, pp. 305–327, 2003.

[10] V. De et al., "Techniques for leakage power reduction," *Des. High-Performance Microprocess. Circuits*, pp. 48–52, 2001.

[11] Z. Chen, L. Wei, A. Keshavarzi, and K. Roy, "IDDQ testing for deep-submicron ICs: challenges and solutions," *IEEE Des. Test Comput.*, vol. 19, no. 2, pp. 24–33, 2002.

[12] D. Duarte, Y. F. Tsai, N. Vijaykrishnan, and M. J. Irwin, "Evaluating run-time techniques for leakage power reduction," In *Proceedings -7th Asia and South Pacific Design Automation Conference, 15th International Conference on VLSI Design, ASP-DAC/VLSI Design 2002*, Bangalore, India, pp. 31–38, 2002.

[13] N. Sirisantana, L. Wei, and K. Roy, "High-performance low-power CMOS circuits using multiple channel length and multiple oxide thickness," In *Proceedings 2000 International Conference on Computer Design*, Austin, TX, USA, pp. 227–232, 2000.

[14] S. Das and S. Kundu, "Simulation to study the effect of oxide thickness and high- K dielectric on drain-induced barrier lowering in N-type MOSFET," *IEEE Trans. Nanotechnol.*, vol. 12, no. 6, pp. 945–947, 2013.

[15] K. Roy and S. C. Prasad, *Low-Power CMOS VLSI circuit design*. India: John Wiley & Sons, 2009.

[16] Y. Taur and T. H. Ning, *Fundamentals of Modern VLSI Devices*. USA: Cambridge University Press, 2002.

[17] K. Nose, M. Hirabayashi, H. Kawaguchi, S. Lee, and T. Sakurai, "VTH-hopping scheme to reduce subthreshold leakage for low-power processors," *IEEE J. Solid State Circuits*, vol. 37, no. 3, pp. 413–419, 2002.

[18] S. Lee and T. Sakurai, "Run-time voltage hopping for low-power real-time systems," In *Proceedings of the 37th Annual Design Automation Conference*, Los Angeles California USA, pp. 806–809, 2000.

[19] C. H. Kim and K. Roy, "Dynamic VTH scaling scheme for active leakage power reduction," *Proceedings of the Conference on Design, Automation and Test in Europe*, Paris, France, pp. 163–167, 2002.

[20] T. Fuse, A. Kameyama, M. Ohta, and K. Ohuchi, "A 0.5 V power-supply scheme for low power LSIs using multi-Vt SOI CMOS technology," In *2001 Symposium on VLSI Circuits. Digest of Technical Papers (IEEE Cat. No.01CH37185)*, Kyoto, Japan, pp. 219–220, 2001.

[21] Y. Taur and T. H. Ning, *Fundamentals of Modern VLSI Devices*. New York: Cambridge University. Press, 1998.

[22] D. P. Foty, *MOSFET Modeling with SPICE: Principles and Practice*. Upper Saddle River, NJ: Prentice Hall PTR, 1997.

[23] Y. Taur, "CMOS scaling and issues in sub-0.25 m systems," *Des. High-Performance Microprocess. Circuits*, pp. 27–45, 2001.

[24] A. Abdollahi, F. Fallah, and M. Pedram, "Leakage current reduction in CMOS VLSI circuits by input vector control," *IEEE Trans. Very Large Scale Integr. Syst.*, vol. 12, no. 2, pp. 140–154, 2004.

[25] A. Agarwal, S. Mukhopadhyay, A. Raychowdhury, K. Roy, and C. H. H. Kim, "Leakage power analysis and reduction for nanoscale circuits," *IEEE Micro*, vol. 26, no. 2, pp. 68–80, 2006.

[26] S. Mukhopadhyay, A. Raychowdhury, and K. Roy, "Accurate estimation of total leakage current in scaled CMOS logic circuits based on compact current modeling," In *Proceedings 2003. Design Automation Conference (IEEE Cat. No.03CH37451)*, Anaheim, CA, pp. 169–174, 2003.

[27] A. Sanyal, A. Rastogi, W. Chen, and S. Kundu, "An efficient technique for leakage current estimation in nanoscaled CMOS circuits incorporating self-loading effects," *IEEE Trans. Comput.*, vol. 59, no. 7, pp. 922–932, 2010.

[28] S. Mukhopadhyay, A. Raychowdhury, and K. Roy, "Accurate estimation of total leakage in nanometer-scale bulk CMOS circuits based on device geometry and doping profile," *IEEE Trans. Comput. Des. Integr. Circuits Syst.*, vol. 24, no. 3, pp. 363–381, 2005.

[29] R.-H. Yan, A. Ourmazd, and K. F. Lee, "Scaling the Si MOSFET: From bulk to SOI to bulk," *IEEE Trans. Electron Devices*, vol. 39, no. 7, pp. 1704–1710, 1992.

[30] M. Bohr, "A 30 year retrospective on Dennard's MOSFET scaling paper," *IEEE Solid-State Circuits Soc. Newsl.*, vol. 12, no. 1, pp. 11–13, 2007.

[31] STMicroelectronics, "An introduction to FD-SOI - YouTube," 2013. [Online]. Available: https://www.youtube.com/watch?v=uvV7jcpQ7UY [Accessed: 28-Sep-2018].

Chapter 14

Low power designs for enhanced CMOS performance

Gowsika Dharmaraj, Ashwin Kumar S,
Chandra Prakash S, and Reba P
PSG Institute of Technology and Applied Research

14.1 INTRODUCTION

As the advancements in technology and fabrication began, with increased integration densities and speed of operation that are common concerns in today's ultra-deep submicron regime, the circuits implemented in CMOS technology started to supersede the older technologies—bipolar transistors, TTL logic, N-channel technology.

It has not been an issue in the development of ICs, especially for consumer electronics concerning power consumption until the 1990s. The 10-year prediction of Moore's law (as shown in Figure 14.1 [1]) in 1965 was accurate in predicting the number of transistors; nonetheless, no prediction was made on power consumption. During 1990–1992, power consumption started to challenge the development of microelectronics. The maximum power dissipation decides the electrical limits of a design. CMOS technologies then dominated the semiconductor industry for their low power operation in addition to meeting new growing requirements such as performance, portability, reliability, and cost [2].

However, with the increased number of transistors per package, as shown in Figure 14.2 [1], and parallelly with the operational frequency because of downscaling the device, the power dissipation in chips increased dramatically. In addition, the failure rate of silicon is magnified by 2, for

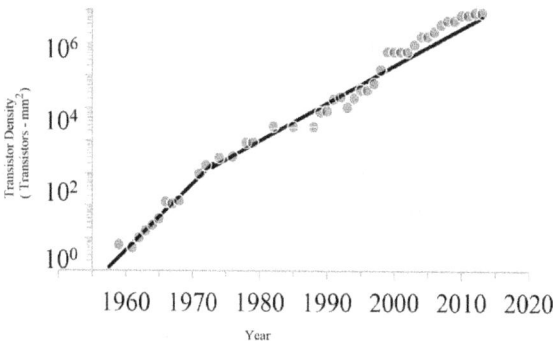

Figure 14.1 Integrated circuit density (transistors-mm²) per year.

DOI: 10.1201/9781003459231-14

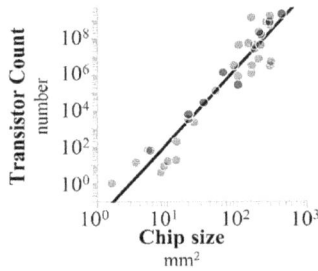

Figure 14.2 A power-law relationship between the chip size (mm²) and the number of transistors.

every 10° increase in temperature in devices that exploit a large amount of power. Thus, the surging intricacies of SoCs, specifically in deep submicron technologies, came the major motivation for the VLSI digital drafting engineers to come up with emerging techniques in order to diminish power consumption at various levels of design stages.

14.2 SOURCES LEADING TO POWER DISSIPATION

Power dissipation [3] in any system is defined as the rate of dissipation of energy over time. This is classified in terms of:

Peak power: It is termed as the maximal instantaneous power dispelled over the particular time period.

Average power: It is the average of instantaneous power dissipated over time period.

The overall power dissipated is given by:

$$p(t) = i(t)v(t) \tag{14.1}$$

where $i(t)$ - current measured instantaneously and $v(t)$ – voltage supplied to bias the circuit.

In CMOS circuits, power dissipation [4] is mainly due to:

- Dynamic component
- Short-circuit and static power components.

Hence, the power dissipation in CMOS circuits considering the above categories can be written as:

$$P_{\text{total}} = P_{\text{switching}} + P_{\text{short}} + P_{\text{leakage}} \cdots \tag{14.2}$$

$$P_{\text{switching}} = \alpha C_L V^2 f \ \dots \tag{14.3}$$

$$P_{\text{short}} = \tau \alpha V I_{\text{short}} \tag{14.4}$$

$$P_{\text{leakage}} = V I_{\text{leakage}} \tag{14.5}$$

where C_L is the capacitance due to switching nodes, f gives the number of times the switching occurs, V denotes the voltage supplied to the circuit, α corresponds to the proportionality factor that gives the probability that a particular node will switch from state 1-> 0 and τ denotes time for the output voltage to settle.

14.2.1 Dynamic power dissipation

This type of power dissipation is due to charging and discharging of unintended and inevitable capacitances, existing in between various parts of a circuit (called parasitic capacitances) present in the nodes of a CMOS circuit. It is consumed only whenever the inputs change, and it represents the maximum power consumption in the circuits. The state of logic elements present in the circuit depends on the applied input values, its frequency of switching, the applied supply voltage and capacitance as described in Figure 14.3.

No dynamic power consumption takes place if there is no switching of nodes. Hence, in CMOS circuits, reducing dynamic power utilization is of foremost importance. The dynamic power dissipation depends on switching activity factor α, design parameters of node capacitance C_L and the supply voltage V. However, it does not depend on transistor size. The switching power is given by equation (14.3). The voltage shows a quadratic relationship in the power equation and reducing the supply voltage proves to be an efficient method for power reduction in CMOS circuits. This is the major focus of the chapter, and techniques to combat the issues related to reduction in the supply voltage are also discussed.

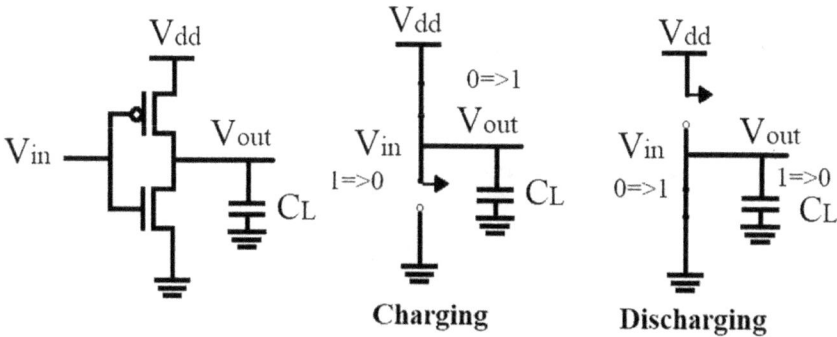

Figure 14.3 Dynamic power dissipation: charging and discharging.

14.2.2 Static/leakage power dissipation

Static/leakage power is consumed in CMOS circuits because of the leakage present in the transistors and also because of current when biased in reverse condition (Figure 14.4).

The equation for leakage power is given by (equation 14.5). Various transistor leakages are subthreshold leakage, gate oxide tunnelling leakage, leakage that occurs in reverse-biased source/drain junctions, gate induced drain leakage (GIDL) as shown in Figure 14.5. Especially in deep submicron technologies, leakage power consumption is a major issue when the devices are in an idle state. So, when scaling down the CMOS circuits, leakage power dominates and various techniques such as stacking, LECTOR (refer [5]) and many others are used to reduce leakage currents.

14.2.3 Short-circuit power dissipation

This category of power dissipation arises because of some finite rise and fall times of the applied input values, as seen in Figure 14.6. Thereby, current (I_{short}) to flows directly from the supply rail towards the ground, making both the transistors in Figure 14.6 turn ON concurrently. This current pathway existing between the applied supply and ground rails for some short time interval when there occurs the transition of inputs, leading to the short-circuit power consumption that is consumed without attributing to the function of circuit. This power consumption is not attributed to

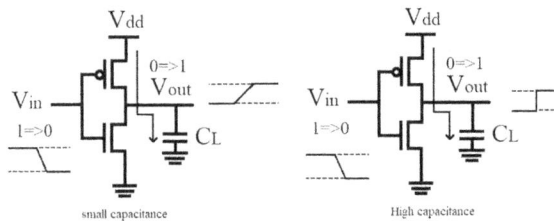

Figure 14.4 Leakage power dissipation due to low and high capacitance.

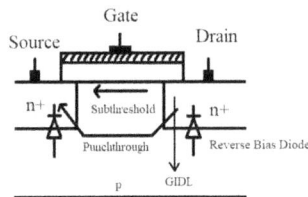

Figure 14.5 Different sources of leakage current in MOS transistor.

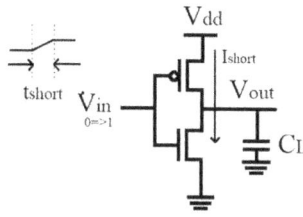

Figure 14.6 The non-zero rise time of the input transition from 0 to 1 cause both the transistors to ON for tshort period, which results in Ishort current.

the behaviour of CMOS structures and is considered as redundant. In the system of circuits having n number of gates, the overall short-circuit power dispelled is given by:

$$P_{short} = V_{dd} \sum_{j=0}^{n} \alpha_j I_{shortj} \tag{14.6}$$

where V_{dd} is the applied supply voltage, α_j is the switching factor of jth circuit and I_{shortj} is the short-circuit current of the jth circuit.

14.3 CIRCUIT-LEVEL IMPLEMENTATION FOR POWER OPTIMIZATION

This section describes power optimization techniques that can be implemented at the circuit level.

14.3.1 Static and dynamic logic

It is widely realized that the dynamic logic implementation requires reduced area and has faster switching speed than static logic. However, the choice in the use of static and dynamic logic is based on various other design aspects such as power utilization, testability and ease of design. The evaluation of power dissipation of a dynamic logic presents a substantial advantage owing to a few reasons such as:

a. The use of fewer number of transistors for implementing the logic and thereby the physical capacitance is lower.
b. They do not consume short-circuit power as there is no direct current path existing between the supply and the ground and,
c. Reduced parasitic node capacitances.

Under the static design, the CMOS logic circuits are implemented considering the steady state behaviour of CMOS structures. It has advantages such as nil static power dissipation and no need for precharge phase. In addition, it does not exhibit any degradation or produce any undesired output voltage because of the charge sharing issue. This issue arises when the charge that is retained in the output node during the precharge stage is divided between the junction capacitance of the transistors happening during the evaluation stage. The following are the considerations in choosing between static and dynamic [6]:

a. **Spurious Transitions:** Because there is non-zero propagation delays between various logic blocks, undesirable effects such as critical races and the dynamic hazards are inbuilt into static logics, and a node can undergo multiple transitions in a clock cycle before it can settle to the correct and desired logic level. This undesired switching mechanism is called spurious transitions. In some cases, it can cause the circuit to consume 30% of the total energy [6] that is actually required to carry out the computations. The behaviour of dynamic logic is such that it does not exhibit this issue as any of the nodes in the circuit can utmost go through one transitional clock that consumes power in a clock cycle; nonetheless, it exposes some power overhead. Usually, this could be eliminated by exploiting low-power consuming self-timed methodology so that, at the functional level, spurious transitions can be suppressed. This method utilizes the combination of the advantageous functionalities of both static and dynamic logic.

b. **Short-Circuit Currents:** In the absence of transients on the applied input values, usually a CMOS inverter does not drive away power. But, if there is a transient present in the input, then for a short duration of time, both the transistors, i.e., NMOS and PMOS, turn ON and conduct. This causes a short-circuit current to flow between the supply and the ground. So, to reduce the effect of short-circuit power expenditure, an inverter can be used as a buffer to make the rise times and the fall times of the applied input values less than or equal to the output rise times or the fall times. However, the dynamic logic ordinarily does not expose the problem of short-circuit currents, flowing as a result of the presence of transients on the applied input values, except in cases when clock skew is predominant or where the static pull-up devices are employed in order have a control over charge sharing.

c. **Parasitic Capacitance:** Comparing to the static logic, the dynamic logic relatively uses small number of transistors for the implementation of the logic function, thereby having low capacitance. This has a direct reflection on the power delay product.

d. **Switching Activity:** The dynamic logic shows a greater disadvantage for its necessity for the precharge operation. As the nodes are evaluated, it is immediately discharged after the pre-charge phase for every clock cycle. Thereby making the switching factor larger. Thereby, making a significant dynamic power consumption.

e. **Power Down Modes:** This is achieved by disabling the clock signal and is inherently applicable to the static logic only. Extra circuitry is required to in order to preserve the state of the dynamic logic, which increases the parasitic capacitance and delays.

14.3.1.1 Pass transistor logic

- Traditional static CMOS logic uses transistors that is twice the number of inputs needed to implement a logic function. This can result in significant implementation area, and the total capacitance of gate terminal is increased due to the presence of a huge number of transistors. This also increases linearly with fan-in.
- The pass transistor logic is the popular alternative to the conventional static CMOS.
- It uses only a single type of MOS, most commonly NMOS, thereby reducing the significant amount of capacitance.
- The full adder circuit that is implemented using pass transistor and the conventional static CMOS logic can be seen in Figures 14.7 and 14.8.
- It is lucid that the number of transistors used for implementing the adder circuit is less using the NMOS-only circuit comparing the conventional circuit.

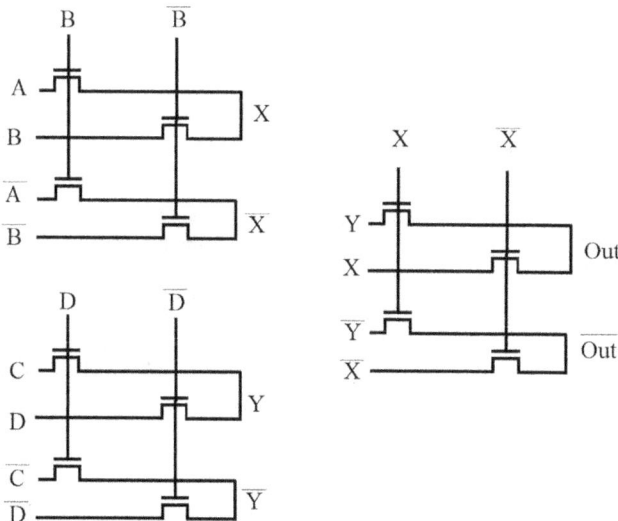

Figure 14.7 The full adder circuit using pass transistor logic. This uses 12 transistors for the implementation.

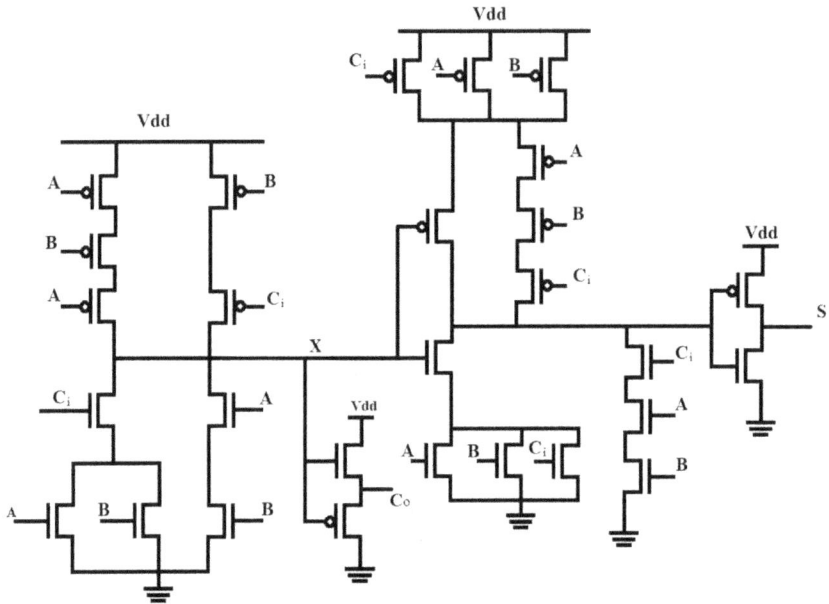

Figure 14.8 The full adder circuit using conventional static CMOS logic. This uses 28 transistors for the implementation.

- Thus, it utilizes only a fewer number of transistors and this property makes its quite attractive.
- The differential/complementary pass logic is employed to produce both true and complementary functions. The circuit for the full adder using CPL/DPL is shown in Figure 14.9.

However, CPL/DPL implementation has basically two complications. First, there can be reduced speed of operation at lower applied power supply, which results from diminished current flow when there is a threshold drop across the pass transistor. This becomes a crucial issue when the low-power-consuming circuits are to be operated at reduced supply voltage levels. Second, since the maximum voltage of the inverters is not V_{DD}, the PMOS transistor present in the inverter circuit is not completely turned OFF. Due to this, a significant amount of static power dissipation arises because of the direct pathway for the current to flow from the applied supply voltage to the ground.

14.3.1.2 Dynamic logic

Dynamic logic is the alternate logic that avoids static power consumption. It exploits a succession of precharge and evaluation stages based on conditions. The elemental construction of the circuit consists of a pull-down

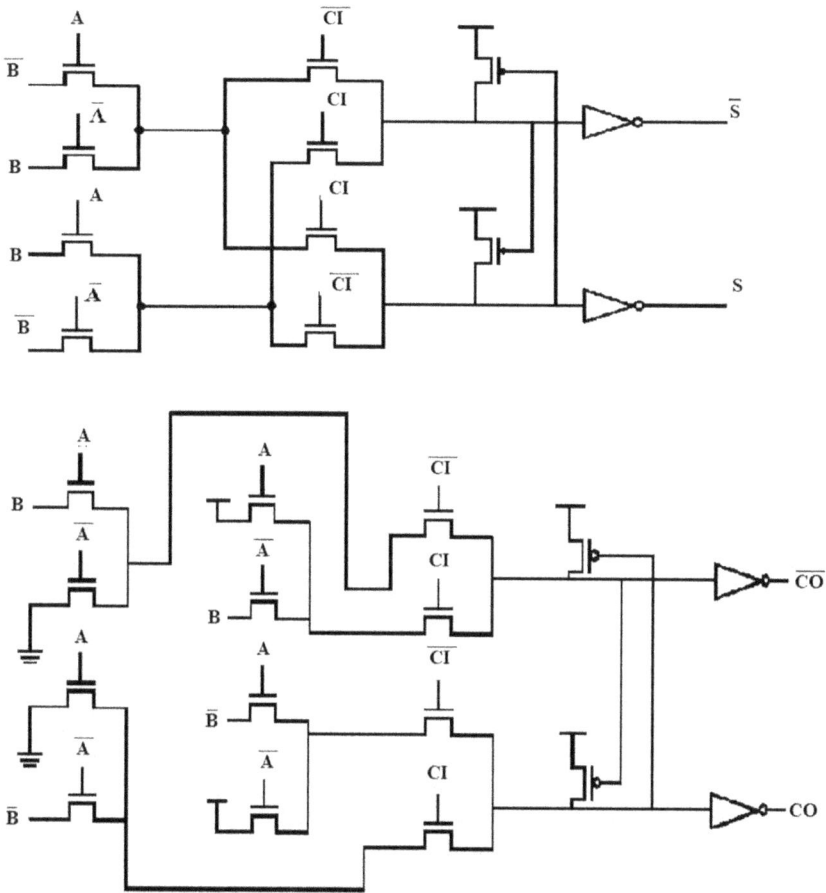

Figure 14.9 The full adder circuit using CPL/DPL logic.

block that can be seen the same as in the complementary CMOS circuit (Figure 14.10) [7]. The main two phases of operation are precharge and evaluation, which are determined by the clock signal. While there are major advantages in dynamic design, there are various issues that must be considered. This includes charge leakage, charge sharing, backgate coupling, and clock feedthrough. Apart from this, the one major issue that challenges the design aspects of the dynamic logic circuits is straight-forwardly cascading the dynamic logic gates so as to implement complicated structures. This actually creates the trouble of reduced final output voltage from the circuit. This is overcome by domino logic. The circuit can do operations faster because of its reduced noise margins relative to static logic. This is improved by scaling down the technology, but it can lead to elevated power utilization.

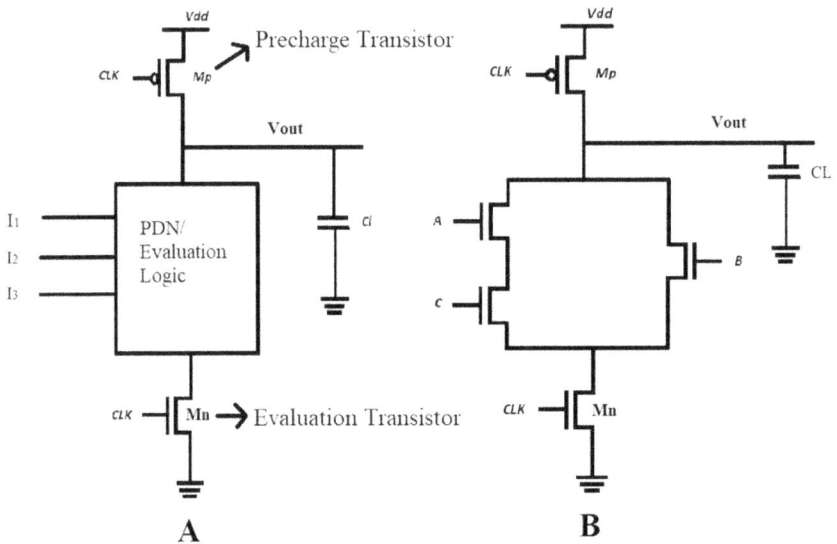

Figure 14.10 (a) Basic dynamic circuit (b) Example of (A+B).C (bar).

14.3.2 Voltage scaling and its issues

From the dynamic power expression (equation 14.2), it is clear that lowering the voltage applied to the circuit is the most efficacious method to diminish power consumption. And so that the electric field can be decreased, the high field effects is subdued. Power delay product (PDP) is defined as the total sum energy spent in each of the transition cycle due to switching. It is given by the equation:

$$\text{Energy per transistion} = P_{\text{total}}/f_{\text{clk}} = C_{\text{total}}V_{\text{dd}}^2 \tag{14.7}$$

where C_{total} is the total effective capacitance during switching, in order to carry out a computation, and it is written as $C_{\text{total}} = p_t C_L$.

As clear from the aforementioned equation, the energy in one transition (as defined as PDP) in "properly designed" CMOS circuits stays in proportion to V_{DD}^2. Thus, it is eminent to decrease the voltage supplied, to show quadratic improvement on PDP (assuming switching frequency remains constant). Given this improvement in PDP, it is not preferred to operate circuits at these low supply voltages as, in spite of diminished dynamic power utilization by reducing voltage supplied to the circuit, the unavoidable trade-off in the design is caused: the increased delay.

The relation between supply voltage and the delay of NMOS and PMOS transistors can be given as [8]:

$$\tau_N = \frac{C_{\text{load}}}{k_n\left(V_{\text{DD}} - V_{\text{Tn}}\right)}\left[\frac{2V_{\text{Tn}}}{V_{\text{DD}} - V_{\text{Tn}}} + \ln\left(\frac{4\left(V_{\text{DD}} - V_{\text{Tn}}\right)}{V_{\text{DD}}} - 1\right)\right] \qquad (14.8)$$

$$\tau_P = \frac{C_{\text{load}}}{k_p\left(V_{\text{DD}} - |V_{\text{Tp}}|\right)}\left[\frac{2|V_{\text{Tp}}|}{V_{\text{DD}} - |V_{\text{Tp}}|} + \ln\left(\frac{4\left(V_{\text{DD}} - |V_{\text{Tp}}|\right)}{V_{\text{DD}}} - 1\right)\right] \qquad (14.9)$$

For the supply voltage V_{DD} decrease, we pay speed penalty, with delay increasing drastically and reducing the performance of the gate. As the primary objective is to suppress the power consumption without compromising performance, the compensation proportionate to the decreased speed at low voltages is essential. So, various approaches such as: (1) lower the threshold voltage; (2) utilize parallel/pipelined architecture; c) lower V_{DD} only for non-critical circuits, and various others are employed.

14.4 V_{th} CONTROL TECHNIQUES

Threshold voltage is the most paramount parameter in technology and parallelly in the circuit design. Because lowering the power supplied to the circuit reduces the throughput, it is effective to decrease the supply voltage. Also, by reducing the threshold voltage, the delay of the circuit decreases. So, scaling down the threshold voltage is crucial so as to mark up to the level of performance. Nonetheless, there arises some issues associated with lowering the threshold:

a. Delay oscillates overpoweringly with fluctuating V_{th}, in the region of low V_{DD}. For instance, delay rises around thrice for $\Delta V_{\text{th}} = +0.15$ at V_{DD} of 1 V [9].
b. Subthreshold leakage increase.
c. Inability for IDDQ test (measure of the extent of defect free CMOS circuits).

To overcome these problems, various Vth control techniques are discussed. This section elaborates on various Vth control techniques: MTCMOS, VTCMOS, DTCMOS, and EVTCMOS.

14.4.1 MTCMOS

Multi-threshold CMOS (MTCMOS) is applicable ubiquitously to CMOS digital circuits. The main feature of MTCMOS is that, it exploits two levels of threshold devices: high and low-threshold voltage level MOFETs in a one sole chip. It also reduces standby leakage current and gate delay, thereby

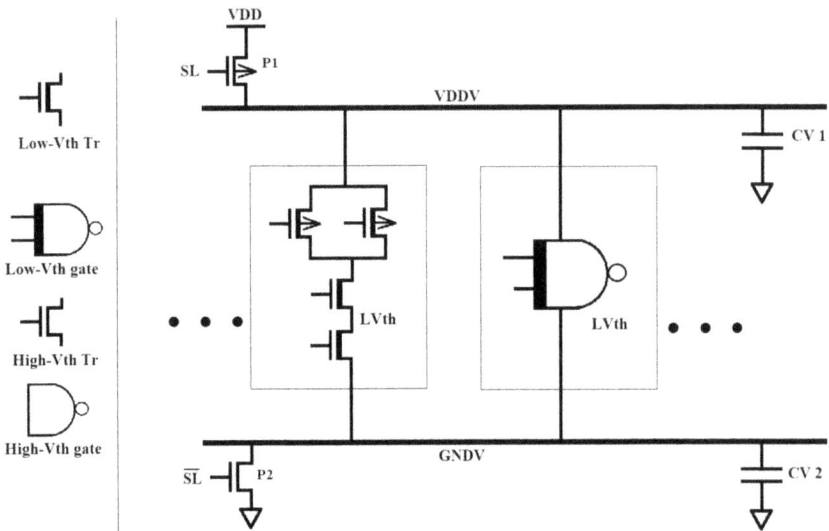

Figure 14.11 MTCMOS circuit.

increasing the performance at low supply voltage, usually operated at 1 V [10]. As seen in Figure 14.11, there are two major features employed: (1) NMOS and PMOS transistors of two different thresholds are used (2) it operates in two modes of operation namely, "active" and "sleep" mode.

A reduced threshold, about 0.2–0.3 V is employed to implement the logic gate. Its terminals are linked to virtual power rails. The real and virtual supply lines are associated using high threshold (0.5–0.6 V) PMOS and NMOS (P1&P2) transistors. *SL* and $\overline{SL}$ are the signals that perform control operation and are supplied correspondingly to the transistors P1 and P2, respectively. These are employed for the active or sleep circuit mode of operation. In operating in active mode: a low value SL is applied so both transistors get turned ON. These transistors offer only a small on-resistance such that the virtual lines function as a true power supply line. Thus, the logic that consists of low threshold MOSFETs operates at low power and high speed.

When functioning in sleep mode of operation, the control signal SL is made high, and thus the transistors P1 and P2 are turned OFF so that the virtual lines will be in floating state. The low-threshold MOSFETs possess subthreshold characteristics that results in a huge amount of leakage current that is almost completely suppressed by P1 and P2. Thereby, power consumption is dramatically reduced during the sleep mode.

Nonetheless, MTCMOS circuit can only suppress standby leakage power but the presence of large MOSFETs can lead to increased area and propagation delay. In addition, if data storage is essential, then a latch circuit is required.

14.4.1.1 Transistor sizing issues

The important parameters affecting the performance of MTCMOS circuits take into account the size considerations of the MOSFETs, including the capacitances of virtual ground rails. If the transistors are sized too large, the area of silicon exploited will turn useless, which also increases the switching energy overhead. Alternatively, if it is smaller, due to increased resistance, the delay of the circuit increases. On considering the effect of virtual capacitances, they should be made large so as to suppress the voltage rise and drop in the virtual lines due to switching of the internal logic.

14.4.1.2 Finite resistance approxiamation

One approach taken, in which the sleep transistor (high V_{th} transistor) can be closely resembled as a linear resistor so the there is a small voltage drop across it (Figure 14.12) [11]. Whenever a discharging current flows, during high to low transition the discharge current flows through the resister and induces a decreased potential (V_x). This basically has two impacts:

a. It brings down the gate voltage from V_{DD} to $V_{DD} - V_X$, and
b. Due to the body/bulk effect, the threshold voltage of pulldown transistor (NMOS) increases.

Combined together, the above two issues lead to reduced discharge current. To maximize capabilities, the resistor must be sized as small as possible, subsequently making the transistor as large as possible. Also, instead of having large capacitances, lowering the effective resistance with correct transistor sizing will be easier to employ in low power, high performance circuits.

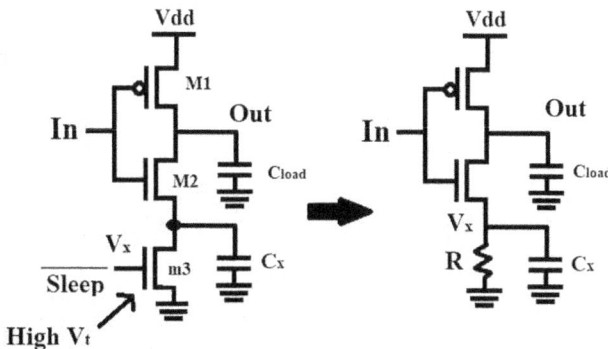

Figure 14.12 Sleep transistor modelled as resistor.

Further moving towards complex MTCMOS circuits, the circuit becomes more susceptible to input vectors, which determine the worst-case performance. It causes a large flow of current in sleep transistors. It is therefore highly challenging to design the sleep transistor of the correct size for better and desired computational performance since delay is dependent on the input vectors and also on the glitching behavior. Even after correctly sizing choices of the transistors, it is paramount to figure out the worst-case pattern criteria of the input vector in order to find the delay.

14.4.1.3 Mutual exclusive discharge patterns

To overcome this, another technique is proposed in which the sleep transistor is built over mutual exclusive discharge patterns, whose performance resembles very close to original CMOS circuit versions for all feasible inputs.

This technique ensures that individual gates meet a local performance in order to satisfy a macro performance [12]. This ascertains that any possible valid combination of gate present in the logic pathway will meet the performance criteria. The MTCMOS circuits do not necessitate identifying the input vector pattern for the worst case. Alternatively, each of the independent gate is allotted with its own high threshold voltage (V_{th}) sleep transistor, and its size can be determined from SPICE software circuit simulations.

Here, initially, the MTCMOS circuits are sized using independent sleep transistors. Then at successive stages, these can be merged as they can be apportioned among the gates that are mutually exclusive and where there is no simultaneous current discharge at the gate. The sleep transistors can then be combined to bring out a sleep transistor for the entire circuit in the last stage. This proves to give best performance within the intended range.

14.4.1.4 Ground bounce and its minimization

It is a potential glitch that is created because of the switching behaviour of currents, which arises at the supply or ground connections. These glitches are induced in proportion to LdI/dt [13]. When the circuit complexity increases, a greater number of signals gets switched concurrently, thereby causing huge leakage currents flowing in finite short time interval. This contributes to the ground bounce phenomenon. The ground bounce is also accentuated by large input and output buffer capacitances. When this value rises above the certain noise margin value of the circuit, it might get switched to an undesired value or switch at the unintended time, causing errors in both static and dynamic logic circuits.

The power-gating circuits [14] that minimizes the ground bounce phenomenon are the ones that have sleep transistors getting turned ON non-uniformly, i.e., in a step-level fashion. These circuits ensure reduced fluctuations in the magnitude of the voltage in power-dividing networks.

This can be done in two ways, as explained in [15]

   a. By effectively changing the value of VGS and thereby VDS, as can be
       seen in Figure 14.13.
   b. By resizing the sleep transistor size that can be seen in Figure 14.14.

In the first technique, during sleep time, until V_{DS} is decreased appreciably,
the sleep transistor is made to conduct with $0 < V_{GS} < V_{DD}$. During active

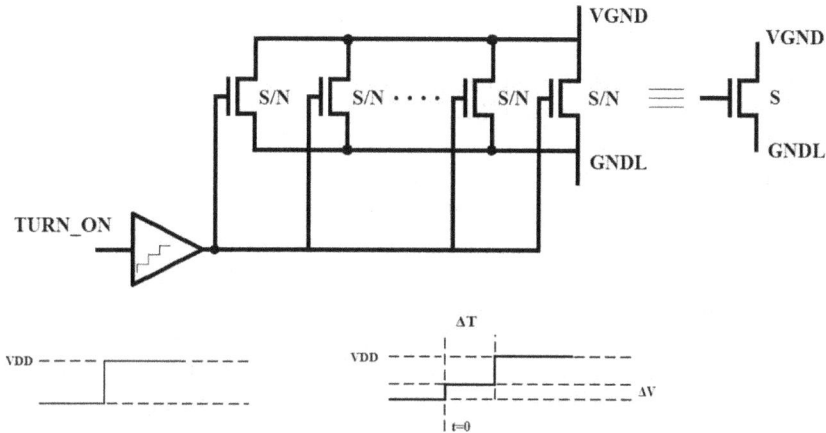

Figure 14.13 The V_{GS} of the sleep transistors increases non-uniformly in a stepwise
          manner.

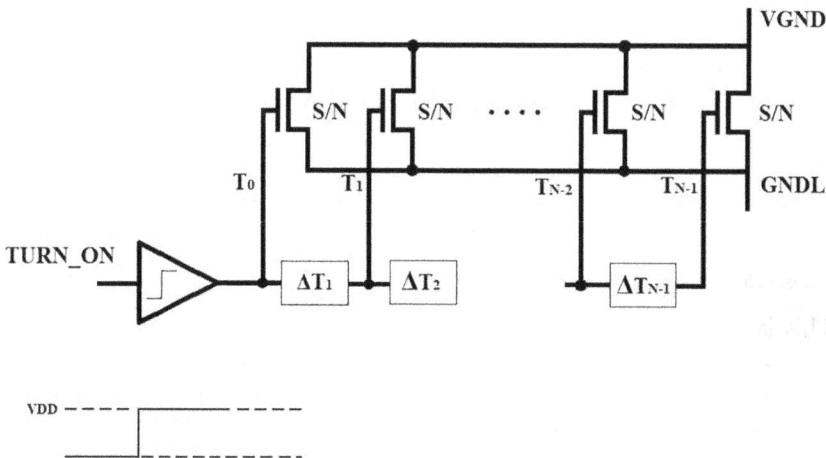

Figure 14.14 The sleep transistors are turned ON non-uniformly in a stepwise manner.

mode, sleep transistor is biased with $V_{GS} = V_{DD}$ and thus made to turn ON completely. During when the V_{DS} of sleep transistor is not very large, the current measures instantaneously will be not much sensitive to the changing V_{GS} of that transistor. Thereby, enabling V_{GS} of transistor to rise in a step-level wise without much increase of the peak current. The second employs resizing the size of transistor.

14.4.1.5 SSCMOS

The super cut-off CMOS is used for the operation at very low supply voltages (<0.7 V), where only a pico-ampere of standby current is observed per logic. Normally, MTCMOS circuit does not function below 0.7 V since, the high V_{th} MOS does not gets turned ON. The SSCMOS pushes the bound of low-voltage operation over MTCMOS [16]. It employs low V_{th} transistors instead of high V_{th} transistors, whose gate is driven by the gate bias generator as seen in Figure 14.15. For a PMOS (NMOS) insertion case, a low PMOS (NMOS) is attached in series along with the low-V_{th} network. The gate is supplied with a 0 V(VDD) during the active circuit mode and the VDDV(VSSV) line is linked to the VDD(VSSV). During the sleep circuit mode of functioning, gate is connected to level of VDD+0.4 V (VSS−0.4 V) so as to completely supress the leakage current. Thus, low-voltage operation is possible and thereby reducing stand-by leakage current. A circuit comparison of MTCMOS and SCCMOS is shown in Figure 14.16. 1-V DSP chip was designed for the mobile phone-based applications has been developed lately by employing MTCMOS.

In the standby mode of operation, because of the large leakage of low threshold voltage (V_{th}) circuits, virtual VDD (VDDV) gets reduced to 0 V, thereby the latches lose the information stored. The degraded capabilities of the sleep transistors are mainly because of the size and the quantity of the current flow.

The traditional MTCMOS latch circuit that preserves the data in sleep mode is shown in Figure 14.17. In sleep mode, the data is preserved in the circuit but it creates problem when operating in high speed. This is because of the presence of the TG1, which is a high threshold MOS transistor present in the critical path. In addition, the circuit takes up large area since the transistors Q1 to Q4 has to supply voltage to the inverters in the data path.

14.4.1.6 Balloon circuit

This is a new approach for storing the data during the sleep mode. Since memory is "blown up" with data at the beginning of the sleep mode, the memory circuit is called balloon circuit (see Figure 14.18). The balloon circuit is affixed to the node A of MTCMOS circuit [17]. The basic operation of the circuit:

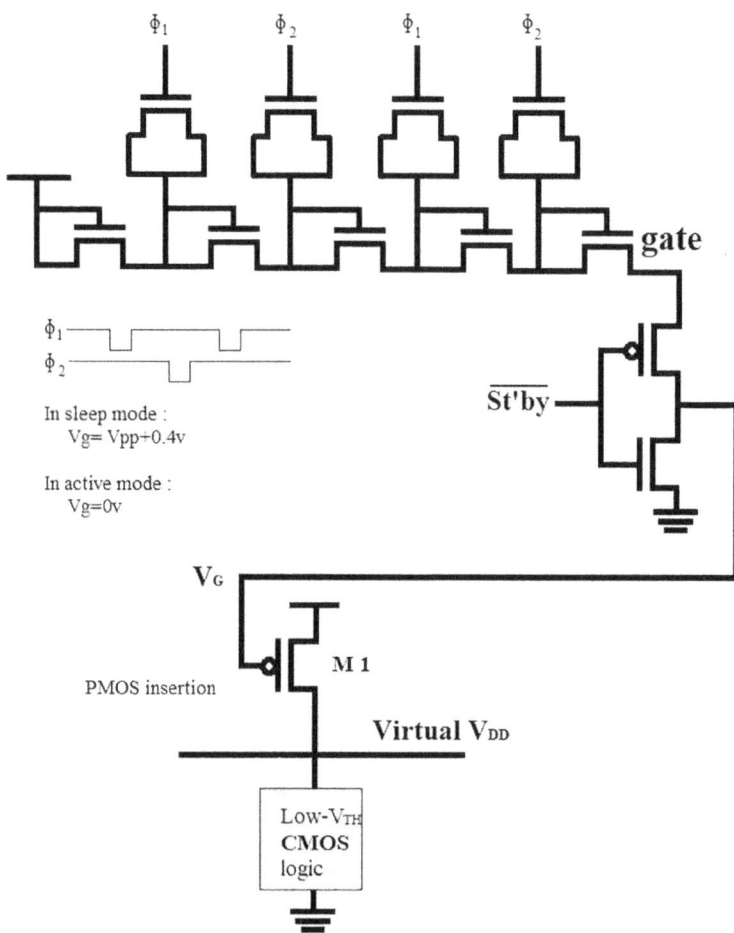

Figure 14.15 Super-cut-off CMOS (SCCMOS).

- In the sleep state, the data is stored in the memory module and parallelly the leakage current is supressed that comes from the memory circuit, using a switch.
- In the active circuit mode, balloon circuit is isolated from load circuit using the switch.
- In the transition period, from active to sleep state, the switch either gets on to read data from the node or reinstate data to the node.

Construction: The balloon circuit is composed of two inverters that contains high-threshold transistors and one transmission gate.

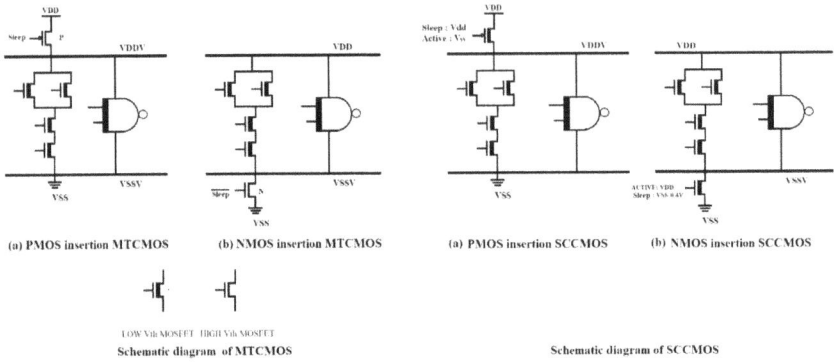

Figure 14.16 (a) Schematic of MTCMOS circuits (b) Schematic of SCCMOS circuits.

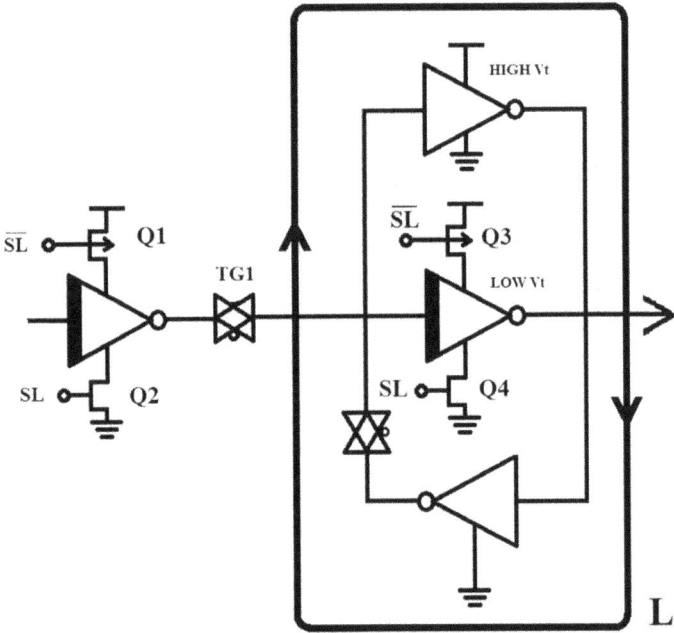

Figure 14.17 Circuit diagram of conventional MTCMOS latch circuit.

- TG2- high threshold TG which isolates memory circuit from MTCMOS logic.
- TG1 and TG3 are a low threshold TG's that are utilised to read and write into the memory circuit.

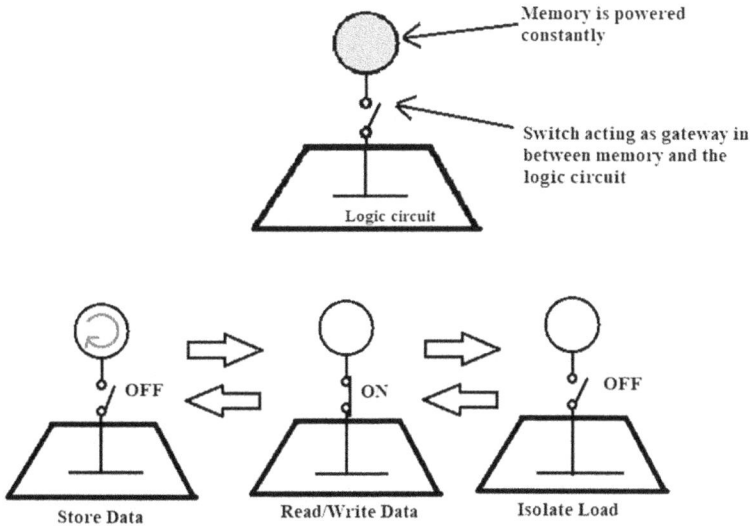

Figure 14.18 Concept of the MTCMOS data-preserving circuit.

Two signals B1 and B2 controls these low-voltage TGs, while sleep signal SL is used to control the high voltage transistor. The circuit goes through four phases of operation (refer Figure 14.19):

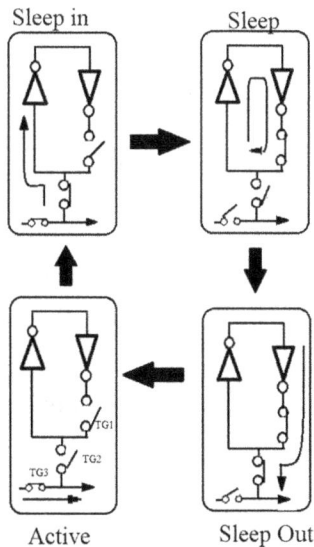

Figure 14.19 Operation of the balloon circuit.

1. Active
2. Sleep
3. Sleep-in
4. Sleep-out, where sleep-in & sleep-out stages are the modes of transition between the sleep and the active stages.

The MTCMOS balloon circuit and the variation of control signals is illustrated in the Figure 14.20. During the sleep state, TG2 is turned OFF, thereby cutting off the standby current from balloon circuit. TG2 is in OFF state even during the active period in order to subdue the load present at the node A, where the balloon circuit is attached.

The circuit inputs the data from the node A during the sleep-in period when TG1 is turned OFF. This gate is in OFF state even during the active circuit period. At this time, the node N will be in floating state as both gates (TG1 & TG2) are turned OFF. To avoid the floating node problem, gate TG1 has to be turned ON. Nonetheless, this leads to increased control signals and area overhead. To mitigate this issue, the gate TG1 is made of low threshold transistors.

TG3 is present between balloon circuit and the output node in order isolate both the circuits. In the sleep-out stage, this gate is turned off that helps to write the data from balloon circuit into node A. Subsequently, it is turned ON later where every node continues its operation, thereby preventing stored data destruction. In the active circuit period, TG3 is in ON state and since it is composed of low-V_{th} transistors, this avoids the reduction of speed performance of the circuit. Usually, balloon circuit is attached to node where the capacitance is lower in order to avoid data destruction because of the charge sharing issue, during the sleep-out stage.

A master-slave D-flip flop (DFF) (see Figure 14.21) is normally utilized for capturing and storing the data. During the high state of the clock signal, the data is latched in the master latch and in the low state of the clock signal, data is stored into the slave latch. However, this circuit will not be able to store back the data during the sleep period.

A circuit that combines the D-flip flop and balloon circuit is exploited in order to avoid the problem of data storage during sleep period. The circuit shown is a clock-independent circuit (see Figure 14.22). A universal clock is fed to the synchronous digital circuits and microprocessors, which contains the pipeline of many registers. This circuit is called as the clock-dependent balloon DFF. Initially designed chip was composed of balloon circuit and a latch. This was fabricated by utilizing 0.5 μm double-metal MTCMOS technology. Although the above techniques overcome the problem of data retention and high-speed operation, there are various setbacks. Extra data-recovery circuitry such as the balloon circuit, inevitably cause an intolerable area penalty and demands an involved control strategy in transitioning from the active state to sleep state.

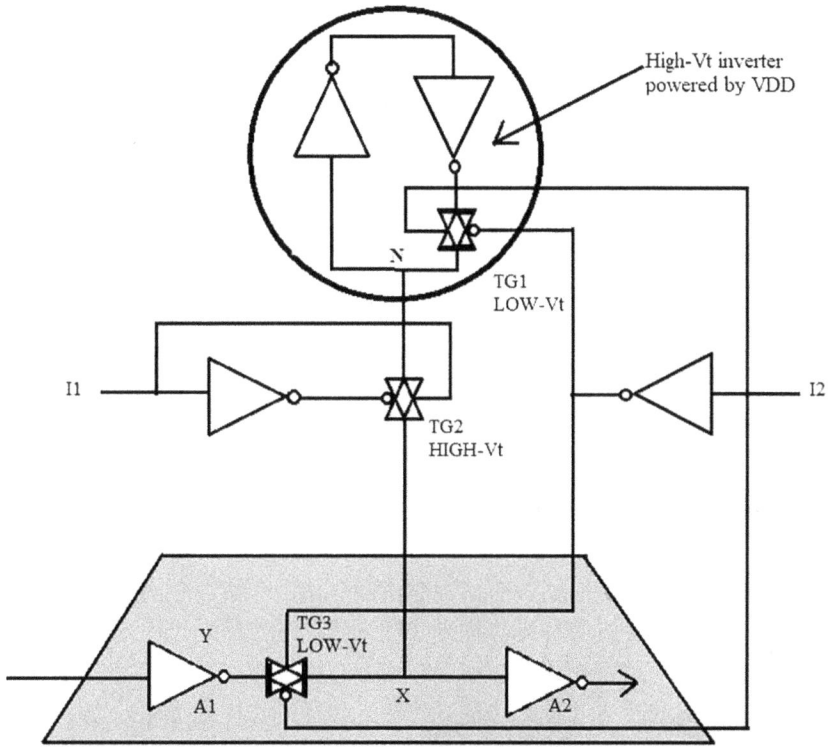

(a). circuit diagram of the Balloon circuit

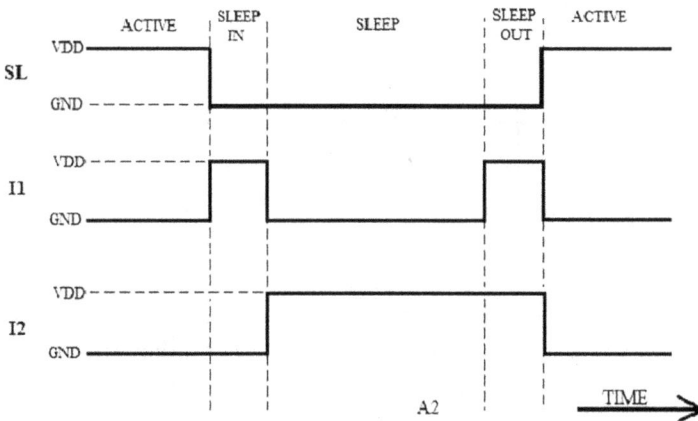

(b). Sequence of control signals for Balloon circuit

Figure 14.20 MTCMOS balloon circuit.

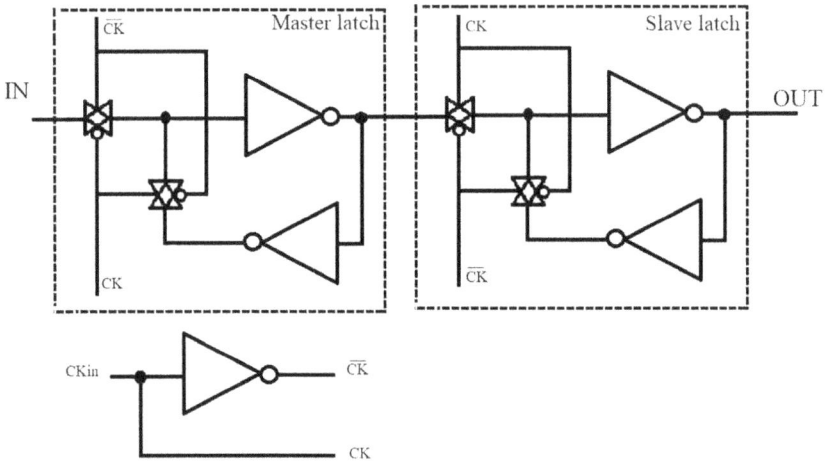

Figure 14.21 Schematic circuit diagram of ordinary CMOS D-flip-flop.

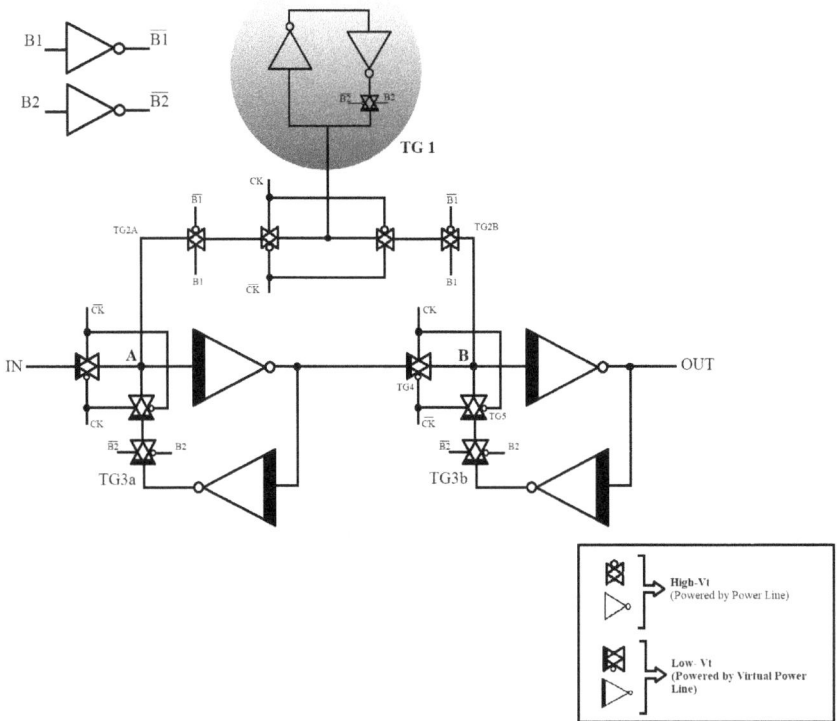

Figure 14.22 Clock-free balloon circuit applied to a D-flip flop.

The empirical outcomes elucidates that the balloon circuit can be employed if the sleep-in and sleep-out stages extends the minimum time to read/write time. Henceforth, if

$$t_{\text{sleep-in/out}} \geq \text{Skew}_{\max} + t_{\text{read/write}} \tag{14.10}$$

The skew that can exist between B1 and B2 can be circumvented.

14.4.1.7 IPS scheme

This power gating supports the intermediate power saving mode and the cut-off mode. While MTCMOS can be built by using a single transistor, in particular NMOS sleep transistors, in IPS scheme additional PFET transistor is added in parallel to the NMOS transistor [18]. There are three modes of operation:

- Cold mode,
- Park mode and
- Run/idle mode.

In RUN/IDLE mode (see Figure 14.23), SL and the HL signals are held high. This forces the NMOS transistor to provide power gating. The n-type FET is utilized to short the virtual ground and the real ground, thereby a full rail VDD supply can be given to the circuit thus enabling high-speed operation.

In COLD mode (see Figure 14.24), also called as non-state retentive mode, SL is kept at the low state and HL is kept at high state so that path for the current to flow is cut-down. Hence, this provides a greater amount of suppression of current arising from both gate and the subthreshold leakage.

In PARK mode (see Figure 14.25), which is also called the retention mode, both SL and HL signals are made low. During this mode of functioning, the voltage of the virtual ground is maintained at the threshold voltage level of the p-type FET, exceeding the ground voltage. Thereby, the potential that is held across the circuit will be $V_{dd} - V_{tp}$. This leads to the reduction in the gate and subthreshold leakage in the circuit.

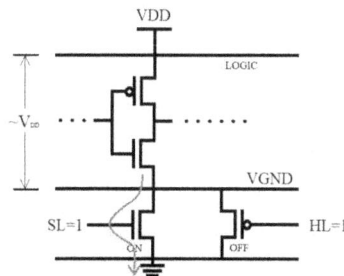

Figure 14.23 Power-gating structure in RUN/IDLE mode indicating the flow of current.

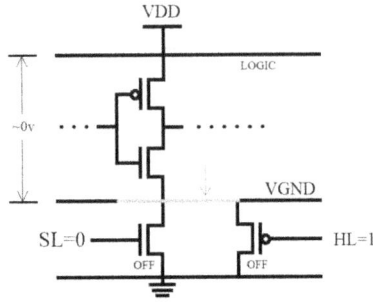

Figure 14.24 Power-gating structure in COLD mode indicating the flow of current.

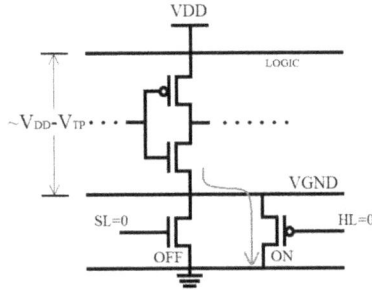

Figure 14.25 Power-gating structure in PARK mode indicating the flow of current.

Because during this mode of functioning, the potential of virtual ground is restricted by V_{tp}, the state is preserved. Additionally, the ground bounce effect due to the transition between modes is comparatively smaller than that in the COLD mode. PARK mode of operation is employed as the intermediate stage when being switched from the COLD to RUN/IDLE mode, in order to reduce ground bounce issue. Henceforth, the series of stage transition starts with the COLD mode followed by the PARK and the RUN/IDLE mode. Experimental latch circuits have been designed and manufactured using 0.13 μm MT-CMOS technology which has a reduced area of 30%, delay decreased by 10%, and 10% reduced active power utilization relative to the traditional MT-CMOS technology [18].

14.4.1.8 VRC

In virtual rails (power/ground) clamp structure, there is a quad-rail configuration where the actual power/ground (VDD/GND) and virtual power/ground (VDDV/VGND) is connected with the help of a NMOS/PMOS transistor along with a diode that is biased forward condition (see Figure 14.26) [19].

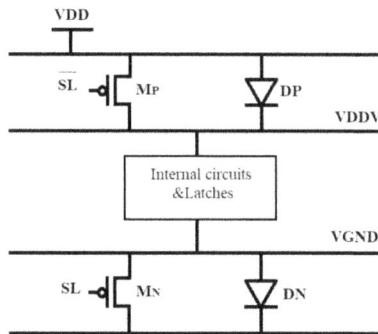

Figure 14.26 The VRC circuit diagram.

These transistors are low Vth transistors, that is identical to the transistors employed in the internal logic circuit. Therefore, the cost of manufacturing and fabricating is reduced (since areas of NMOS/PMOS are smaller compared to that in MTCMOS). In sleep mode of functioning, SL signal is made low and the NMOS/PMOS transistors are turned off. The potential drops/rises from VDD/GND level. However, the potential change of VDDV and VGND are clamped due to the in-built potential of the diode. Therefore, it is feasible to store the data because of the voltage difference between virtual supply (VDDV) and ground (VGND) rails during sleep mode. This also diminishes the noise arising because of ground bound effect by restricting the potential of virtual ground. Experimentally, a 24-bit multiplier employing a 0.25 μm CMOS double-layer fabrication technology was used that gave up to 98% leakage current reduction and data storage capability without speed degradation [18].

14.4.2 VTCMOS

While the MTCMOS technique has been introduced and various approaches have been taken to reduce several drawbacks, this is not useful application to the memory elements unless the circuit is modified which leads to increased area and speed peanlities. Variable threshold scheme (VTCMOS) proves to overcome these shortcommings. Furthur, (1) it reduces the reduction of speed in the worst-case condition in the low V_{dd} regime because of the dynamic variation in V_{th}; (2) it also reduces the power dissipation when in the sleep mode for small V_{th}.

The conventional techniques introduced to reduce V_{th} fluctuation is: self-adjusting threshold voltage scheme (SATS). In this scheme, fluctuation in the potential is reduced by the use of self-substrate biasing (refer Figure 14.27). This scheme is composed of a sensor that helps to sense the leakage and a self-substrate bias (SSB) circuit [20]. The current that arises because of the leakage from the MOSFET is sensed by the sensor. This in turn produces

Self-Adjusting V_{th} Circuit

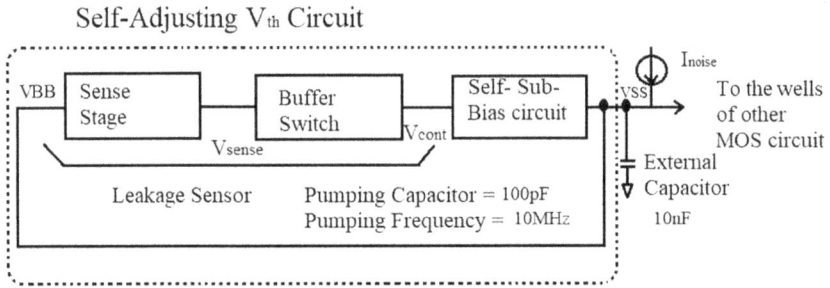

Figure 14.27 The block diagram of self-adjusting V_{th} scheme (SATS).

a proportional output voltage V_{cont}, which is supplied to the SSB circuit. This voltage is conditionally controlled in such a way that the SSB circuit is made to operate in the case only when the leakage exceeds a particular pre-defined value. When SSB is triggered, body bias voltage reduces. Subsequently, this increases the V_{th} thereby supresses the leakage current. Hence, V_{th} is maintained at the lowest feasible value that satisfies the power requirements. This uses an internal bias generator circuit.

The stand-by power reduction SPR scheme overcomes the issue of increased power in the low V_{th} regime during the sleep mode of operation [21]. In standby mode, the substrate is supplied so as to diminish current because of subthreshold leakage by increasing the threshold potential. During active mode of operation, the body bias is not supplied in order to ensure high-speed operating conditions. This scheme is composed of a circuit that performs level shifting along with the voltage-switch part (refer Figure 14.28). In active mode, S is asserted high state and also N-well bias equals to V_{DD} $(V_{N\text{-WELL}} = V_{DD})$. The P-well bias potential, $V_{P\text{-WELL}}$ is made to equal V_{SS}. In standby mode, $V_{N\text{-WELL}}$ becomes V_{NBB} and $V_{P\text{-WELL}}$ becomes -2 V. The SPR circuit utilizes an external power supply in order to bias substrate.

The variable threshold (VT) voltage technology does not rely on an external supply so as to bias the substrate. It also does not create any setback in speed and area of the chip. VT is applicable to both the logical gates and also to the memory elements. In this scheme, with the help of substrate bias conditional control, the threshold voltage of the transistor is varied. This is done by employing a VT voltage circuit. The principle of working of VTCMOS is controlling the body bias voltage V_{BB} in order to compensate dynamic variation in V_{th}. The threshold, V_{th} gets dynamically varied through the body bias V_{BB}. In active mode of operation, V_{th} is lowered in order to improve the on-current, benefitting from both low power dissipation and high switching speed. The following methodology of VT scheme was employed in a $4\,mm^2$, 8×8 discrete cosine transform (DCT) processor [22].

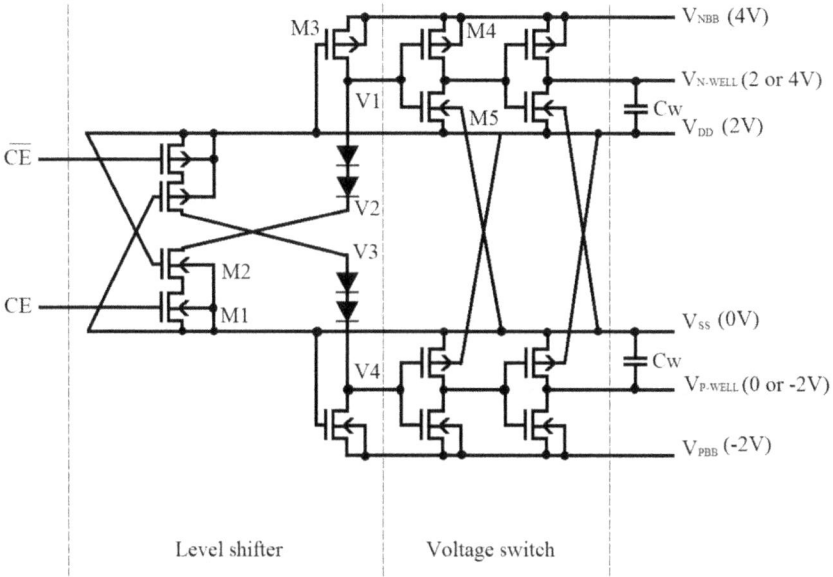

Figure 14.28 The SPR circuit.

During the sleep mode of operation, VT circuit induces a larger substrate bias so as to increase V_{th}. Hence, it can be ensured that the power dissipated is reduced to a large extent as the current because of the subthreshold leakage is supressed by making the threshold voltage large.

The major blocks of VTCMOS scheme are composed of: four of identical leakage current mirrors (LCM's), SSB circuit (SBB) and a substrate charge injector (SCI) as shown in Figure 14.29. The SBB circuit consists of a pump circuit two ring oscillators with high and low operational frequencies in the order of Mhz. The functioning of SBB and SCI are controlled by comparing VBB with ranges of voltages specified in four LCM's. as—$V_{active(+)}$, V_{active}, $V_{active(-)}$, and $V_{standby}$, where $V_{active(+)} > V_{active} > V_{active(-)} > V_{standby}$. These potentials can be assigned in the four LCM's by slightly varying the sizes of the transistors present in the bias circuit. SSB will draw current from the bulk as such to low V_{BB} and SCI will inject current in to increase VBB. The four of the voltage ranges are compared as follows:

1. After power on, $V_{BB} > V_{active(+)}$: SSB draws small amount of current from the substrate and thereby lowering body bias, V_{BB} by employing a ring oscillator.
2. When $V_{BB} < V_{active(+)}$ the frequency that drives the pump gets reduced and draws some amount of current in order to have more precise control on body bias, V_{BB}.

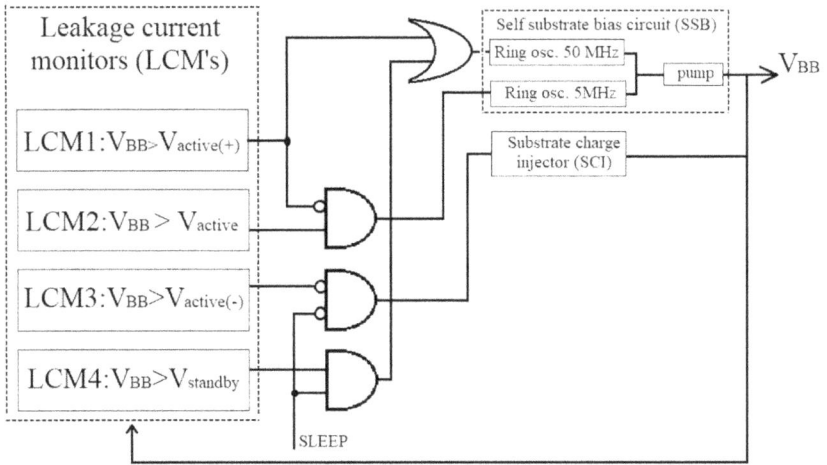

Figure 14.29 Block diagram of VT scheme.

3. When $V_{BB} < V_{active}$: SSB circuit relinquishes. But, whenever there is a leakage current in the MOS, V_{BB} increases. At some voltage, when it reaches V_{active}, SSB circuit will turn on. By this mechanism, V_{BB} is set under control by ON-OFF working of SSB.

4. When $V_{BB} > V_{active(-)}$ the SCI injects currents into the substrate, so as to increase V_{BB}.

   Whenever VBB exceeds beyond $V_{active(+)}$ and $V_{active(-)}$, V_{BB} is recovered back to SSB and SCI. Henceforth, even if the voltage exceeds beyond say, because of a power line bump, then it is swiftly got back with the help of SSB and SCI.

14.4.3 DTCMOS

Dynamic threshold CMOS (DTCMOS) is proposed for improving performance in ultra-low volatges. DTCMOS can be accomplished by binding the gate and the body together. This is known for the improved speed performance of Silicon On Insulator (SOI) digital circuits that operate under low power supply conditions. SOI CMOS VLSI technology is suitable for integartion of VLSI structures that drives on low power utilizing reduced supply voltage. The demand for supporting environment for deep submicron technologies had led to the growth of SOI CMOS technology. The SOI devices are different from the bulk devices in such a way that SOI devices have a distinct buried oxide layer, that helps to separate body from bulk/substrate (refer Figure 14.30). It is found that the speed performance of can be improved at 25% when a bulk CMOS circuit is replaced by SOI circuit, at reduced power consumption [23]. The operation is elaborated as:

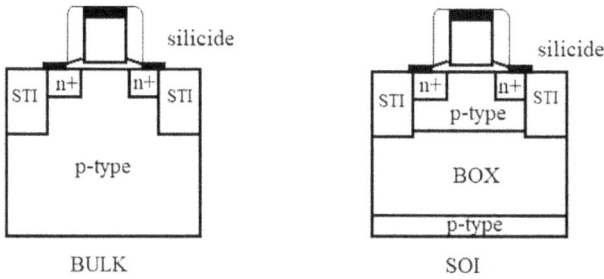

Figure 14.30 Cross section of bulk and SOI MOS devices.

1. In standard MOSFET operation: The terminals- source, gate, drain and body contacts are separate. And the body is either floated or grounded.
2. In DTCMOS operation: The body is tied to the gate terminal.

The N-type channel and p-type channel SOI devices employing DTCMOS technology is shown in the Figure 14.31. This technique binds the gate and body of the MOSFET [24]. The crossectional structure of SOI DTCMOS is shown in Figure 14.31.

On considering NMOS behaviour, the junction between the body and source is biased in forward direction, thereby forcing the threshold voltage to drop below the voltage at zero body bias. This does not create an additional cost of higher leakage current and in fact, have the same leakage as that of the standard device at zero body bias. Reduced threshold voltage compared to the voltage at zero body bias is achieved by a subthreshold of providing a 60 mV/dec swing. A similar improvement can be observed when gate of DTCMOS is raised above threshold when, the junction between the body and source is biased in forward direction. An another advantage of DTCMOS is that the mobility of the carriers is increased, due to the reduction of the depletion charge along with the effective perpendicular field present in the channel [25].

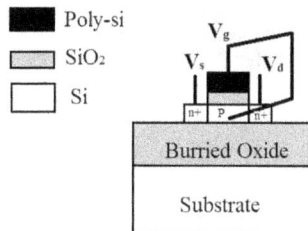

Figure 14.31 Structure of SOI DTCMOS.

Consider the case of 90 nm SOI technology—that consists of high threshold (large aspect ratio) and low threshold devices (small aspect ratio) [26]. High threshold devices have small on-current and low leakage while low-threshold devices induces greater amount of on-current and reduced leakage current arising during when the device is in OFF condition. SOI DTCMOS technique can be adopted at low voltage with the gate of the main (large) transistor controlling gate of auxiliary device (small) and source is binded to body as illustrated in Figure 14.32. The SOI DTCMOS technique do not pave way to rise the leakage current in subthreshold regime, since main transistor controls the small transistor and thus employing DTCMOS technique, power utilization is considerably reduced.

14.4.4 Dual-V_{th} CMOS

The logic of any digital circuit is composed of low V_{th} transistors in order for low power utilization. Nonetheless, this arrives at the cost of high leakage currents. Consequently, the challenge is to maintain performance while having low power consumption. To mitigate this issue, dual- V_{th} devices are realised.

In this technique, for a logic circuit, high threshold device are used for the paths (non-critical) that exhibits to shorter delay in the circuit, in order to minimize the current arising due to leakage. On the other hand, devices operating on low threshold are employed for critical paths, achieving high performance and low power simultaneously. This is illustrated in the Figure 14.33. Critical Paths are referred as the any longest path in a circuit that results in the longest delay. It determines the maximal operating rate (speed) of the circuit. However, its not quite possible to assign every of non-critical paths with a device having a high threshold. It is because of the complexity of the circuit. So, in order to attain maximal power conservation under low power restrictions, heuristic algorithm have been proposed. This is used for choossing and allocating the oprtimum high threshold potential.

The lower limit of the low-V_{th} is identified from the noise margins and high threshold potential ranges from low V_{th} to 0.5 V_{DD}. Since, not all non-critical

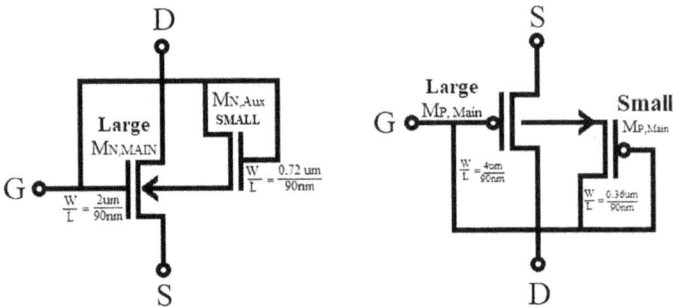

Figure 14.32 N-channel and P-channel SOI devices employing DTCMOS technique.

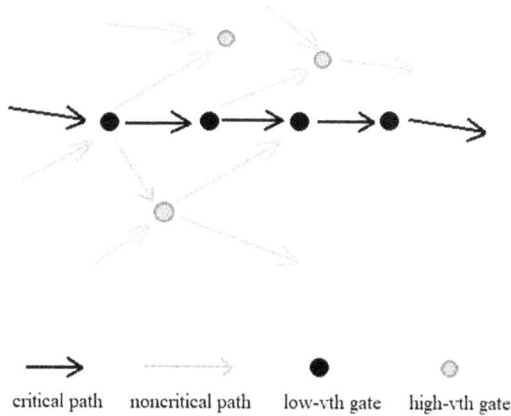

critical path   noncritical path   low-vth gate   high-vth gate

Figure 14.33 Dual-V_{th} CMOS scheme.

paths can be assigned with high-V_{th} devices, the choice of the optimal one is determined by breadth-first search algorithm. The working of this alogorithm is beyond the scope of the chapter. This techniques proves to reduce leakage power by more than 50% [27]. And it is found that, given a low $V_{th}=0.2\ V_{DD}$, the high V_{th} can lie between 0.3 V_{DD} and 0.4 V_{DD}.

14.5 DYNAMIC LOGIC DESGN FOR POWER OPTIMIZATION

The above tecniques discussed, aims to minimize power utilization in circuits operated at reduced voltage. Apart from various issues in dynamic logic design, such as charge leakage, charge sharing, backgate coupling and clock feedthrough, there exists a critical issue that is challenging. It is the design of the cascading gates in the dynamic logic. This problem arises because of the applied inputs, given to the next stage gets precharged to high state i.e., 1. This causes the accidental discharge during the beginning of evaluation phase. In order to mitigate this, domino logic was proposed— where the inputs can make a transition (0->1) only during evaluation period. The principle of operation of domino logic is:

- During precharge state,the output is charged upto V_{DD} and consequently, the result from the inverter turns to be 0.
- In evaluation stage, logic gate dispenses charges and makes the output of the inverter to make a transisition from 0->1.

Thereby ensuring no change in output is produced during the precharge stage. However, the significant amount of power consumption limits its performance. Though in order to minimize power utilization, the applied supply voltage can be scaled down, but the following issues are raised.

• Reducing the supply voltage increases the delay.
• In order to compensate for the delay, down scaling of the threshold voltage is performed along with reduced supply voltage.
• Minimization in threshold voltage leads to improved speed however, leakage due to subthreshold current is increased.
• The precharge node in the domino circuits may get discharged due to this leakage current.

Thereby, degarding the performance of the domino circuits and as when operating at high frequencies. So, few techniques that are the modifications of the domino logic are proposed in [28,29] such as:

• Footerless domino logic.
• Footed domino logic.
• Current mirror footed domino logic.
• High-speed clocked delay domino logic.
• Modified high-speed clocked delay domino logic.
• Conditional evaluated domino logic.
• Conditional stacked keeper domino logic.

14.6 COMPARATIVE ANALYSIS

The brief summary of operation of various modifications of domino logic are provided in Table 14.1:

The research [28] proposed different topologies above (except for the foot-driven stack transistor domino logic) using a TSMC (Taiwan Semiconductor Manufacturing Company Limited) 65 nm technology with V_{DD} of one volt and a temperature, $T = 27°C$ for the load capacitance, $C_L = 100$ fF. The results were proven to show that CEDL technique cunsumes low power while the CMFDL occupies less area.

14.7 CONCLUSION

Power consumption is the bottleneck of circuit's performance when extending to deep submicron technologies. This chapter thus provides insights into low power CMOS circuit design techniques, by incorporating the fundamentals of power consumption considerations. The various techniques in reducing the supply voltage level, at reduced threshold voltage such as the power-gating structures (MTCMOS), VTCMOS, DTCMOS, and Dual-V_{th} CMOS provides the knowledge of basic low power circuit techniques. Finally, the analysis gives a comparative study between the variations in domino logic, that is widely used.

Domino logic

Operation	

Footer less domino logic (FLDL) (Figure 14.34):

During low phase of the clock, (precharge phase), M2 turns ON, and dynamic node N_D gets charged to V_{DD}. During high phase of the clock (evaluation period) the outcomes from the circuit changes depending on the inputs applied to evaluation logic block. During this time, M2 gets ON thereby, links the dynamic node to V_{DD} preventing unintended dispense of charges from the node.

Advantage:
Robustness of the circuit is boosted by effective enlargement in the size of the keeper transistor

Drawback:
In the evaluation period, when all of the applied inputs are made low, there arises a leakage because of the subthreshold and tunelling current. Also, as increasing size of keeper transistor boosts the toughness of circuit, that also creates penalty of increase in power consumption and delay.

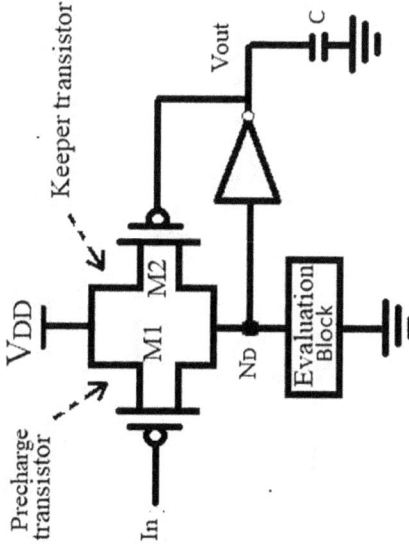

Figure 14.34 Footerless domino logic (FLDL).

(Continued)

Domino logic

	Operation
Footed domino logic (Figure 14.35):	This circuit performs the same operation in the coarse of low and high cycle of the clock as the footerless domino logic. But, a transistor (M3) is inserted in the footer of this circuit.	**Advantage:** The current arising because of the leakage is minimzed significantly by introducing a transistor in the footer. **Drawback:** The footer transistor creates delay and also curtails the speed. Additionaly, the robustness of the circuit gets reduced for high fan-in gates.

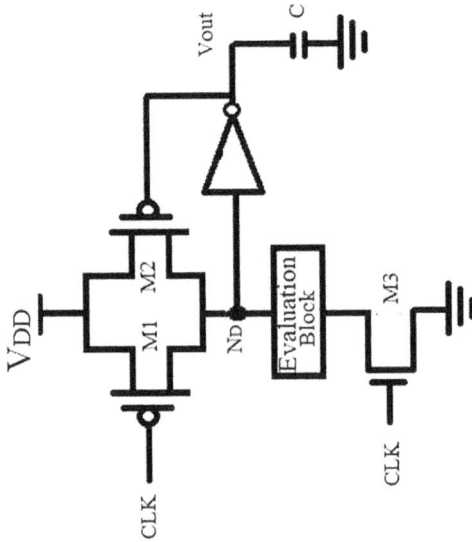

Figure 14.35 Footed domino logic (FDL).

(Continued)

Domino logic

	Operation
Current mirror footed domino logic (CMFDL) (Figure 14.36):	In order to reduce delay in the FDL logic, current mirrors M4 and M5 are inserted in the PDN. The transistor M6 is inserted that provides a path for the feedback present in between the gate of current mirror and the output. This M6 avoids the discharging of the dynamic node.	**Advantage:** During the evaluation phase, when clock signal is in high state along with having all the applied inputs low, the stacked transistors, M3 and M4 reduces the subthreshod current. The stacked transistor M4 provides noise immunity to the circuit. **Drawback:** The current mirrors reduces delay, however increases the discharging current in the circuit.

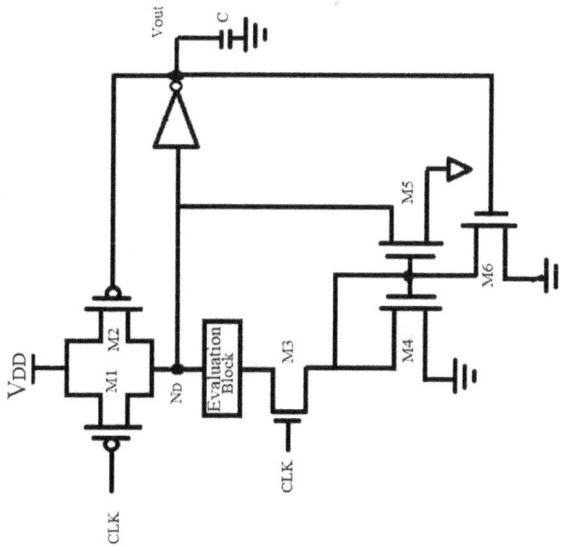

Figure 14.36 Current mirror footed domino logic (CMFDL).

(Continued)

Domino logic

Operation

High-speed clocked delay domino logic (HSCD) (Figure 14.37):

In start of precharge mode, M3 is ON. Node G_N will be maintained at a reduced voltage and switches OFF N2. Then following a delay of two inverters M3 is switched OFF. The transistors of PDN are sized such that voltage at G_i is low relative to V_{th} of M4, so that M4 is turned OFF during precharge state. During the evaluation stage, M3 switched OFF making the voltge at N high. This voltage will bias the transistors in the PDN and this decreases the leakage current in the logic.

Advantage: The rate of operation can be maximized by enlarging the size of N1, N2 or the PDN transistor.

Drawback: But by enlarging the size of M3, the potential at N gets reduced, and this rises by enlarging the size of evaluation logic. As width of the evaluaton transistors is enlarged, delay of the circuit increases.

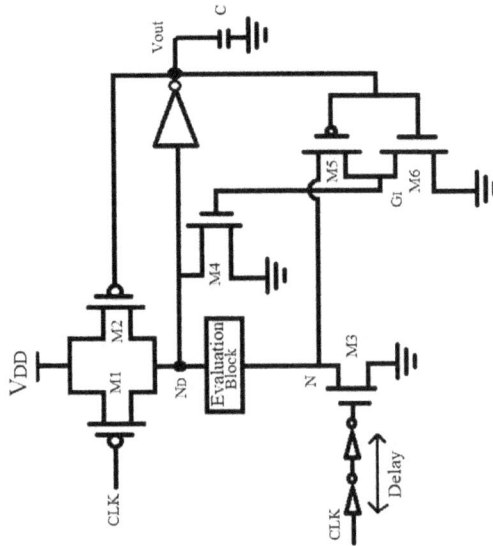

Figure 14.37 High-speed clocked delay domino logic (HSCD).

(*Continued*)

Domino logic	Operation	
Modified HSCD (M-HSCD) (Figure 14.38) 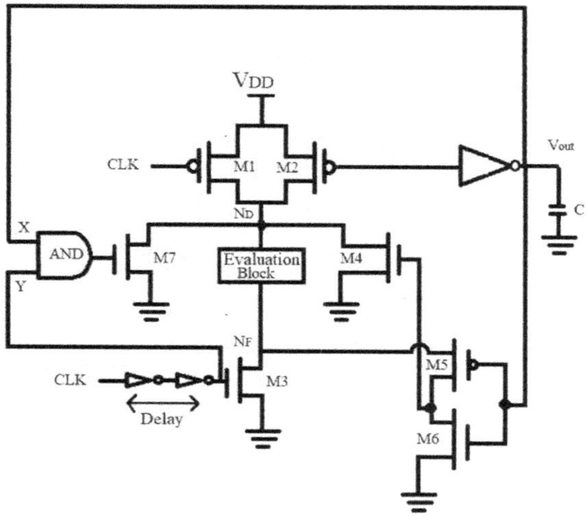	It is a modified HSCD circuit that is composed of one AND gate along with a NMOS transistor M7. During the evaluation phase, the dynamic node N_D discharges to 0 whenever one or more inputs turns high. During this interval, the applied input X to gate will turn high and the input Y will be turn on after two inverters delay. So, this biases the transistor M7 and thus speed increases.	**Advantage:** The speed of the circuit is increased because of AND gate and MD trnasistor. **Drawback:** During the precharge state, the of M3 stays in the high inpedance state and thus it leads to power consumption.

Figure 14.38 Modified high speed clocked delay domino logic (M-HSCD).

(Continued)

Domino logic	Operation	
Conditional evaluated domino logic (CEDL) (Figure 14.39):	This circuit contains a stacked transistors M4 and M5 that is present between dynamic node and the bottom ground. These two of the transistors switches ON conditionally based on ON/OFF condition of M3. During the evaluation phase, the footer transistor M3 swicthes OFF following the delay equivalent to the 2 inverter delay, causing high voltage at N. This value can be changed based on evaluation and footer transistor sizes. Increasing the size of M3, lowers the voltage at N, while by enlarging the size of evaluation transistors, potential at N rises.	**Advantage:** The gate of M4 that is floating in the prechrage stage in M-HSCD is resolved by inserting a NMOS transistor in stack with M4 tuned OFF. **Drawback:** There is an optimum value for the footer width, above which the delay of the circuit increases.

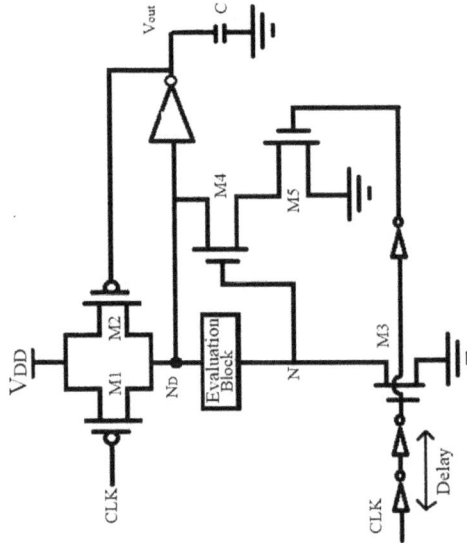

Figure 14.39 Conditional evaluated domino logic (CEDL).

(Continued)

Domino logic

	Operation
Conditional stacked keeper domino logic (CSK-DL) (Figure 14.40):	Here, the node N is provided as feedback for the transistors M7 and M8 for dispensing charge in the dynamic node. During high phase of the clock, for a delay of two inverters, the X and Y nodes will turn 0 and thereby, M4, M5 turns ON while the M6 turns OFF. Node N charges to a voltage depending on the sizes of evaluation and footer transistors. Thereby, M8 turns ON and the dynamic node discharges to zero. M7 is also in ON state and it lowers the voltage at node P to $V_{DD} - V_{th}$. This increases the current in the keeper transistor.

Advantage:
Robustness of the circuit is increased because of the increase in the current through keeper transistor. **Drawback:** Degarded noise performance.

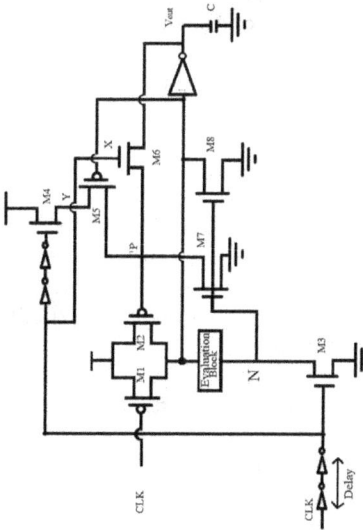

Figure 14.40 Conditional stacked sleeper domino logic (CSK-DL).

REFERENCES

[1] Burg, D., & Ausubel, J. H. (2021). Moore's law revisited through Intel chip density. *PLoS One*, 16(8), e0256245.

[2] Padmavathi, B., Geetha, B. T., & Bhuvaneshwari, K. (2017). Low power design techniques and implementation strategies adopted in VLSI circuits. In *2017 IEEE International Conference on Power, Control, Signals and Instrumentation Engineering (ICPCSI)* (pp. 1764–1767). Chennai, India: IEEE.

[3] Varadharajan, S. K., & Nallasamy, V. (2017). Low power VLSI circuits design strategies and methodologies: A literature review. In *2017 Conference on Emerging Devices and Smart Systems (ICEDSS)* (pp. 245–251). Mallasamudram, India: IEEE.

[4] Rabaey, J. M., & Pedram, M. (Eds.). (2012). *Low Power Design Methodologies* (vol. 336). New York: Springer Science & Business Media.

[5] Verma, P., & Mishra, R. A. (2011). Leakage power and delay analysis of LECTOR based CMOS circuits. In *2011 2nd International Conference on Computer and Communication Technology (ICCCT-2011)* (pp. 260–264). Allahabad: IEEE.

[6] Chandrakasan, A. P., Sheng, S., & Brodersen, R. W. (1992). Low-power CMOS digital design. *IEICE Transactions on Electronics*, 75(4), 371–382.

[7] Kuroda, T. (2002). Low-power, high-speed CMOS VLSI design. In *Proceedings. IEEE International Conference on Computer Design: VLSI in Computers and Processors* (pp. 310–315). Germany: IEEE.

[8] Sakurai, T., Kawaguchi, H., & Kuroda, T. (1997). Low-power CMOS design through VTH control and low-swing circuits. In *Proceedings of the 1997 International Symposium on Low Power Electronics and Design* (pp. 1–6), Monterey, CA, USA.

[9] Mutoh, S. I., Douseki, T., Matsuya, Y., Aoki, T., Shigematsu, S., & Yamada, J. (1995). 1-V power supply high-speed digital circuit technology with multithreshold-voltage CMOS. *IEEE Journal of Solid-State Circuits*, 30(8), 847–854.

[10] Kao, J., Chandrakasan, A., & Antoniadis, D. (1997). Transistor sizing issues and tool for multi-threshold CMOS technology. In *Proceedings of the 34th Annual Design Automation Conference* (pp. 409–414), Anaheim California, USA.

[11] Kao, J., Narendra, S., & Chandrakasan, A. (1998). MTCMOS hierarchical sizing based on mutual exclusive discharge patterns. In *Proceedings of the 35th Annual Design Automation Conference* (pp. 495–500), San Francisco California USA.

[12] Chang, Y. S., Gupta, S. K., & Breuer, M. A. (1997). Analysis of ground bounce in deep sub-micron circuits. In *Proceedings. 15th IEEE VLSI Test Symposium (Cat. No. 97TB100125)* (pp. 110–116). Monterey, CA, USA: IEEE.

[13] Kim, S., Kosonocky, S. V., & Knebel, D. R. (2003). Understanding and minimizing ground bounce during mode transition of power gating structures. In *Proceedings of the 2003 International Symposium on Low Power Electronics and Design* (pp. 22–25), Seoul, Korea (South).

[14] Abdollahi, A., Fallah, F., & Pedram, M. (2007). A robust power gating structure and power mode transition strategy for MTCMOS design. *IEEE Transactions on Very Large Scale Integration (VLSI) Systems*, 15(1), 80–89.

[15] Kawaguchi, H., Nose, K. I., & Sakurai, T. (1998). A CMOS scheme for 0.5 V supply voltage with pico-ampere standby current. In *1998 IEEE International Solid-State Circuits Conference. Digest of Technical Papers, ISSCC. First Edition (Cat. No. 98CH36156)* (pp. 192–193). San Francisco, CA, USA: IEEE.

[16] Shigematsu, S., Mutoh, S. I., Matsuya, Y., Tanabe, Y., & Yamada, J. (1997). A 1-V high-speed MTCMOS circuit scheme for power-down application circuits. *IEEE Journal of Solid-State Circuits*, 32(6), 861–869.

[17] Kim, S., Kosonocky, S. V., Knebel, D. R., & Stawiasz, K. (2004). Experimental measurement of a novel power gating structure with intermediate power saving mode. In *Proceedings of the 2004 International Symposium on Low Power Electronics and Design* (pp. 20–25), Newport Beach, CA, USA.

[18] Kumagai, K., Iwaki, H., Yoshida, H., Suzuki, H., Yamada, T., & Kurosawa, S. (1998). A novel powering-down scheme for low Vt CMOS circuits. In *1998 Symposium on VLSI Circuits. Digest of Technical Papers (Cat. No. 98CH36215)* (pp. 44–45). Honolulu, HI, USA: IEEE.

[19] Kobayashi, T., & Sakurai, T. (1994). Self-adjusting threshold-voltage scheme (SATS) for low-voltage high-speed operation. In *Proceedings of IEEE Custom Integrated Circuits Conference-CICC'94* (pp. 271–274). San Diego, CA, USA: IEEE.

[20] Seta, K., Hara, H., Kuroda, T., Kakumu, M., & Sakurai, T. (1995). 50% active-power saving without speed degradation using standby power reduction (SPR) circuit. In *Proceedings ISSCC'95-International Solid-State Circuits Conference* (pp. 318–319). San Francisco, CA, USA: IEEE.

[21] Kuroda, T., Fujita, T., Mita, S., Nagamatsu, T., Yoshioka, S., Suzuki, K., ... & Sakurai, T. (1996). A 0.9-V, 150-MHz, 10-mW, 4 mm/sup 2/, 2-D discrete cosine transform core processor with variable threshold-voltage (VT) scheme. *IEEE Journal of Solid-State Circuits* 31(11), 1770–1779.

[22] Kuo, J. B., & Lin, S. C. (2004). *Low-Voltage SOI CMOS VLSI Devices and Circuits*. New York: John Wiley & Sons.

[23] Assaderaghi, F., Sinitsky, D., Parke, S. A., Bokor, J., Ko, P. K., & Hu, C. (1997). Dynamic threshold-voltage MOSFET (DTMOS) for ultra-low voltage VLSI. *IEEE Transactions on Electron Devices*, 44(3), 414–422.

[24] Chang, L., Ieong, M., & Yang, M. (2004). CMOS circuit performance enhancement by surface orientation optimization. *IEEE Transactions on Electron Devices*, 51(10), 1621–1627.

[25] Lin, W. C. H., & Kuo, J. B. (2010). Low-voltage SOI CMOS DTMOS/MTCMOS circuit technique for design optimization of low-power SOC applications. In *Proceedings of 2010 IEEE International Symposium on Circuits and Systems* (pp. 3833–3836). Paris, France: IEEE.

[26] Chen, Z., Diaz, C., Plummer, J. D., Cao, M., & Greene, W. (1996). 0.18 um dual Vt MOSFET process and energy-delay measurement. In *International Electron Devices Meeting. Technical Digest* (pp. 851–854). San Francisco, CA, USA: IEEE.

[27] Moradi, F., Cao, T. V., Vatajelu, E. I., Peiravi, A., Mahmoodi, H., & Wisland, D. T. (2013). Domino logic designs for high-performance and leakage-tolerant applications. *Integration*, 46(3), 247–254.

[28] Garg, S., & Gupta, T. K. (2018). Low power domino logic circuits in deep-submicron technology using CMOS. *Engineering Science and Technology, an International Journal*, 21(4), 625–638.

[29] Wei, L., Roy, K., & De, V. K. (2000). Low voltage low power CMOS design techniques for deep submicron ICs. In *VLSI Design 2000. Wireless and Digital Imaging in the Millennium. Proceedings of 13th International Conference on VLSI Design* (pp. 24–29). Calcutta, India: IEEE.

Chapter 15

Recent advances in carbon nanotubes-based sensors

Amandeep Kaur and Jitender Kumar
University of Delhi

Avtar Singh
Adama Science and Technology University

15.1 WHAT IS A SENSOR?

A sensor is a device that reacts to changing situations in the environment. Sensors are embedded in our bodies, automobiles, airplanes, cellular, telephone, radar and chemical plants and have countless other applications. Without the use of sensors, there would be no automation. In the past few years, sensor technology has grown a lot and has drawn everyone's attention. The basic function of the sensor is to measure important quantities such as intensity of light, sound and pressure which get converted in terms of an electrical signal, usually a voltage or current by the sensor [1].

15.2 CHARACTERISTICS OF A GOOD SENSOR

Any good sensor should have the following qualities:

1. **High Sensitivity:** able to detect even small traces of vapor in ppm.
2. **High Selectivity:** able to identify a specific vapor in mixture of vapor.
3. **Stability:** with a given input one always gets the same output.
4. **Response Time:** time to acquire an equilibrium state. It should be as low as possible (~ up to few seconds).
5. **Recovery Time:** time taken by the sensors to return to its original state after the VOCs are removed. It should be as low as possible (~ generally few seconds).
6. **Reproducibility:** able to produce the same results on different trials

DOI: 10.1201/9781003459231-15

15.3 CARBON NANOTUBE AS A SUITABLE MATERIAL FOR SENSOR

The discovery of CNTs in 1991 by Ljima has gained increased attention due to their wide range of potential applications. CNTs have unique structural, mechanical, optical, thermal and electrical properties. A CNT has an exceptional one-dimensional atomic structure with very high aspect ratio (=length/diameter). It is one of the strongest and stiffest materials in nature having tensile strength of ~ 11–150 GPa (much greater than that of steel 0.38–1.55 GPa) and Young's modulus of 270–950 GPa (as compared to 200 GPa for stainless steel). The bondings in CNTs are in sp 2 form, where each atom is bonded with three neighboring atoms, as in graphite. This structural bond is stronger than the sp 3 bond of diamond, which gives CNT a unique strength. The emission spectra single-walled carbon nanotubes (SWCNT) lie in the NIR region and is very sensitive to the local ambient, thus facilitating the biological detection. Moreover, CNTs are extensively used as an active channel in transistors and conductors due to their high mobilities (up to 10,000 cm 2 V -1 s -1) and high electrical conductivities at room temperature. These nanotubes also exhibit superior thermal properties twice as high as diamond and are thermally stable up to 2,800°C in a vacuum or inert atmosphere. Its electric-current-carrying capacity is 1,000 times higher than copper wires ~ 10^9 A/cm^2. Due to these amazing properties, CNT is suitable for a wide variety of applications such as sensors, actuators whose electronic and optical properties can be harnessed. CNTs are ideal for sensing application because their electronic properties are highly sensitive to the local chemical environment. CNTs have high chemical stability and foreign species or chemical groups can be easily attached to their surface through chemical treatment. These nanotubes are usually produced by three methods such as arc discharge, CVD and laser ablation. CVD is a widely used technique as it allows good control over the length and the structure of the nanotubes in comparison to the other methods. The laboratories for the growth of CNTs are well established with outstanding control over the yield at low cost. Ease of its growth techniques and availability of large surface area for adsorption of vapor molecules makes it a strong candidate for sensing applications as shown in Figure 15.1. The basic principle of CNT-based vapor sensor is transfer of charge carriers between CNT and vapor molecules leading to increase in the resistance of sensor. CNT-based sensors have an added advantage of high sensitivity toward low vapor pressure species such as nerve agents, blister agents and explosives, which are impossible to detect with available conventional sensors. Also, the structure of CNT is hollow and open-ended from both sides, which allows the vapor molecules to permeate easily into the nanotubes [2].

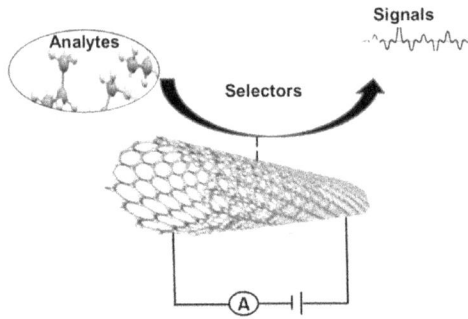

Figure 15.1 Carbon nanotube as sensor.

15.4 STRUCTURE OF CNTs

CNTs are graphene sheets of covalently bonded carbon atom rolled into cylinders with both ends open or normally capped by fullerene-like structure as shown in Figure 15.2. There are mainly two types of CNTs: single-walled carbon nanotubes (SWCNTs) which contain one graphene sheet rolled into cylinder. Its diameter is 1–5 nm. The other variant is SWCNTs, which has many graphene sheets rolled into concentrically nested cylinders (with interlayer spacing of 304 Å) together. The pure SWCNT structure, which is made solely of carbon atoms, can be represented as a rolled-up tubular shell of graphene sheet made of carbon atoms arranged in hexagonal rings of the benzene type. A single atomic layer of crystalline graphite is represented by graphene sheets, which are seamless cylinders created from a honeycomb lattice. A MWCNT is a stack of concentric cylinder-shaped rolls of graphene. As a single molecule containing millions of atoms, each nanotube can have a diameter as small as 0.7 nm and a length of tens of micrometers. The SWCNTs typically have a tube thickness of just one atom and a circumference of only 10 atoms. Nanotubes can be thought

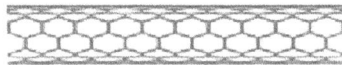

SWCNTs: Single Walled Carbon Nanotubes

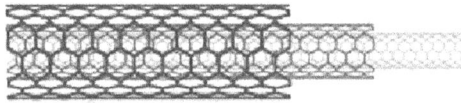

MWCNTs: Multi Walled Carbon Nanotubes

Figure 15.2 Structure of carbon nanotube, SWCNTs and MWCNTs.

of as having a practically one-dimensional structure due to their typical enormous length-to-diameter ratio (aspect ratio) of about 1,000. Larger MWCNTs are made up of numerous single-walled tubes stacked on top of one another. Only nanostructures having an outside diameter of less than 15 nm are eligible for the designation MWCNT; otherwise, the structures are referred to as carbon nanofibers. CNTs are different from carbon fibers, which are made up of strands of stacked graphite sheets rather than a single molecule. There are three conceivable varieties of carbon nanotubes in addition to two different basic forms. Armchair carbon nanotubes, zigzag carbon nanotubes, and chiral carbon nanotubes are three different forms of CNTs. The way the graphite is "rolled up" throughout the production process determines how these different forms of carbon nanotubes are produced. Different forms of SWCNTs are possible depending on the rolling axis' relationship to the hexagonal network of the graphene sheet and the closing cylinder's radius.

The number of unit vectors along each of the two directions in the graphene's honeycomb crystal lattice serves as the indices and for the chiral vector, respectively. When the nanotube is referred to as "zigzag," "armchair" and all other configurations are referred to as chiral. The three main forms of SWCNTs—chiral, zigzag and armchair [3].

CNT can be further classified into three categories, namely, armchair, zigzag and chiral. This classification is based on the wrapping angle of the graphene sheet. The way the grapheme sheet is rolled up is denoted by a pair of indices (n, m) known as the chiral vector. These integers n and m denote the number of unit vectors along the two directions of hexagonal lattice of graphene sheet. If $m=0$, the nanotubes are called "zigzag," and if $m=n$, then nanotubes are called "armchair"; otherwise tubes are called chiral. CNTs can be either metallic or semiconducting depending on the chirality. The chirality of nanotubes has major impact on its transport properties, mainly the electronic properties [3].

15.5 LIMITATION OF CNT-BASED SENSOR

Despite having limitless advantages of CNT, vapor sensors based on pristine CNT have certain limitations. They exhibit low sensitivity, low selectivity, irreversibility and long recovery time in comparison to other available sensors. To overcome these problems, several research groups have incorporated CNTs in polymer matrix and metal matrix. The advantage of making polymer composite is that it can be easily processed, can take a variety of shapes, can withstand high service temperature and composite becomes mechanically strong provided the CNTs are adequately dispersed within the matrix. Incorporation of CNTs in polymer has brought about a new reflux of research in lightweight high-performance reinforced polymer composite. CNT polymer composites can be processed in different ways that will be discussed in the subsequent section. Composites are processed easily without damaging CNTs properties, thus reducing the manufacturing cost. These composites have found great importance in various industries such as automotive, transportation, aerospace, consumer products and constructions. CNT polymer composites have improved mechanical strength and also result in the enhancement of electrical conductivity, thermal conductivity and dimensional stability. Conductive polymer composites can be formed at low loading of CNT due to their high aspect ratio and high conductivity. These composite film structures can be incorporated into or placed close to the VOC source, providing an opportunity for early identification. To locate any leaks inside the car, composite tapes can be fixed to fuel tanks or pipes. By absorbing any unwelcome VOCs in the environment, composites can also cover packaging materials to extend shelf life [2].

15.6 METHODS TO PREPARE POLYMER CNT COMPOSITE

The polymer CNT composite can be made by following techniques:-

15.6.1 Melt mixing

It is the best technique for creating CNT: Polymer nanocomposite at a large scale. When making thermoplastic polymer and CNT composites, this approach is typically used. The benefit of this method is that CNTs can be dispersed without the use of a solvent. Extruders, injection machines and twin screw mixers, among other specialized equipment, are used. High shear pressures and high temperatures are used in this procedure. To get a uniform dispersion of CNT in polymer, counter-rotating motors are used. Nanotube clumps are helped to disentangle by these shear forces. Twin screw mixers are used to melt and mix thermoplastic polymers. Above its melting point, polymer is melted. In the mixer, CNTs and polymer are introduced while many parameters, including the processing temperature, screw speed and mixing duration, are controlled. This method is best suitable for existing industrial operations. The degree of contact between the CNTs and polymer has improved. Melt mixing, however, is a violent process since it severely reduces the aspect ratio of the CNTs. High shear forces can potentially harm polymer structures. The main drawbacks of this method include its high cost and significant CNT degradation during the shear mixing procedure. Its use is similarly limited to thermoplastic matrices with low filler concentrations.

15.6.2 In-situ polymerization

The homogenous dispersion of CNT in the thermosetting polymer may be achieved using this approach, which is effective. In this procedure, CNT and monomers are combined initially, either with or without the aid of a solvent. By polymerizing the appropriate monomer using the appropriate processes, the CNT: polymer composite is created. Quite a strong CNT polymer interface is created, leading to improved mechanical and electrical conductivity. If the monomer has a low viscosity, this approach has the benefit of not requiring a solvent. However, using solvent is required to provide greater dispersion when the monomer's viscosity is high. This method is typically favored over others because nanotubes might take part in the polymerization process, forming covalent bonds between CNT and polymer in the process. This very practical manufacturing method enables the creation of nanocomposite materials with high nanotube loading. This method's main flaw is that it requires careful control over a number of variables, including the temperature and length of the polymerization, the concentration of the initiator, the solvent content, and the amount of agitation needed to disperse the nanotubes.

15.6.3 Mixing the solution

The easiest method for preparing composites on a lab scale is this one. In this method, the appropriate solvent, such as nitromethane, toluene, xylene, DI water, etc., is mixed with the CNT and polymers. The advantage of utilizing a solvent is that it helps the nanotubes stay in their unaggregated state. The solvent, CNTs, and polymer are combined in typical concentrations. The CNTs are then spread out using ultrasonication. There are two ways to apply ultrasonication: gently or vigorously. The nanotubes' length can be reduced by using high-power ultrasonication for an extended period of time, so these parameters must be tuned. This phase is carried out so that the CNTs are evenly distributed throughout the polymer matrix and the bundles of entangled CNTs can be successfully broken down without harming or breaking them. Controlling the intensity and duration of ultrasonication is necessary for this aim. Solution is centrifuged at a specific speed after ultrasonication to help undispersed CNTs settle. Then, 70% of the dispersed solution that is on top is decanted off and used to create the composite film. Techniques for processing solutions provide dispersions of high quality that are simple to use [4].

15.7 CNTs FUNCTIONALIZATION

Due to inadequate interfacial contact, which causes poor homogeneous dispersion, the full potentiality of CNTs when placed in polymer matrix has been severely constrained. Due to strong Vanderwaal attraction among the neighboring nanotubes, they tend to remain in bundle form and thus lack solubility in any common organic solvent as shown in Figure 15.3. This results in CNT agglomerates, which further impede the performance of

Figure 15.3 Plain and functionalized carbon nanotubes.

the devices based on them. Therefore, CNT bundles need to be separated into individual nanotubes to take their full advantage (i.e., for adsorption of vapor molecules). It is possible by treating CNTs with certain chemicals at higher temperature. This process is known as functionalization of nanotubes. Functionalization adds reactive groups (e.g.–OH, –COOH, amine, etc) on the surface of CNT. Depending on the chemical treatment performed, specific groups are attached to the CNT surface. This process also leads to the destruction of the CNT surface changing the hybridization from sp 2 to sp 3. These functionalized nanotubes show great stability in various organic solvents by transforming them from hydrophobic to hydrophilic.

Moreover, functionalized nanotubes are easily dispersed in organic solvents, which improves interfacial bonding between CNTs and polymer matrices. It results in stronger nanotube-polymer interaction, which will further enhance in nanocomposite mechanical properties.

There are mainly three ways for CNT functionalization, namely, covalent (chemical), covalent functionalization involving direct cycloaddition to π electrons of the CNTs and non-covalent (physical) as shown in Figure 15.4. As demonstrated in Figure 15.4, non-covalent functionalization involves wrapping CNT with polymer while making use of the nanotubes' interfacial properties without impairing their intrinsic qualities. It involves the individualization of nanotubes in an aqueous or organic solvent by functionalizing CNT with aromatic chemicals and surfactants that can wrap around the nanotubes' structure. This method is quite effective at making nanotubes more soluble in common solvents. In this case, the physical structure of CNT is not damaged, but the extent of dispersion of CNTs in the polymer obtained by this technique is not up to mark. However, this technique is rarely employed because it becomes difficult to remove the

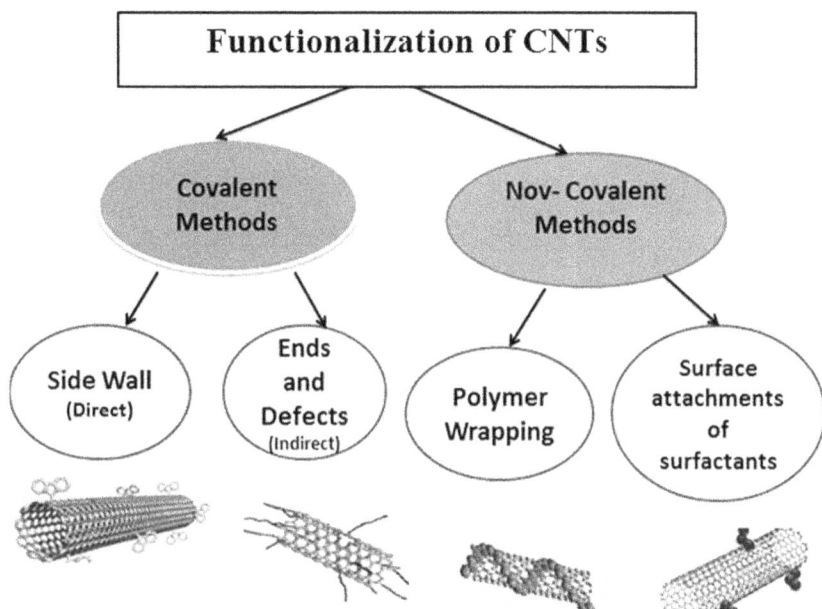

Figure 15.4 Types of functionalization.

wrapped structure in nanocomposite preparation. Covalent functionalization of CNTs can be achieved by the introduction of functional groups on the surface of CNT. It is achieved by treating CNTs with strong acids such as H_2SO_4 and HNO_3, which tend to debundle the nanotube, leading to better dispersion. Depending upon the chemical treatment performed, particular functional group is attached to the surface of CNT such as carboxylic acid, alcohol and ester groups. These groups may undergo further reactions to render them compatible with the polymer matrix. However, this method has a few drawbacks. During the functionalization process, ultrasonication is performed for long duration, which not only damages the CNT's physical structure but also introduces the defects on the CNT surface and sidewalls. In some cases, nanotubes are broken down into small pieces. This results in severe degradation in mechanical properties of CNT as well as disruption of π electrons in nanotube. Therefore, suitable functionalization route should be selected that modifies the CNT surface in such a way that it leads to the homogeneous dispersion of CNTs in the polymer matrix with minimum damage of its original structure. Another class of reactions, generating covalent functionalization, involves direct cycloaddition to the -electrons of the CNTs. These reactions are performed under mild conditions that do not induce CNT shortening and cause little damage to the CNT surface [5] (Figure 15.4).

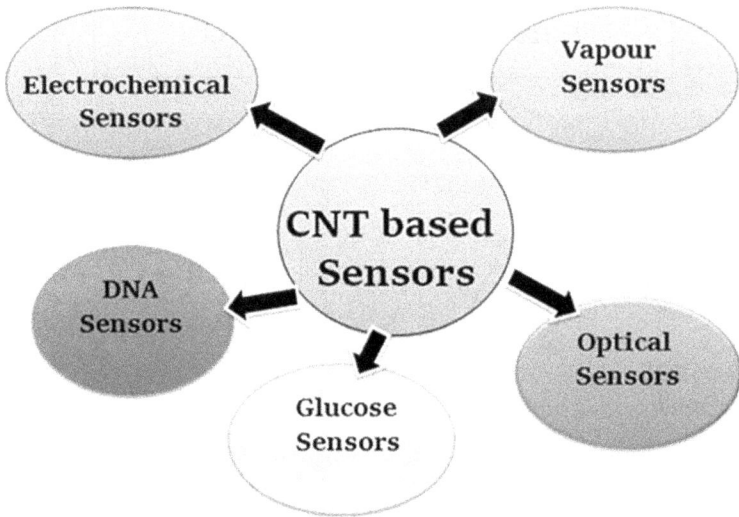

Figure 15.5 CNT-based sensors.

15.8 CNT-BASED VAPOR SENSOR

A composite made of polymer and CNT is frequently employed as a sensor because of its exceptional qualities. However, the degree of CNT dispersion in the polymer matrix has a significant impact on how well the nanocomposite performs. Realizing a high degree of nanotube dispersion is therefore crucial. Due to their strong Vanderwaal interactions, CNTs have a high inclination to form aggregates. When high shear pressures are used to separate the nanotubes, the nanotubes are also harmed, changing their characteristics. CNT dispersion in the polymer matrix is improved through functionalization; however, this is not a long-term fix. Therefore, a variety of techniques have been used to generate high-quality nanotube mixing in the polymer matrix. There are various factors that influence the performance of CNT-based sensors (Figure 15.5).

Number of nanotubes present: The concentration of nanotubes in the polymer must be tuned for improved sensing in order to create great sensors. If the CNT doping level in the polymer matrix is just above the percolation threshold, the sensor has been found to operate exceptionally well. A 3D network is established perfectly at this concentration.

Functionalized Nanotube: In the polymer matrix, a functionalized nanotube is evenly distributed. This enhances the composite sensor's overall performance. The physical structure of CNTs is typically severely damaged by traditional functionalization procedures, which also tend to change the majority of their sp2 bonds into sp3-linked carbon atoms. As a result, there are fewer vapor adsorption sites accessible on the CNT surface, which worsens the sensitivity. In some instances, the functionalization of nanotubes is accomplished through direct cyclo addition to CNT electrons, which doesn't affect the nanotubes' physical structure. Better responsiveness is displayed by such sensors [2].

Effect of physicochemical properties: The selectivity of the composite polymer nanotube sensor is affected by the physicochemical parameters of the analyte. When the sensor was subjected to various organic vapors, such as acetone, methanol, and ethanol. In each scenario, their response, response time, and recuperation time vary. It was discovered that the electrical and structural characteristics of both species, i.e., the sensing material and the vapor molecule, mutually determine the selectivity. In addition, the adsorption and chemical characteristics of the analyte have a significant impact on the ability to distinguish between various organic vapors. The sensor's maximum electronegativity and smallest molecular size were found to be reasons for its strong methanol vapor selectivity [6].

15.9 CNT-BASED BIOSENSORS

Biosensors are tools that can transform biological stimuli into signals for study and detection. By utilizing important analytes that cause disease, these devices help researchers increase their understanding of biological circuits. The knowledge gathered through the use of biosensors may enable earlier disease detection and more effective treatment.

Numerous biosensors are available, including those based on cells or tissues, immune systems, DNA, magnetic, thermal, piezoelectric, and optical sensors. They have incredible properties like stability, great target selectivity, biocompatibility, and little nonspecific binding. Biosensors based on CNTs are in demand. It can act as both the electrodes for electrochemical methods and as a transducer in a chemiresistor or chemFET. For developing practical sensors, the range of the sensors must correspond to concentrations of the targeted biomarker. There are a variety of biosensors developed and presently used in the healthcare industry [7].

15.9.1 CNT-based optical biosensors

Due to their high sensitivity, quick rate of detection, non-invasiveness and non-destructive mode of operation, optical biosensors have attracted a lot of interest in recent years. Over the past two decades, research has been focused on extending this branch of biosensors due to the aforementioned properties as well as optical biosensors' capacity to detect a wide range of analytes. The idea of pushing optical biosensors into the near-infrared (nIR) region of the electromagnetic spectrum has gained a lot of attention recently.

The low end of the visible spectrum or even the ultraviolet (UV) region must often be used to excite many optical biosensors or fluorescent probes, which can shorten sensor lifetimes by photobleaching organic molecules. Using UV light on biological material raises safety issues related to phototoxicity. Due to the light's reflection, absorption, and scattering by the tissue, water, and blood, biological tissues also absorb and block UV and visible light. The depth of tissue penetration for visible light has been calculated to be larger than 0.1 cm as a result; nevertheless, as wavelengths move toward the nIR spectrum, the depth of tissue penetration increases. The nIR-I (780–900 nm) and nIR-II (900–1,700 nm) regions of the spectrum allow light to reach tissue depths of up to 1 cm and 3 cm, respectively. The increased tissue penetration depth observed with nIR wavelengths is due to the light's off resonance frequency, which is lower in comparison to water, blood, and tissue and causes less absorption and scattering. CNTs have undergone extensive research and have been determined to be ideal for the creation of DNA sensors among the nIR optical biosensors. The nIR-II range (900–1,700 nm) of fluorescence results from the band gap between semiconducting SWNT electrons being on the order of 1 eV, while excitation typically occurs in the nIR-I (780–900 nm) range or in the longer visible range (>600 nm). Another advantageous aspect that motivates the development and application of SWNT as biosensors is the Stoke-shift between the excitation and emission wavelengths of SWNT, which results in a reduced autofluorescence when employed for imaging in biological samples [8].

15.9.2 Electrochemical CNT biosensors

Electrochemical sensors are the most common because of their portability and low cost. The large surface areas of nanotubes and wonderful electrical properties make them suitable for biosensors. Nanotube improves the immobilization of biological recognition elements and also enhances electron transfer which increases the sensitivity and selectivity of these sensors. Electrochemical enzyme sensors are even more popular than electrochemical due to the high demand for glucose sensing. They have found application in the recognition of the ripening of fruits and freshness of fish.

There are several approaches to immobilize nanotubes on the electrode, including adsorption, drop-casting, and direct production. However, if nanotubes are not evenly disseminated, the performance of the sensor would suffer. As a result, highly sensitive sensors are developed using functionalized nanotubes.

15.9.3 Electrochemical immunosensors

Devices known as immunosensors generate signals in response to interactions between antibodies and antigens during binding. There are several approaches to identify the binding of the corresponding pair member, creating a stable complex. On a sensor surface, either the antigen or the antibody can be immobilized. Nanotube-based immunosensors can be produced using sandwich assay technology. This method involves immobilizing antibodies at the electrode—either directly to the CNTs or to another binding component at the electrode surface—and functionalizing the electrode with nanotubes to facilitate the binding of the analyte proteins. The binding of the analyte proteins is then detected by the addition of secondary antibodies containing reporter molecules. If these reporter molecules are present at the electrode surface and they bind to the immobilized analyte, the electrical signal will change [7].

15.10 CNT-BASED GLUCOSE SENSOR

The most popular test for determining the blood glucose level in a human body is the glucose sensor. A variety of disorders, including diabetes, can be diagnosed and managed with the aid of concentration monitoring. In recent years, glucose oxidase (GOx) has been used as the selection in the development of electrochemical and chemFET sensors. Geetha et al developed a composite of CNTs and CuO for the detection of glucose as an electrolyte in artificial sweat. The CNT-CuO nanocomposite was made using the complicated precipitation process. Chronoamperometry and cyclic voltammetry were employed to examine the CNT CuO NC's capacity to detect glucose. Electrochemical tests reveal that the so-developed sensor showed high sensitivity and selectivity. Besides this, it offered a rapid response and good stability. To identify and reduce diabetes risk, prevent heart disease, diabetic retinopathy, renal failure, and nerve degeneration, CNT-CuO NC is an appropriate material to include in wearable sensor systems. This is a result of CNT-CuO NC's ability to sense glucose [9].

15.11 CNT-BASED DNA SENSOR

The identification of DNA is crucial for the detection of infectious agents and pathogens, agents of biowarfare, diagnosis and treatment of genetic abnormalities, and the discovery of new drugs. High levels of sensitivity, selectivity, and reproducibility are offered by CNT-based DNA sensors. Due to their extremely high surface area and biocompatibility, Au nanoparticles are currently used for the development of DNA sensors with CNT. Because of their high surface-to-volume ratio, suitable electrical characteristics, and quick electron transfer rate, nanotubes are thought to be a good material for DNA sensor platforms. For the electrochemical detection of double-stranded DNA, CNTs are mixed with polymer on a high-density polyethylene substrate and employed as conducting channels in chemiresistors [10].

REFERENCES

[1] Patel, B. C., Sinha, G. R., & Goel, N. (2020). Introduction to sensors. In *Advances in Modern Sensors: Physics, Design, Simulation and Applications.* IOP Publishing.

[2] Kaur, A., Singh, I., Kumar, J., Madhwal, D., Bhatnagar, P. K., Mathur, P. C., ... & Paiva, M. C. (2013). An environment friendly highly sensitive ethanol vapor sensor based on polymethylethacrylate: Functionalized-multiwalled carbon nanotubes composite. *Advanced Science, Engineering and Medicine,* 5(10), 1062–1066.

[3] Eatemadi, A., Daraee, H., Karimkhanloo, H., Kouhi, M., Zarghami, N., Akbarzadeh, A., ... & Joo, S. W. (2014). Carbon nanotubes: Properties, synthesis, purification, and medical applications. *Nanoscale Research Letters,* 9(1), 1–13.

[4] Mohd Nurazzi, N., Asyraf, M. M., Khalina, A., Abdullah, N., Sabaruddin, F. A., Kamarudin, S. H., ... & Sapuan, S. M. (2021). Fabrication, functionalization, and application of carbon nanotube-reinforced polymer composite: An overview. *Polymers,* 13(7), 1047.

[5] Kaur, A., Singh, I., Kumar, J., Bhatnagar, C., Dixit, S. K., Bhatnagar, P. K., ... & da Conceicao Paiva, M. (2015). Enhancement in the performance of multi-walled carbon nanotube: Poly (methylmethacrylate) composite thin film ethanol sensors through appropriate nanotube functionalization. *Materials Science in Semiconductor Processing,* 31, 166–174.

[6] Kaur, A., Singh, I., Kumar, A., Rao, P. K., & Bhatnagar, P. K. (2016). Effect of physicochemical properties of analyte on the selectivity of polymethylmethacrylate: Carbon nanotube based composite sensor for detection of volatile organic compounds. *Materials Science in Semiconductor Processing,* 41, 26–31.

[7] Ferrier, D. C., & Honeychurch, K. C. (2021). Carbon nanotube (CNT)-based biosensors. *Biosensors,* 11(12), 486.

[8] Hofferber, E. M., Stapleton, J. A., & Iverson, N. M. (2020). Single walled carbon nanotubes as optical sensors for biological applications. *Journal of the Electrochemical Society,* 167(3), 037530.

[9] Geetha, M., Maurya, M. R., Al-maadeed, S., Muthalif, A. A., & Sadasivuni, K. K. (2022). High-precision nonenzymatic electrochemical glucose sensing based on CNTs/CuO nanocomposite. *Journal of Electronic Materials,* 51(9), 4905–4917.

[10] Nouri, M., Meshginqalam, B., Sahihazar, M. M., Sheydaie Pour Dizaji, R., Ahmadi, M. T., & Ismail, R. (2018). Experimental and theoretical investigation of sensing parameters in carbon nanotube-based DNA sensor. *IET Nanobiotechnology,* 12(8), 1125–1129.

Chapter 16

Low power and low area multiplier and accumulator block for efficient implementation of FIR filter

J. L Mazher Iqbal
Vel Tech Rangarajan Dr.Sagunthala R&D
Institute of Technology and Science

G. Narayan
Vellore Institute of Technology

T. Manikandan
Rajalakshmi Engineering College

M. Meena
Vels Institute of Science, Technology & Advanced Studies

Jose Anand
KCG College of Technology

16.1 INTRODUCTION

Fixed logic ASICs provide high productivity benefits in a wide range of application domains. They offer benefits such as high performance and low cost in high volume. There are some drawbacks related to these kinds of devices such as high nonrecurring engineering (NRE) cost, low flexibility, difficult and long design cycle and long and expensive time to market. It takes quite a few months to years to design and validate depending upon the complexity and size and of the fixed logic device. If the circuit does not work as anticipated or if design requires changes, it must be redesigned. On the other hand, the benefits of reconfigurable logic devices are low development cost (No upfront NRE cost), short time to market in turn revenue, fast design cycle and ease for designers. With parallel computing, extensive variety of logic capacity, I/O capabilities, operating speed and power parameters of reconfigurable logic devices, designers are able to rapidly model and assess their designs on an operational circuit. The same device may be embedded into the final system. Amendment to the design

DOI: 10.1201/9781003459231-16

is possible after the complete system has been transported to the customer. Digital FIR filters may be implemented using highly sophisticated digital signal processing (DSP) processors. The multiply accumulate (MAC) unit is the basic computational unit of DSP processor. FPGAs and DSPs each have their own advantages and disadvantages. The disadvantage of using DSP processors are limited number of instructions/clock, limited memory & device connectivity, sequential execution, i.e. performing one operation on a single set of data at a time, the fixed number of operations, bandwidth limitations, higher power consumption and low operational speed. Hence, the implementation of programmable digital FIR filter on digital signal processor is not a very efficient approach. Hence, efficient programmable and dedicated FIR filters, parallel and pipelined FPGA-based hardware accelerator without full multipliers are implemented to enhance the overall performance. Therefore, FPGA-based digital FIR filters completely optimize in terms of reconfigurable, parallelism, high number of instructions/clock, high bandwidth flexible I/O & memory connectivity, power characteristics, rapid prototyping, short time to market, no NRE cost, and bandwidth requirement.

In this chapter, we have developed the reconfigurable FPGA-based implementation to perform operations like FIR filtering Fourier transform. Also, we have developed a new FIR filter algorithm for a set of fixed coefficients, and FIR filter using optimized look-up table (LUT). The reconfigurable processing element architecture use pipelining and parallel dispensation. The processing element (PE) is based on bit serial arithmetic, and the constant shift method (CSM) is developed. The CSHM categorizes common computation and uses it again between different multiplications. A new algorithm considering the adder graph technique for a set of fixed coefficients is developed for dedicated architecture where the multipliers can be grouped in multiplier blocks (MB). The proposed algorithm chooses common adder graphs that can be maximally shared with the remaining coefficients, while previous dependence-graph algorithms consider one coefficient at a time and do not take into account the effects on the remaining coefficients while synthesizing the coefficient. The low complex memory-based FIR filter architecture using optimized LUT multiplier is presented. The reconfigurable FIR filter architectures and algorithms offer configurability, low complexity, low area, low power, and high speed.

Filtering shows a significant role in very large scale integrated circuit (VLSI) signal processing. The methods of filtering are extensively used in various electronic devices to eliminate chunk of signal that harms the signal. The key requirement of using the filter in VLSI signal processing is to increase the quality of a signal by decreasing the noise to obtain or to associate the preferred slice of signal. Filters are used for two purposes: one for separation of signal and for restoration of signal. Record of audio signal which is associated with deprived equipment is filtered to provide the recovering sound signal using filters. When filters are realized, the coefficients are the furthermost composite module. Finite impulse response (FIR) digital filter is typically used in DSP applications that are loudspeaker and channel equalization, noise cancelation, and DSP applications. The cost of employment of an FIR filter can be decreased by decreasing the complexity of the coefficients. The reconfigurable FIR filters are excessively flexibility in modeling the magnitude response. The basic FIR filters are shown in Figure 16.1. Cascade apprehension is usually used for FIR filter design and implementation. The filter is realized by divided into numerous low-order modules. The cascade realization is usually used for upper-order filter implementation; the cascade realization is shown in Figure 16.2.

The FIR filters are realized using reconfigurable computing hardware such as FPGA. The key benefit of the FPGA-based cascade apprehension is its modularity. In the context of FIR filters, a multiplier and accumulator (MAC) module is the method of multiplying a coefficient with the input signal with delayed samples and accumulating the result. For the realization of FIR filter complex computational methods are vital.

The estimation of area and speed for various types of altered distributed arithmetic (DA) specified for the FIR filter has been proposed. Patel and Patel [1] presented various methods in terms of power, delay, area, and platform for implementation. Lou, Yu, and Meher [2] presented a detailed exploration and important pathway for FIR filters. Alam [3] presented a method to decrease the complexity by decreasing the multipliers count in the cascaded filter for the chosen description. Though the method needs complex delay components, it results in less memories and additional

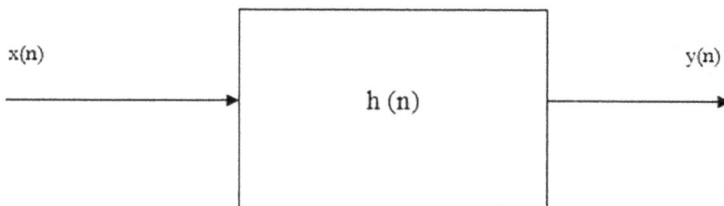

Figure 16.1   FIR filter block diagram.

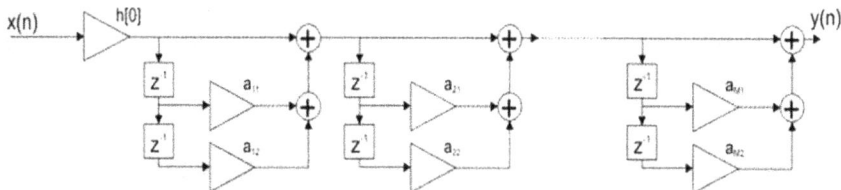

Figure 16.2   FIR filter cascade realization.

energy-effective memory structures when time-multiplexed. Pan and Meher [4] presented various memory structures and their access arrangements and comparisons in terms of their effectiveness when using memories. Dam, Cantoni, and Nordholm [5] discussed the model of variable digital filters (VDF) used for numerous DSP applications. The VDF realizes effective hardware execution with the least-square principle, an effective branch-cutting structure is offered to obtain the optimal solution. Ye and Yu [6] presented genetic algorithm (GA)-based multiplier-less FIR filters with single-stage and cascade arrangement. The limitation in the methodology is GAs frequently fail to discover viable solutions, it needs separate search space. Ye and Yu [7] discussed the sparse filter method at bit-level multiplier-less FIR filters. The multiplier-less filter scheme decreases the search space. Iqbal and Varadarajan [8] presented an algorithm considering the condensed adder graph method for a set of stationary coefficients. The condensed adder graph method is helpful in numerous filter applications such as FIR or infinite impulse response (IIR) filters, where the multipliers can be clustered. The algorithm selects common adder graphs which are exceedingly shared with the remaining coefficients. Iqbal and Varadarajan [9] discussed the comparison of various realizations of dynamically reconfigurable FIR filter architecture using computation-sharing multiplier. The shift and add unit produces {1, 3, 5, 7, 9, 11, 13, 15} and {0, 1, 2, 3, 4, 5, 6, 7} binary common sub-expression (BCS). Individually, these "8" BCSs are then fed to the data selector unit. Iqbal and Varadarajan [10] presented the parallel and cascade architectures of FIR architectures with shift and add unit which accomplishes the multiplication. The shift and add unit produces {1, 3, 5, 7, 9, 11, 13, 15} BCS. Iqbal and Varadarajan [11] presented FIR filler with memory-based multiplication. LUT stores pre-computed multiplication and summation performed in FIR filter and the stored value is retrieved through memory indexing. Iqbal and Narayan [12] discussed the least mean square (LMS) adaptive filter architecture using reversible PE architecture with less power dissipation because of reversible methodology adapted in the processing elements. Iqbal and Manikandan [13] discussed the FPGA-based reconfigurable common sub-expression elimination (CSE) architectures using LUT and serial processing for DSP applications. Iqbal and Kishore [14] discussed the employment of sensors using Programmable System on Chip (PSoC) microcontroller and printed circuit board (PCB) strip. Karthik and Iqbal [15] discussed the speech recognition in the existence of numerous noises. The paper proposes the convolutional Encoder and Decoder to eliminate the noise.

16.2 RECONFIGURABLE DEVICES CONFIGURATIONS

Reconfigurable computing is developed using either schematic entry tools or hardware description languages (HDL). The design steps in the design flow are shown in Figure 16.3. The description of functional design such as word width, register allocation, arithmetic operations, logical operations, and control flow is called register transfer level (RTL) description. HDL such as VHDL and Verilog are used to express RTL. Synthesis includes power optimizations, slack optimization, and generic optimization [16]. The synthesis is followed by partitioning, placement, and routing [17]. The alteration of net list to an FPGA-structured bit stream is called employment. Partitioning, placement, and routing are the main phase of implementation, which allocate FPGA resources such as CLBs and interconnections [18]. Large circuit is partitioned into smaller sub-circuit. Placement allocates physical positions of circuits on the FPGA although routing interconnects all the sub-circuits by blustering the fuses [19,20]. The output of the design implementation is the configuration bit stream.

The requirements of DSP computational building blocks are hardware optimization, flexibility i.e. each building block could be arranged to develop numerous diverse applications. The scheme should be adequately all-purpose so that it can be effortlessly combined with additional blocks to develop complete DSP systems. One of the key processes in realizing DSP

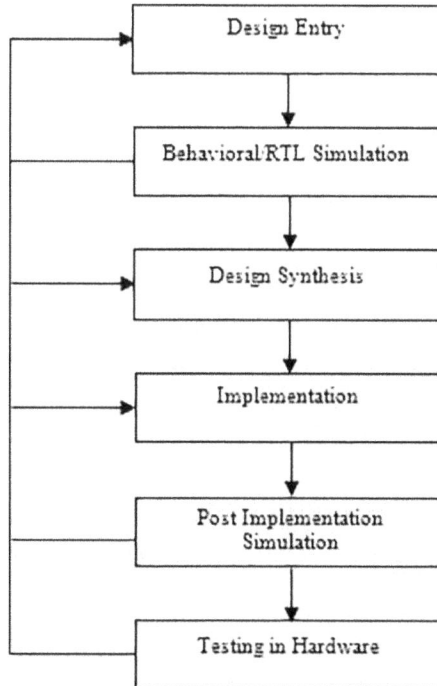

Figure 16.3 High-level view of a possible FPGA design flow.

functions is multiplication and addition. The multipliers are the most costly blocks in numerous DSP function. We have designed and implemented various MB as PE to implement reconfigurable FPGA-based hardware accelerator to perform various DSP functions on reducing hardware complexity, accuracy, speed, and power dissipation in this dissertation.

Utmost DSP applications such as transforms and filters require hardware unit that performs the addition and multiplications. The configuration of such hardware unit is known as multiply and accumulator (MAC) unit. The operation performed by MAC unit is called a MAC or MAC operation. Add/subtract unit and a register is needed to implement accumulation at the output of the multiplier. We have designed and implemented the reconfigurable FPGA-based hardware accelerator (PE-based architecture) for FIR filter. MAC operation is performed by the basic reconfigurable FIR filter architecture module. New dynamically reconfigurable MB are used in the analysis of the proposed reconfigurable PE architecture that multiplies two n-bit number x and y and gives the product 2n bits wide. The output of the multiplier block is added/subtracted from the contents of the accumulator in the add/subtract unit. The outcome is saved in the accumulator. The MAC unit performs multiplication and accumulation simultaneously. The accumulator accumulates the product of the previous multiplications while multiplier is calculating a product. The accumulator is idle during the first multiplication since there is nothing to accumulate. The multiplier is idle during the last accumulation since all the N products have been calculated. The typical feature of many viable DSP devices is the operation of the multiplier and accumulator works in parallel to efficiently accomplish the MAC process for each cycle.

16.3 PROPOSED OPTIMIZED FIR FILTER

We have designed and implemented an optimized multiplier and adder block with high performance such as low power, low area, and high speed. The proposed multiplier and adder blocks are applied to FIR filter in order to decrease the computational density. The FIR filter work with better performance in terms of low power, area and high speed of operation compared to conventional FIR filter.

16.3.1 Proposed CSM multiplier block

Proposed algorithm for the development of efficient multiplier block

- Step 1: Get the input Xin
- Step 2: The coefficient H corresponding to the filter description is loaded into LUT

- **Step 3:** Perform right shift operation for Xin if it is even or Xin+1 for ODD
- **Step 4:** Based on the position of "1"s in Xin perform the corresponding left shifting on H.
- **Step 5:** Accumulate all partial products using the carry save adder block
- **Step 6:** Perform left shift on the accumulated result to get final multiplied output.

Figure 16.4 shows the proposed multiplier block. It takes inputs as (15:0) for both "x" and "h" and converts both inputs into binary numbers. Perform 1 right shift operation for h. After performing the right shift operation h is set

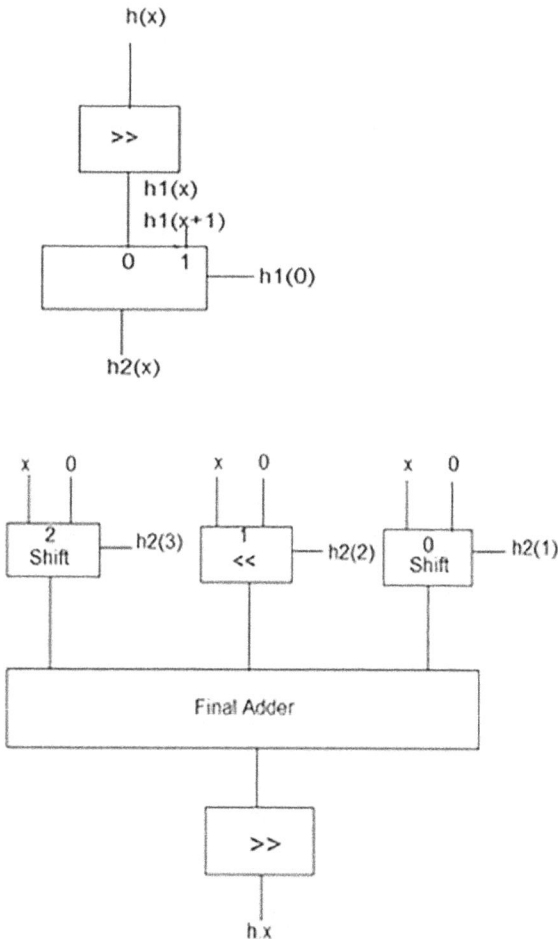

Figure 16.4 Block diagram of modified multiplier.

down to (15:1). Check for 1's in h and perform the left shift operation based on the positions of 1's in h. Then perform the addition operation. Finally, perform the right shift operation for the final addition output.

16.3.2 Adder block diagram

Carry save adder (CSA) is a category of digital adder; it is used in DSP architecture such as filters to add the sum of two or additional n-bit numbers in binary it is different from other digital adders. One is an arrangement of partial sum bit and additional which is a categorization of carry bit. In this, we are using basic arithmetic operations, and we will calculate from right to left as shown in Figure 16.5.

16.4 RESULTS AND ANALYSIS

In this section, simulation results are presented to demonstrate the efficiency of the technique for designing optimized low power, delay, and high-speed FIR filter. Frequently used FIR filters from literatures are used as standards. Altogether design illustrations are realized in VHSIC-HDL and synthesized using Vivado Design suite and compared the results of speed, power, delay, device utilization with conventional FIR filter and modified FIR filter. On comparing the modified and conventional multiplier, the device utilization,

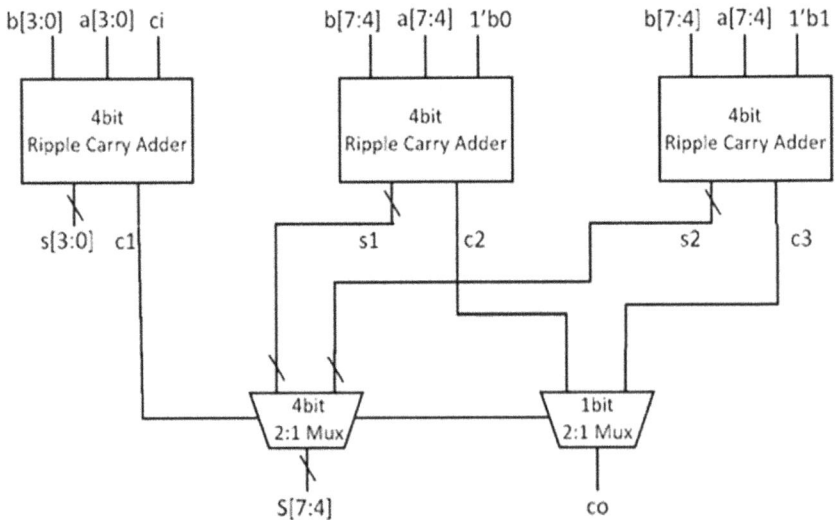

Figure 16.5 Adder block.

speed, and time of the proposed multiplier is less than the conventional multiplier in both LUT and IO sources. The device utilization summary is shown in Table 16.1. The power analysis of the proposed multiplier is shown in Table 16.2. The on-chip power is reduced (0.242 W) compared to conventional multiplier (8.047 W). It gives the details of dynamics, signals, logic, and device static. Table 16.3 shows the device utilization summary (LUT, LUTRAM, IO, logic signals, and on-chip power) of the proposed multiplier and conventional multiplier. Table 16.4 shows the dynamic,

Table 16.1 Device utilization summary

Source	Proposed multiplier utilization	Conventional multiplier utilization	Available
LUT	57	71	3,03,600
IO	31	32	600

Table 16.2 On chip power

Source	Proposed multiplier utilization	Conventional multiplier utilization
Dynamic	13.401 W (97%)	13.446 W (97%)
Signals	0.540 W (4%)	0.604 W (4%)
Logic	0.399 W (3%)	0.498 W (4%)
I/O	12.462 (93%)	12.344 (92%)
Device static	0.427 W (3%)	0.427 W (3%)
Total on-chip power	13.827 W	13.874 W

Table 16.3 Device utilization summary

Source	Proposed multiplier utilization	Conventional multiplier utilization	Available
LUT	-	21	3,03,600
LUTRAM	-	1	1,30,800
FF	-	29	6,07,200
IO	16	19	600
BUFG	-	1	32

Table 16.4 On chip power

Source	Proposed multiplier utilization	Conventional multiplier utilization
Dynamic	0.00 W (0%)	7.717 W (96%)
Signals	-	0.309 W (4%)
Logic	0.00 W (0%)	0.154 W (2%)
I/O	0.00 W (0%)	7.254 (96%)
Device static	0.242 W (100%)	0.330 W (4%)
Total on-chip power	0.242 W	8.047 W

signal, logic, IO, device static, and total on-chip power of the proposed and conventional multiplier. Figure 16.6 shows the power analysis of the modified multiplier. It will show the details of signals, power, device static, IO, and on-chip power. Figure 16.7 shows the simulation waveform for the modified multiplier. Figure 16.7 shows the schematic diagram of the modified multiplier after the synthesis of code. Figure 16.8 shows the device summary of the conventional multiplier after synthesizing the code.

Figure 16.9 shows the power analysis of the conventional multiplier. It shows the details of signals, power, device static, IO, and on-chip power.

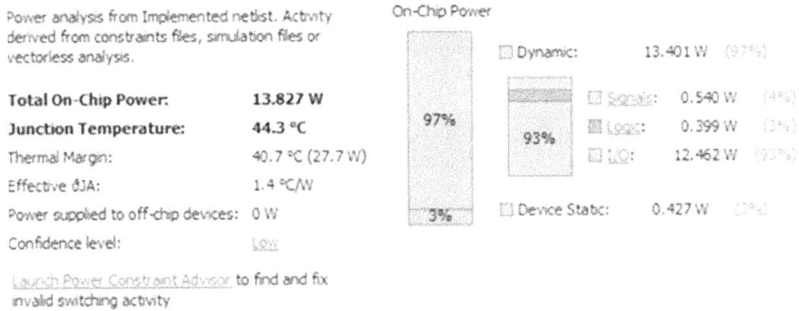

Power analysis from Implemented netlist. Activity derived from constraints files, simulation files or vectorless analysis.

On-Chip Power

Total On-Chip Power:	13.827 W
Junction Temperature:	44.3 °C
Thermal Margin:	40.7 °C (27.7 W)
Effective θJA:	1.4 °C/W
Power supplied to off-chip devices:	0 W
Confidence level:	Low

Dynamic: 13.401 W (97%)

Signals: 0.540 W (4%)
Logic: 0.399 W (3%)
I/O: 12.462 W (93%)

Device Static: 0.427 W (3%)

97%
93%
3%

Launch Power Constraint Advisor to find and fix invalid switching activity

Figure 16.6 Power analysis of modified multiplier.

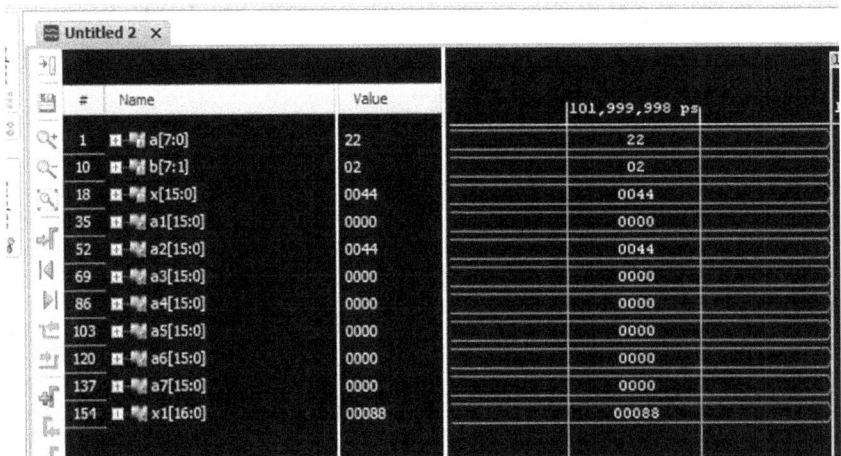

Untitled 2 X

#	Name	Value	101,999,998 ps
1	a[7:0]	22	22
10	b[7:1]	02	02
18	x[15:0]	0044	0044
35	a1[15:0]	0000	0000
52	a2[15:0]	0044	0044
69	a3[15:0]	0000	0000
86	a4[15:0]	0000	0000
103	a5[15:0]	0000	0000
120	a6[15:0]	0000	0000
137	a7[15:0]	0000	0000
154	x1[16:0]	00088	00088

Figure 16.7 Simulation waveform for modified multiplier.

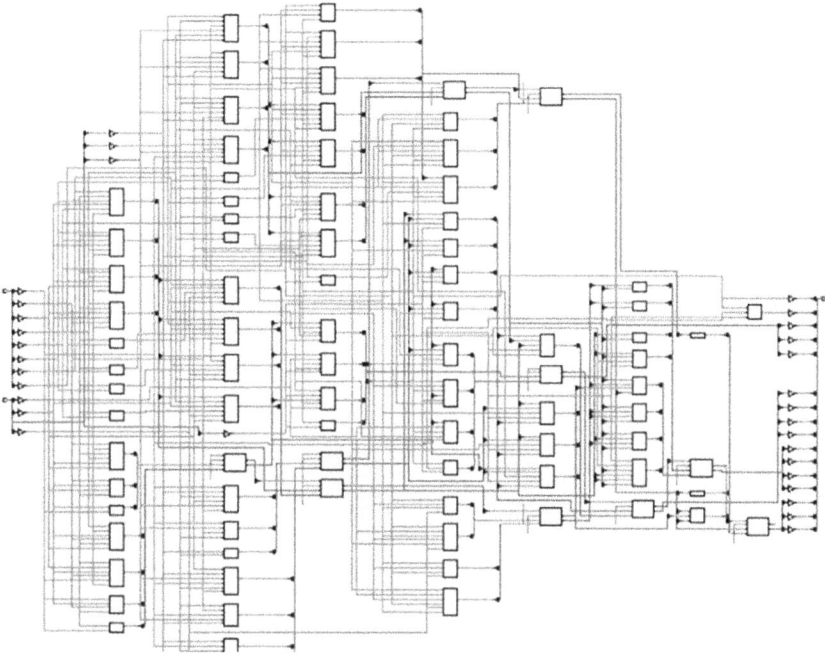

Figure 16.8 Schematic diagram for modified multiplier.

Power analysis from Implemented netlist. Activity derived from constraints files, simulation files or vectorless analysis.

On-Chip Power

Total On-Chip Power:	**13.874 W**
Junction Temperature:	**44.4 °C**
Thermal Margin:	40.6 °C (27.7 W)
Effective ϑJA:	1.4 °C/W
Power supplied to off-chip devices:	0 W
Confidence level:	Low

97%

92%

3%

Dynamic: 13.446 W   (97%)

Signals: 0.604 W   (4%)
Logic: 0.498 W   (4%)
I/O: 12.344 W (92%)

Device Static: 0.427 W   (3%)

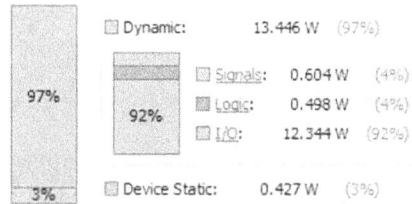

Figure 16.9 Power analysis for conventional multiplier.

Figure 16.10 shows the power analysis of the modified FIR filter. It shows the details of signals, power, device static, IO, and on-chip power.

Figure 16.11 shows the utilization of the modified FIR filter. It gives the details of LUT and IO implementations.

Figure 16.12 shows the schematic diagram of the modified FIR filter after generating the bit stream

Power analysis from Implemented netlist. Activity derived from constraints files, simulation files or vectorless analysis.

On-Chip Power

Total On-Chip Power: **0.242 W**

Junction Temperature: **25.3 °C**

Thermal Margin: 59.7 °C (41.1 W)

Effective ϑJA: 1.4 °C/W

Power supplied to off-chip devices: 0 W

Confidence level: High

☐ Dynamic: 0.000 W   (0%)

Logic: 0.000 W   (0%)

I/O: 0.000 W   (0%)

☐ Device Static: 0.242 W   (100%)

100%

Launch Power Constraint Advisor to find and fix invalid switching activity

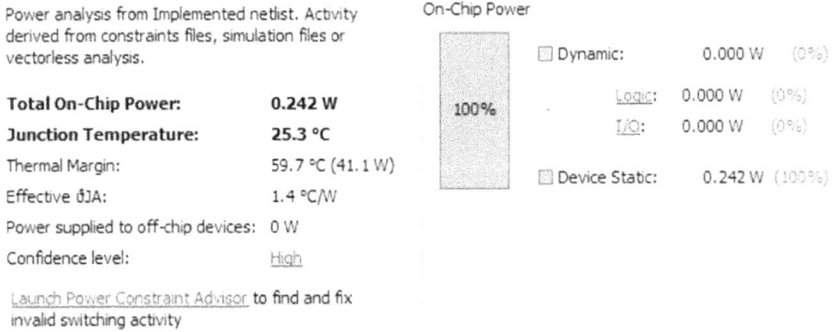

Figure 16.10 Power analysis for modified FIR filter.

Resource	Utilization	Available	Utilization %
IO		16	600	2.67

Figure 16.11 Utilization for modified FIR filter.

Figure 16.13 shows the simulation waveform for the modified FIR filter. It also shows the input and output wires of the given simulated code.

Figure 16.14 shows the power analysis of conventional FIR filter. It shows the details of signals, power, device static, IO, and on-chip power.

Figure 16.15 shows the utilization of modified multiplier of the modified multiplier. It gives the details of LUT and IO implementations.

16.5 CONCLUSION

In this chapter, an effective FIR filter is designed using modified multiplier and accumulator block. The proposed architecture is implemented using Xilinx Vivado and the parameter such as the LUT count, slices; flip-flops, power, and speed are analyzed. The modified multiplier and accumulator blocks reduce the computational complexity, power, and area and increase the speed of operations of the FIR filter as compared to the conventional FIR filter. The multiplication and division are performed using shift operation. The results express the noteworthy decrease in both arithmetic operations and hardware essential to implement those maneuvers collective with acceptable runtimes. Contrast with associated work established on the obtainable data for the proposed method shows improved results in FIR filter. Thus, the proposed novel filter architecture is implemented with high speed, less area, and less power.

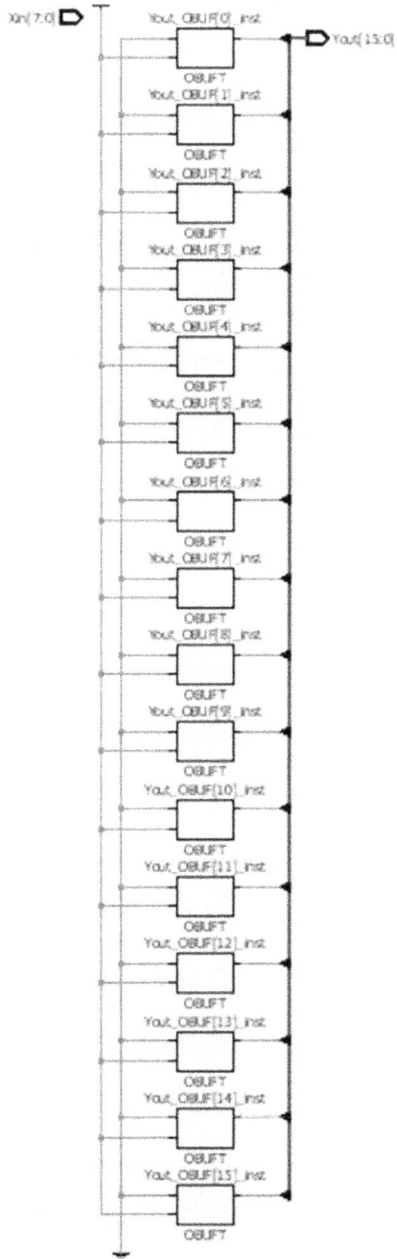

Figure 16.12 Schematic diagram for modified FIR filter.

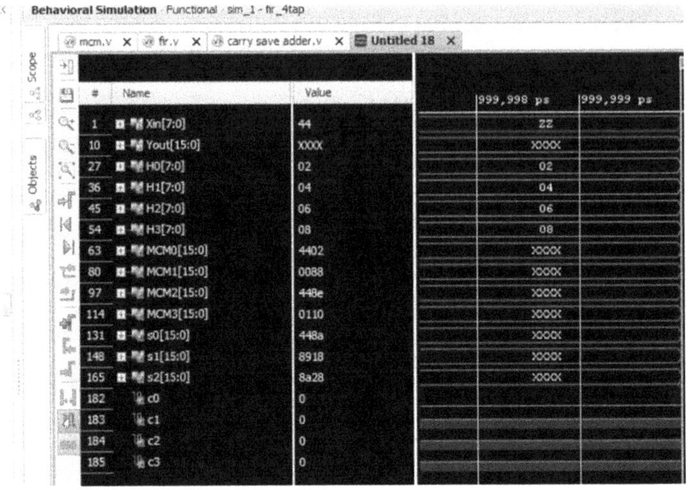
Behavioral Simulation · Functional · sim_1 · fir_4tap
mcm.v
fir.v
carry save adder.v
Untitled 18
Scope
Objects
Name
Value
999,998 ps
999,999 ps
1 Xin[7:0] 44 zz
10 Yout[15:0] XXXX XXXX
27 H0[7:0] 02 02
36 H1[7:0] 04 04
45 H2[7:0] 06 06
54 H3[7:0] 08 08
63 MCM0[15:0] 4402 XXXX
80 MCM1[15:0] 0088 XXXX
97 MCM2[15:0] 448e XXXX
114 MCM3[15:0] 0110 XXXX
131 s0[15:0] 448a XXXX
148 s1[15:0] 8918 XXXX
165 s2[15:0] 8a28 XXXX
182 c0 0
183 c1 0
184 c2 0
185 c3 0

Figure 16.13 Simulation waveform for modified FIR filter.

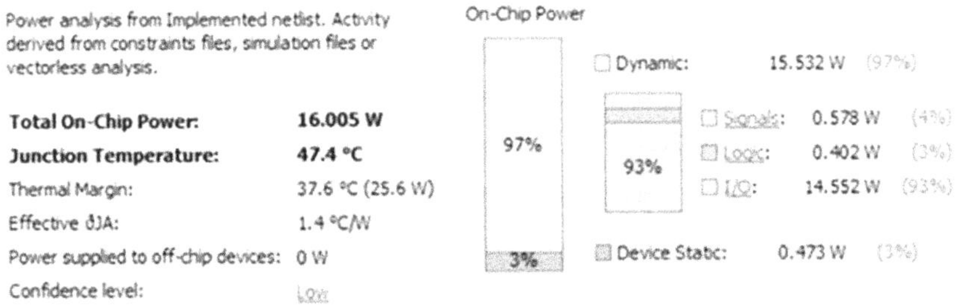
Power analysis from Implemented netlist. Activity derived from constraints files, simulation files or vectorless analysis.
Total On-Chip Power: 16.005 W
Junction Temperature: 47.4 °C
Thermal Margin: 37.6 °C (25.6 W)
Effective θJA: 1.4 °C/W
Power supplied to off-chip devices: 0 W
Confidence level: Low
On-Chip Power
97%
3%
Dynamic: 15.532 W (97%)
93%
Signals: 0.578 W (4%)
Logic: 0.402 W (3%)
I/O: 14.552 W (93%)
Device Static: 0.473 W (3%)

Figure 16.14 Power analysis for conventional FIR filter.

Resource	Utilization	Available	Utilization %
LUT	21	303600	0.01
LUTRAM	1	130800	0.01
FF	29	607200	0.01
IO	19	600	3.17
BUFG	1	32	3.13

Figure 16.15 Utilization for conventional FIR filter.

REFERENCES

[1] Sejal D. Patel and M. C. Patel, "Research trends in area optimized FIR filter implementation on FPGA," *Int. J. Sci. Eng. Technol. Res. (IJSETR)*, 6(3), 1–6, 2017.

[2] X. Lou, Y. J. Yu and P. K. Meher, "Fine-grained critical path analysis and optimization for area-time efficient realization of multiple constant multiplications," *IEEE Trans. Circuits Syst.-I*, 62(3), 863–872, 2015.

[3] Syed Asad Alam, "Techniques for efficient implementation of FIR and particle filtering," Dissertations, No 1716, Linköping University Electronic Press, pp. 1–109, 2016.

[4] Y. Pan and P. K. Meher, "Bit-level optimization of adder-trees for multiple constant multiplications for efficient FIR filter implementation," *IEEE Trans. Circuits Syst.-I*, 61(2), 455–462, 2014.

[5] Hai Huyen Dam, Antonio Cantoni, Kok Lay Teo, and Sven Nordholm, "FIR variable digital filter with signed power-of-two coefficients," *IEEE Trans. Circuits Syst.-I*, 54(6), 1348–1357, 2007.

[6] W. B. Ye and Y. J. Yu, "Single-stage and cascade design of high order multiplier less linear phase FIR filters using genetic algorithm," *IEEE Trans. Circuits Syst.-I*, 60(11), 2987–2997, 2013.

[7] W. B. Ye and Y. J. Yu, "Bit-level multiplier less FIR filter optimization incorporating sparse filter technique," *IEEE Trans. Circuits Syst. I*, 61(11), 3206–3215, 2014.

[8] J L Mazher Iqbal and S. Varadarajan, "A new algorithm for FIR digital filter synthesis for a set of fixed coefficients," *Eur. J. Sci. Res.*, 59(1), 104–114, 2011.

[9] J. L. Mazher Iqbal and S. Varadarajan, *Performance Comparison of Reconfigurable Low Complexity FIR Filter Architectures* (vol. 250, pp.637–642). Berlin Heidelberg: Springer LNCS-CCIS, 2011. https://doi.org/10.1007/978-3-642-25734-6-151.

[10] J L Mazher Iqbal and S. Varadarajan, "High performance reconfigurable FIR filter architecture using optimized multipliers," *Circuits Syst. Signal Process.*, 32, 663682, 2013. https://doi.org/10.1007/s00034-012-9473-3.

[11] J. L. Mazher Iqbal and S. Varadarajan, "Memory based and memory less computation for low complexity reconfigurable digital FIR filter" *WSEAS Trans. Syst.* 12(3), 142–153, 2014.

[12] J. L. Mazher Iqbal and G. Narayan, "Design and implementation of efficient adaptive filter using high performance reversible adder" *J. Adv. Res. Dyn. Control Syst.*, 10, 1494–1499, 2018.

[13] J. L. Mazher Iqbal and T. Manikandan, "FPGA based reconfigurable architectures for DSP computations", *Adv. Intell. Syst. Comput.*, 1163, 587–594, 2020. https://doi.org/10.1007/978-981-15-5029-4.

[14] J. L. Mazher Iqbal, Munagapati Siva Kishore, Arulkumaran Ganeshan and G. Narayan, "Design and implementation of SOC-based noncontact-type level sensing for conductive and nonconductive liquids, hindawi," *Adv. Mater. Sci. Eng.*, 2021, Article ID 7630008, 2021, https://doi.org/10.1155/2021/7630008.

[15] A. Karthik and J. L. Mazher Iqbal, "Efficient speech enhancement using recurrent convolution encoder and decoder, *Wirel. Pers. Commun.*, 2021. https://doi.org/10.1007/s11277-021-08313-6.

[16] J. Anand, J. Raja Paul Perinbam, and D. Meganathan, "Design of GA-based routing in biomedical wireless sensor networks", *Int. J. Appl. Eng. Res.*, 10(4), 9281–9292, 2015.

[17] P. Prem Kumar, K. Duraiswamy, and Jose Anand, "An optimized device sizing of analog circuits using genetic algorithm" *Eur. J. Sci. Res.*, 69(3), 441–448, 2012.

[18] Anand J., J. Raja Paul Perinbam, and D. Meganathan, "Performance of optimized routing in biomedical wireless sensor networks using evolutionary algorithms", *Comptes rendus de l'Academie bulgare des Sciences, Tome*, 68(8), 1049–1054, 2015.

[19] T. Thomas Leonid, M. Mary Grace Neela, and Jose Anand, "Signed pipelined multiplier using high speed compressors," *Int. J. Res. Comput. Appl. Robot.*, 1(6), 29–38, 2013.

[20] K. Sivachandar, V. Amudha, B. Ramesh, J. Anand, M. ShanmugaSundari, and S. Jerril Gilda, "MIMO-IDMA system performance for SUI and LTE frequency selective channels", *Adv. Parallel Comput. Algorithms, Tools and Paradigms*, 41, 422–428, 2022.

Chapter 17

One-sided Schmitt-Trigger-based 10T SRAM cell with expanded read/write stabilities and less leakage power dissipation in 10-nm GNRFET technology

Erfan Abbasian
Babol Noshirvani University of Technology

Mahdieh Nayeri
Islamic Azad University

Shilpi Birla
Manipal University Jaipur

17.1 INTRODUCTION

Static random access memory (SRAM) is potentially considered as a cache memory for microprocessors due to its fast operation and consumes a huge portion of the area and power [1]. Contemporary power-efficient battery-powered devices require stable, low-power, and small-area SRAMs. In advanced technologies, leakage power dissipation dominates total power consumption [2]. This is because in an SRAM memory, only the involved SRAM bitcell(s) perform read/write operation and the other remaining bit-cells are in hold mode and should maintain data [3]. A look-forward solution to reduce leakage power dissipation is to decrease power supply voltage (V_{DD}) because there is a linear relation between leakage power and V_{DD} [4]. However, with V_{DD} reduction and technology scaling, the variations in process, voltage, and temperature (PVT) parameters highly increase and therefore degrade the performance of SRAM cells [5]. Since SRAM cells with the minimum feature size are required to increase the density, the complementary metal-oxide-semiconductor (CMOS) is not the most efficient device because it suffers highly from the increased leakage, and gate losing control of channel, difficulty to choose suitable oxide material, and other known short-channel effects (SECs) in the nanometer region [6,7].

DOI: 10.1201/9781003459231-17

Therefore, researchers seek to find potential alternatives to CMOS, which provide better properties and can overcome the aforementioned problems.

Field-effect transistors (FET) made of carbon nanotubes (CNTs) and graphene nanoribbons (GNRs), known as CNTFET and GNRFET, exhibit nearly similar excellent properties including higher ON-to-OFF currents ratio (I_{ON}/I_{OFF}), higher and same mobilities for both types of transistors (N and P), and mitigated SECs, and therefore, they can be considered as potential substitutes for traditional CMOS [8,9]. GNRFET, unlike CNTFET, is compatible with existing CMOS manufacturing technologies [10]. Chirality-related issues and higher resistance at metal-contact junctions degrade the CNTFET performance [11]. On the other hand, it is possible to fabricate GNRFET through transfer-free, simple silicon-compatible, and in-situ processes [12]. In GNRFET devices, lower resistance at metal-contact junctions results in reduced delay and heat dissipation and improved power is in contrast to traditional silicon-based CMOS and CNTFET [13,14]. These prove that GNRFET is a superlative choice to combat power dissipation with increased scaling, and therefore, GNRFET is the most appropriate alternative to CNTFET.

However, the commercial 6T SRAM cell designed with GNRFET devices still shows unacceptable read and write stabilities in low V_{DD}. This is owing to that there are natural fights between different transistors contributed to the specific operations [15]. And so, the traditional 6T structure must be modified by taking into account techniques at the circuit level. In this regard, a one-sided Schmitt-trigger 10T (OSST10T) SRAM cell is designed and suggested in this chapter. The suggested OSST10T enhances read stability through a robust back-to-back structure made of normal and Schmitt-trigger inverters and increases writability by using a dynamic feedback-cutting technique. Moreover, the amount of power consumed in the idle mode is lessened by employing only one bitline and multiple stacked transistors. The remaining parts of the chapter are structured as below. The traditional 6T structure, along with its limitations and difficulties, is investigated in Section 17.2. Section 17.3 presents an overview of GNRFET. The structure and working of the suggested OSST10T design are demonstrated in Section 17.4. Section 17.5 gives the obtained outcomes and discusses them. A conclusion is derived in Section 17.6.

17.2 THE TRADITIONAL 6T SRAM DESIGN: STRUCTURE, LIMITATIONS, AND DIFFICULTIES

Transistor-level structure of the traditional 6T bitcell is illustrated in Figure 17.1 [16]. It utilizes two back-to-back inverters, so that the left inverter and the right inverter are made by (M1 and M2) and (M3 and M4), respectively. The data accumulated in the cell are sustained by Q and

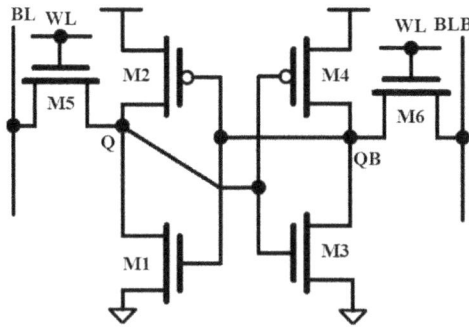

Figure 17.1 The transistor-level structure of the traditional 6T design.

QB. The two bitline (BL) and bitline bar (BLB) are connected to the latch core via M5 and M6, respectively, gated by read/write-wordline (WL), to execute a particular operation. As observed in Figure 17.1, the transistors presented in the pull-down network (M1 and M3) and the transistors presented in the pull-up network (M2 and M4) are in the conflict with the transistors presented in the access path (M5 and M6) during the read and write operations, respectively. This issue has occurred owing to the employment of one shared access path for accomplishing a particular operation. Hence, static noise margins (SNMs) for reading mode, known as RSNM, and for writing mode, known as WSNM, which gauge read-stability and write-ability, respectively, are not improved at the same time. In other words, an increase in RSNM leads to a decrease in WSNM and vice versa. Due to the subthreshold conditions and PVT variations induced by the grave increase in threshold voltage (V_{th}), the 6T SRAM cell suffers highly from instability in different operations. Furthermore, The traditional 6T design has low amounts of RSNM and WSNM and is less reliable at severe low V_{DD} [17].

Transistor sizing plays a very important role in the traditional 6T SRAM cell due to its intrinsic conflicts mentioned earlier. Hence, the proper size of transistors should be selected for the correct read/write operation in the traditional 6T cell.

Let us consider a read '0' operation in the traditional 6T design, so that node Q (QB) accumulates a '0' ('1') at the beginning. Hence, both the transistors M4 and M1 are at ON-state. Pair of BL and BLB should be kept at V_{DD} at the beginning by a precharge circuit (not shown), and then, the WL is raised to V_{DD} to enable the transistors M5 and M6. The path created by M5 and M1 pulls the BL down. Simultaneously, the Q node's voltage rises because of the current flowing from drain to source terminals in M5, and a write '1' operation may take place. Therefore, to avoid the read upset, the width of the transistor M1 must be larger than that of the transistor M5. During the read operation, the transistors are sized according to the necessitated $\beta_{\text{ratio}} = \dfrac{\beta_{\text{pull-down}}}{\beta_{\text{access}}}$, where $\beta = \mu C_{\text{ox}} \dfrac{W}{L}$. In the traditional 6T design, by adjusting the β_{ratio} between 1.2 and 3, a complete read operation with no fails is performed [18].

Now let us consider a write '1' operation in the 6T SRAM cell. According to the above-mentioned supposition, M4 and M1 are at ON-state. Pair of BL/BLB are placed at V_{DD}/GND, by means of a write driver (not shown). BL will be unable to raise the Q node's voltage through M5 because of the β_{ratio} value. Therefore, the write operation will be performed through the path comprising BLB and M6. At the same time, the transistor M4 is at ON-state and prevents flipping the storage node QB. Hence, the width of the transistor M6 must be larger than that of the transistor M4, which makes the initial high voltage at the storage node QB change to '0'. During the write operation, the transistors are sized according to the necessitated $\gamma_{\text{ratio}} = \dfrac{\beta_{\text{pull-up}}}{\beta_{\text{access}}}$. In the traditional 6T SRAM cell, by selecting γ_{ratio} less than 1.8, a complete write operation with no fails is performed [18].

17.3 GRAPHENE NANORIBBON FIELD-EFFECT TRANSISTOR (GNRFET)

GNRFETs are the good replacement for silicon-based MOSFETs in next-generation nanoelectronics. This is because of the stupendous attributes including high mobility, high $I_{\text{ON}}/I_{\text{OFF}}$, and faster switching [6,14]. Graphene, in nature, has no bandgap, hence to open up the bandgap, it must be converted into 1-D GNR form with a width of less than 10 nm [9]. GNRs are split into zigzag and armchair, known as ZGNR and AGNR, respectively, in accordance with the shape of the edge. ZGNR indicates metallic attributes and AGNR can indicate metallic and semiconductor properties together [9]. Based on FET's design, GNRFETs are split into two types of MOS and Schottky-Barrier (SB), known as MOS-GNRFETs and SB-GNRFETs, respectively. In MOS-GNRFETs, channel and drain and source contacts are based on GNR, which reservoirs (drain and source) highly doped by acceptor and donor doping, while in SB-GNRFETs, drain and source contacts are based on metal and channel is based on GNR [9,10]. In the MOS-GNRFET, the current condition is based on thermionic condition, but in the SB-GNRFET, it is based on SB tunneling due to the formation of SBs in grapheme-metal junctions. MOS-GNRFETs have several advantages over SB-GNRFETs such as faster switching (shorter delay), higher trans-conductance, and higher $I_{\text{ON}}/I_{\text{OFF}}$ ratio [6,19,20].

Figure 17.2 illustrates the structure of the MOS-GNRFET device with six parallel AGNRs [21]. The strength of driving is improved by the parallel

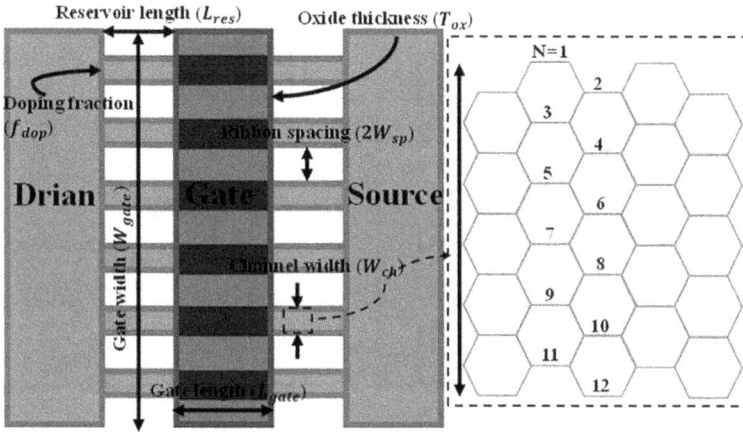

Figure 17.2 An illustration of the MOS-GNRFET structure.

connections of AGNR in the device. In this figure, the length of channel is symbolized by L_{ch}, width of gate by W_{gate}, length of reservoir by L_{res}, two adjacent devices space by $2W_{sp}$, and doping level fraction in the reservoir by f_{dop}. The channel width (W_{ch}) is calculated by the number of dimer lines (N), which is expressed by equations (17.1 and 17.2) [9].

$$W_{ch} = \sqrt{3}d_{cc}\frac{(N+1)}{2} \tag{17.1}$$

$$W_{gate} = \left(W_{ch} + 2W_{sp}\right) \times n_{rib} \tag{17.2}$$

in which the distance of C-C (C: Carbon) bond and the number of AGNRs used are symbolized by $d_{cc} = 142$ nm and n_{rib}, respectively. The bandgap is affected by the N with three-period impact. The bandgap for $N = 3p$, $N = 3p+1$, and $N = 3p+2$ is moderate, large, and small, respectively [9].

17.4 THE SUGGESTED OSSTI0T DESIGN: STRUCTURE AND WORKING

17.4.1 Bitcell structure

The transistor-level structure of the suggested OSST10T design is depicted in Figure 17.3. The suggested cell utilizes a robust back-to-back inverters structured by a normal inverter, formed by M1-M3, and an ST inverter, formed by M4-M8. The data accumulated in the cell are sustained by Q and QB. The suggested design performs the operations using one bitline (BL) to reduce leakage power dissipation and dynamic power consumption.

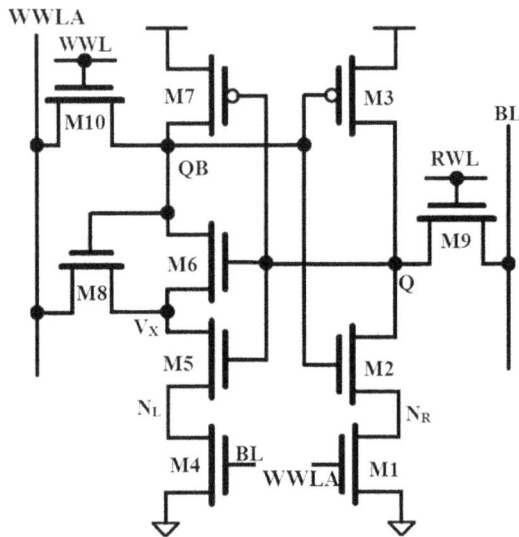

Figure 17.3 The suggested OSST10T SRAM cell's structure at transistor level.

Transistor M9, controlled by read-wordline (RWL), is enabled for both read and write operations, and transistor M10, gated by write-wordline (WWL), is enabled for write operation only. Dynamic feedback-cutting transistors M1, gated by write-wordline-A ($WWLA$), and M4, gated by BL, are used for improving the suggested cell's writability. Various signals employed in the suggested OSST10T SRAM bitcell with their status level in all different modes of operation are reported in Table 17.1.

17.4.2 Hold mode

Both RWL and WWL are pulled down to ground to disable M9 and M10, which remove read and write paths, respectively, thereby decoupling BL from the cell core. $WWLA$ is raised to V_{DD} to enable the feedback mechanism of the ST inverter through transistor M8. Moreover, transistors M1 and M4 are at ON-state because BL and $WWLA$ are kept at V_{DD}. Therefore, the datum and its complement are maintained by the robust back-to-back structure formed by transistors M1-M8.

17.4.3 Single-ended read operation

BL is connected to V_{DD} at the beginning by a precharge circuit (not shown). Afterwards, RWL is raised from low-to-high level ('0' to 'V_{DD}') to begin the read cycle by turning on transistor M9. Transistor M10 is turned off by pulling down WWL and transistor M1 is turned on by asserting $WWLA$.

Table 17.1 The signals used in the suggested OSST10T design, along with their status level

Signals	Hold mode	Read operation	Write '1' ('0') operation
RWL	'0'	'1'	'1'
WWL	'0'	'0'	'1'
WWLA	'1'	'1'	'0' ('1')
BL	'1'	Precharge	'1' ('0')

In this situation, the transistor M8 presented in ST inverter enables the feedback mechanism. When storing node Q maintains a logic '0', BL is pulled down to the ground via the path created by M9-M2-M1. Therefore, the logic '0' accumulated in the cell's node Q is read. When storing node Q maintains a logic '1', BL remains as it is because transistor M2 is at OFF-state, and therefore, a logic '1' is read from the cell. In the suggested design, the read-disturb problem is resolved by the employment of the robust back-to-back structure formed by the normal inverter (M1-M3) and the ST inverter (M4-M8).

17.4.4 Write operation with pseudo-differential structure

Writing cycle is started by pulling up RWL and WWL, which establish the write paths by turning on transistors M9 and M10, respectively. $BL/WWLA$ is either kept at V_{DD}/GND or GND/V_{DD} based on the new data, which will update the cell's content. Suppose that the Q node's content, which initially is the logic '0', is renewed by the logic '1'. In this situation, BL and $WWLA$ are set to V_{DD} and GND, respectively. By pulling down $WWLA$, transistor M1 is turned off, which decouples the node Q from the ground. From the other point of view, the transistor M8 presented in the ST inverter disables the feedback mechanism, thereby, reducing the switching voltage of the ST inverter. The path formed by BL-M9 and the path formed by M10-$WWLA$ charges the '0' storing node Q and discharges the '1' storing node QB, respectively. Using a similar process, the '1' storing node Q can be updated to '0'.

17.5 SIMULATION SETUP, RESULTS, AND ANALYSES

The suggested OSST10T design performance is evaluated and compared with that of the conventional 6T design [16] from the points of view of stabilities, operations delay, and power dissipation. Both the conventional 6T and proposed OSST10T designs are designed and simulated using MOS-GNRFET devices [9] with 10 nm technology node (SPICE code is available in [22]). The MOS-GNRFET device parameters characteristics are reported in Table 17.2.

Transistors used for designing the conventional 6T SRAM bitcell should be sized properly. Therefore, transistors M1 and M3 are sized to have 30 nm gate width and the other transistors are sized to have 20 nm gate width [19]. This strategy has been used for transistor sizing in the suggested OSST10T SRAM bitcell, as reported in Table 17.3.

The outcomes achieved at $V_{DD}=0.5$ V have been summarized in Table 17.4. The suggested OSST10T design improves RSNM by 1.49× and WSNM by 4.71×, increases read delay by 2.54× and write delay by 1.04×. The leakage power of the presented design is 0.45× as that of the conventional 6T design. The following subsections justify these achievements.

17.5.1 Stability

In the SRAM bitcell, the read/write stabilities is gauged by RSNM/WSNM, and measured from the butterfly curves, as mentioned in [23–26]. The read-butterfly curves for the suggested OSST10T and traditional 6T designs at $V_{DD}=0.5$ V are shown in Figure 17.4a. The suggested OSST10T SRAM bit-cell mitigates the read-disturbance issue by using the robust back-to-back inverters structured by the normal and ST inverters, thereby, increasing RSNM by 1.49×. The proposed OSST10T and 6T designs are compared

Table 17.2 MOS-GNRFET device parameters characteristics [9]

Parameters	Values
Channel length (L_{gate})	10 nm
No. of dimmer lines (N)	12
Distance between two adjoining AGNRs ($2W_{sp}$)	1.9806 nm
No. of AGNRs (n_{ribbon})	6 and 9
Gate width (W_{gate})	20 and 30 nm
Thickness of gate oxide (T_{ox})	0.95 nm
Percentage of line-edge roughness (P_r)	0
Gate dielectric constant (SiO_2)	3.9
Doping level fraction	0.001
V_{DD}	0.7V
Temperature	25°C

Table 17.3 Transistor sizing (nm) for comparison SRAM cells

SRAM Cells	M1	M2	M3	M4	M5	M6	M7	M8	M9	M10
6T [16]	30	20	30	20	20	20	-	-	-	-
OSST10T (Prop.)	30	30	20	30	30	30	20	20	20	20

Channel length is set to 10 nm.

Table 17.4 Performance comparison between proposed OSST10T and 6T designs at $V_{DD}=0.5\,V$

SRAM cells	RSNM (V)	WSNM (V)	Read delay (ns)	Write delay (ps)	Leakage power (nW)
6T [16]	0.0950	0.0652	0.089	6.344	0.940
OSST10T (Prop.)	0.1414	0.3069	0.226	6.574	0.420

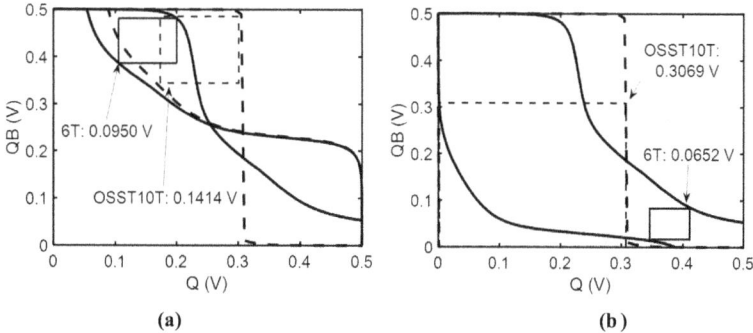

Figure 17.4 Static noise margin comparison at 0.5 V. (a) RSNM and (b) WSNM.

with each other from the point of view of WSNM at $V_{DD}=0.5\,V$, as depicted in Figure 17.4b. The suggested OSST10T SRAM bitcell employs the dynamic feedback-cutting technique for writability improvement, and therefore, offers a 4.71× higher WSNM in contrast to the traditional 6T SRAM bitcell.

Figures 17.5 and 17.6 show the values of RSNM and WSNM, respectively, against V_{DD} variations for suggested OSST10T and traditional 6T SRAM bitcells. As observed, the suggested cell shows the highest values of RSNM and WSNM at all V_{DD} values considered for simulations.

17.5.2 Speed performance

The time required for executing a specific operation in an SRAM cell, which can be reading or writing, characterizes how fast that cell is. The read delay definitions for the traditional 6T design with the differential reading structure and the proposed OSST10T design with the single-ended reading structure are different. To estimate the read delay, the definitions specified in [3,15,27] have been used. Figure 17.7 compares the read delay of the investigated SRAM bitcells versus V_{DD} values. The suggested cell indicates a 2.54× higher read delay at $V_{DD}=0.5\,V$ because of its single-ended read operation.

The writing delay is considered as how long it takes to charge (discharge) the Q node's voltage, which initially stores '0' ('1'), up to 90% (10%) of V_{DD}

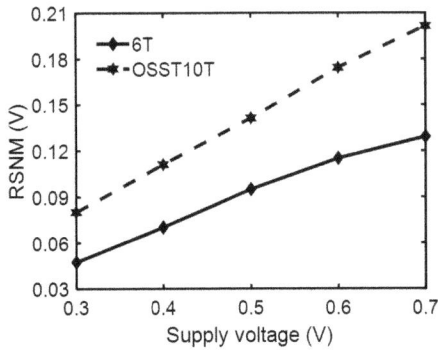

Figure 17.5 RSNM comparison against V_{DD} values.

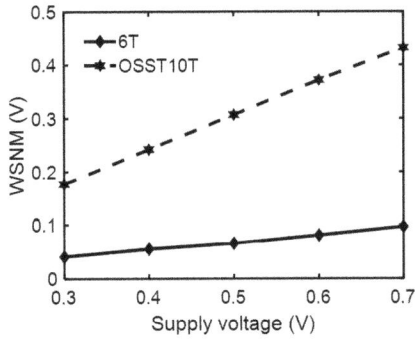

Figure 17.6 WSNM comparison versus V_{DD} values.

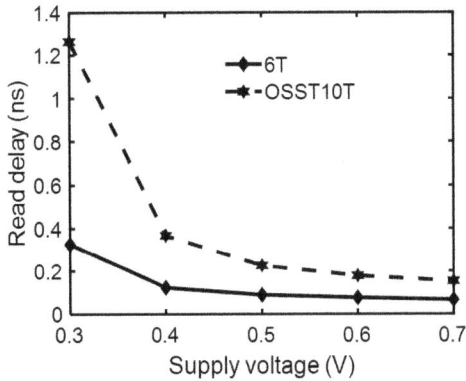

Figure 17.7 Read delay comparison versus V_{DD} values.

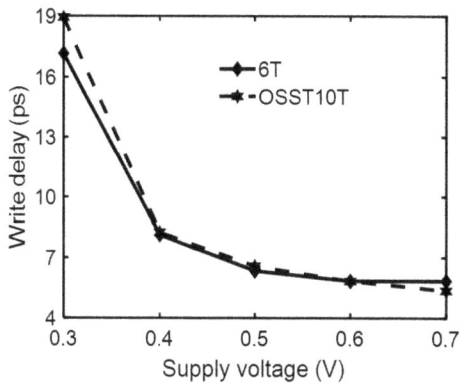

Figure 17.8 Write delay comparison versus V_{DD} values.

for writing '1' ('0') [3,15,28]. In Figure 17.8, the average write delay of the investigated designs versus V_{DD} values are compared. The suggested design shows a 1.04× higher average write delay at $V_{DD}=0.5$ V. This increase in write delay is owing to the employment of stacked transistors M1 and M4 in the normal inverter and the ST inverter, respectively. Moreover, Since $WWLA=V_{DD}$ during the write '0' operation, the feedback mechanism of the ST inverter is enabled via transistor M8, thereby prolonging the writing process.

17.5.3 Leakage power dissipation

In modern technology, a huge percentage of the SRAM's whole power is associated with leakage power dissipation [2,29–31]. And so, designing a low-leakage power SRAM bitcell is very challenging. The suggested OSST10T SRAM bitcell improves the dissipated leakage power using only one bitline and employing multiple stacked transistors in both the normal and ST inverters. In Figure 17.9, the investigated SRAM designs are compared with each other from the leakage power point of view at a V_{DD} range of 0.3–07 V. The proposed design dissipates 0.45× of leakage power in the traditional 6T design at $V_{DD}=0.5$ V.

17.6 CONCLUSION

A ten-transistor SRAM cell, namely OSST10T, with expanded read/write stabilities and reduced leakage power dissipation is introduced in this work. It employs the single-bitline structure to perform operations. In the suggested OSST10T design, the latch core is structured by the robust back-to-back inverters made of normal and Schmitt-trigger inverters. This

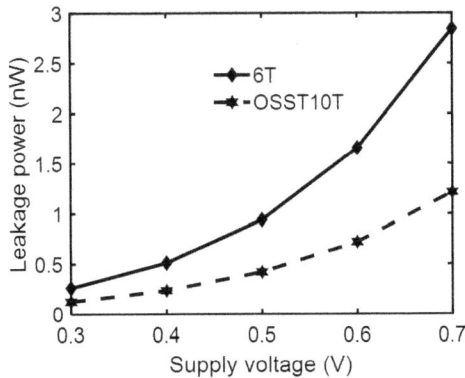

Figure 17.9 Leakage power comparison versus V_{DD} values.

mitigates the read-disturb problem, thereby, enhancing RSNM. To improve WSNM, a data-dependent feedback-cutting technique is employed. Since the back-to-back inverters principally contribute to the leakage power dissipated by an SRAM cell, the suggested SRAM cell reduces this kind of power using single-bitline structure and employing multiple stacked transistors. In the suggested design, the $RSNM/WSNM/T_{RA}/T_{WA}/P_{leakage}$ is 1.49×/4.71×/2.54×/1.04×/0.45× as that of the conventional 6T.

REFERENCES

[1] E. Abbasian and M. Gholipour, "Single-ended half-select disturb-free 11T static random access memory cell for reliable and low power applications," *International Journal of Circuit Theory and Applications*, vol. 49, pp. 970–989, 2021.

[2] G. Pasandi and S. M. Fakhraie, "A 256-kb 9T near-threshold SRAM with 1k cells per bitline and enhanced write and read operations," *IEEE Transactions on Very Large Scale Integration (VLSI) Systems*, vol. 23, pp. 2438–2446, 2014.

[3] E. Abbasian, F. Izadinasab, and M. Gholipour, "A reliable low standby power 10T SRAM cell with expanded static noise margins," *IEEE Transactions on Circuits and Systems I: Regular Papers*, vol. 69, no. 4, pp. 1606–1616, 2022.

[4] J. P. Kulkarni and K. Roy, "Ultralow-voltage process-variation-tolerant Schmitt-trigger-based SRAM design," *IEEE Transactions on Very Large Scale Integration (VLSI) Systems*, vol. 20, pp. 319–332, 2012.

[5] S. Ahmad, M. K. Gupta, N. Alam, and M. Hasan, "Low leakage single bitline 9T (SB9T) static random access memory," *Microelectronics Journal*, vol. 62, pp. 1–11, 2017.

[6] E. Abbasian and M. Gholipour, "A variation-aware design for storage cells using Schottky-barrier-type GNRFETs," *Journal of Computational Electronics*, vol. 19, pp. 987–1001, 2020.

[7] E. Mahmoodi and M. Gholipour, "Design space exploration of low-power flip-flops in FinFET technology," *Integration*, vol. 75, pp. 52–62, 2020.

[8] P. G. Sankar and K. Udhayakumar, "MOSFET-like CNFET based logic gate library for low-power application: A comparative study," *Journal of Semiconductors*, vol. 35, p. 075001, 2014.

[9] Y.-Y. Chen, A. Sangai, A. Rogachev, M. Gholipour, G. Iannaccone, G. Fiori, et al., "A SPICE-compatible model of MOS-type graphene nano-ribbon field-effect transistors enabling gate-and circuit-level delay and power analysis under process variation," *IEEE Transactions on Nanotechnology*, vol. 14, pp. 1068–1082, 2015.

[10] M. Gholipour, Y.-Y. Chen, A. Sangai, N. Masoumi, and D. Chen, "Analytical SPICE-compatible model of Schottky-barrier-type GNRFETs with performance analysis," *IEEE Transactions on Very Large Scale Integration (VLSI) Systems*, vol. 24, pp. 650–663, 2015.

[11] Z. Zhang, M. A. Turi, and J. G. Delgado-Frias, "SRAM leakage in CMOS, FinFET and CNTFET technologies: Leakage in 8T and 6T SRAM cells," In *Proceedings of the Great Lakes Symposium on VLSI*, New York, NY, United States, pp. 267–270, 2012.

[12] M. C. Lemme, L.-J. Li, T. Palacios, and F. Schwierz, "Two-dimensional materials for electronic applications," *Mrs Bulletin*, vol. 39, pp. 711–718, 2014.

[13] Y. Matsuda, W.-Q. Deng, and W. A. Goddard, "Contact resistance properties between nanotubes and various metals from quantum mechanics," *The Journal of Physical Chemistry C*, vol. 111, pp. 11113–11116, 2007.

[14] M. Nayeri, P. Keshavarzian, and M. Nayeri, "Approach for MVL design based on armchair graphene nanoribbon field effect transistor and arithmetic circuits design," *Microelectronics Journal*, vol. 92, p. 104599, 2019.

[15] E. Abbasian, "A highly stable low-energy 10T SRAM for near-threshold operation," *IEEE Transactions on Circuits and Systems-I: Regular Papers*, vol. 69, no. 12, pp. 5195–5205, 2022.

[16] M. R. Jan, C. Anantha, and N. Borivoje, *Digital Integrated Circuits: A Design Perspective*, Upper Saddle River, NJ: Prentice Hall, 2003.

[17] E. Abbasian and M. Gholipour, "Robust transmission gate-based 10T sub-threshold SRAM for internet-of-things applications," *Semiconductor Science and Technology*, vol. 37, no. 8, p. 085013, 2022.

[18] S. Ahmad, B. Iqbal, N. Alam, and M. Hasan, "Low leakage fully half-select-free robust SRAM cells with BTI reliability analysis," *IEEE Transactions on Device and Materials Reliability*, vol. 18, pp. 337–349, 2018.

[19] E. Abbasian, M. Gholipour, and F. Izadinasab, "Performance evaluation of GNRFET and TMDFET devices in static random access memory cells design," *International Journal of Circuit Theory and Applications*, vol. 49, pp. 3630–3652, 2021.

[20] E. Abbasian, M. Nayeri, and S. Sofimowloodi, "Simulation-based recommendations for digital circuits design using Schottky-Barrier-type GNRFET," *ECS Journal of Solid State Science and Technology*, vol. 11, p. 071001, 2022.

[21] E. Abbasian, T. Mirzaei, and S. Sofimowloodi, "A stable low leakage power SRAM with built-in read/write-assist scheme using GNRFETs for IoT applications," *ECS Journal of Solid State Science and Technology*, vol. 11, no. 12, p. 121002, 2022.

[22] SPICE, Model of graphene nanoribbon FETs, [Online], Available: https://nanohub.org/resources/17074.

[23] E. Abbasian, S. Birla, and M. Gholipour, "Ultra-low-power and stable 10-nm FinFET 10T sub-threshold SRAM," *Microelectronics Journal*, vol. 123, p. 105427, 2022.

[24] E. Abbasian, M. Gholipour, and S. Birla, "A single-bitline 9T SRAM for low-power near-threshold operation in FinFET technology," *Arabian Journal for Science and Engineering*, vol. 47, no. 11, pp. 14543–14559, 2022

[25] A. Dolatshah, E. Abbasian, M. Nayeri, and S. Sofimowloodi, "A sub-threshold 10T FinFET SRAM cell design for low-power applications," *AEU-International Journal of Electronics and Communications*, vol. 157, p. 154417, 2022.

[26] M. Karamimanesh, E. Abiri, K. Hassanli, M. R. Salehi, and A. Darabi, "A write bit-line free sub-threshold SRAM cell with fully half-select free feature and high reliability for ultra-low power applications," *AEU-International Journal of Electronics and Communications*, vol. 145, p. 154075, 2021.

[27] E. Abbasian and M. Gholipour, "A low-leakage single-bitline 9T SRAM cell with read-disturbance removal and high writability for low-power biomedical applications," *International Journal of Circuit Theory and Applications*, vol. 50, no. 5, pp. 1537–1556, 2022.

[28] E. Abbasian, S. Birla, and M. Gholipour, "A 9T high-stable and low-energy half-select-free SRAM cell design using TMDFETs," *Analog Integrated Circuits and Signal Processing*, vol. 112, no. 1, pp. 141–149, 2022.

[29] E. Abbasian and M. Gholipour, "Design of a highly stable and robust 10T SRAM cell for low-power portable applications," *Circuits, Systems, and Signal Processing*, vol. 41, no. 10, pp. 5914–5932, 2022.

[30] E. Abbasian, S. Birla, and M. Gholipour, "A comprehensive analysis of different SRAM cell topologies in 7-nm FinFET technology," *Silicon*, vol. 14, pp. 6909–6920, 2022.

[31] E. Abbasian and M. Gholipour, "Improved read/write assist mechanism for 10-transistor static random access memory cell," *International Journal of Circuit Theory and Applications*, vol. 50, no. 10, pp. 3642–3660, 2022.

Chapter 18

Advances in low-power devices
FinFETs, nanowire FETS, and CNT FET

Savitesh Madhulika Sharma
Siddartha Institute of Science and Technology

Avtar Singh
Adama Science and Technology University

18.1 INTRODUCTION TO LOW-POWER DEVICES

The increased complexity brought on by having multiple applications on a single IC chip requires VLSI designers to redesign for IC area, execution, cost, and reliability; power considerations, which were typical of simple power once had a purely symbolic meaning, but since they have been given a more concrete meaning identical speed and load for the region. The IC uses low-cost, high-performing components. Micron devices will give way to submicron devices as a result of technological advancements [1]. It is IC that speed and power consumption need to be balanced because they are inherently incompatible. In other words, increased speed causes higher power scattering, and vice versa. Convenient modern portable devices demand quick calculations and complex functionality with little power consumption.

18.2 FINFETs

18.2.1 Introduction to FinFETs

FinFET is a multigate transistor with a fin-like structure as shown in Figure 18.1 results quasi-planar structure. The fin shape actually allows the gate to wrap around the channel region to increase the electrostatic integrity. Due to multiple gates various types of multigate structures can be derived such as double, triple, and quadruple gate FETs such as wrapped around FETs, and surrounding gate FETs. The FinFET offers better immunity from short channel effects (SCE), low power consumption, and high volume integration.

DOI: 10.1201/9781003459231-18

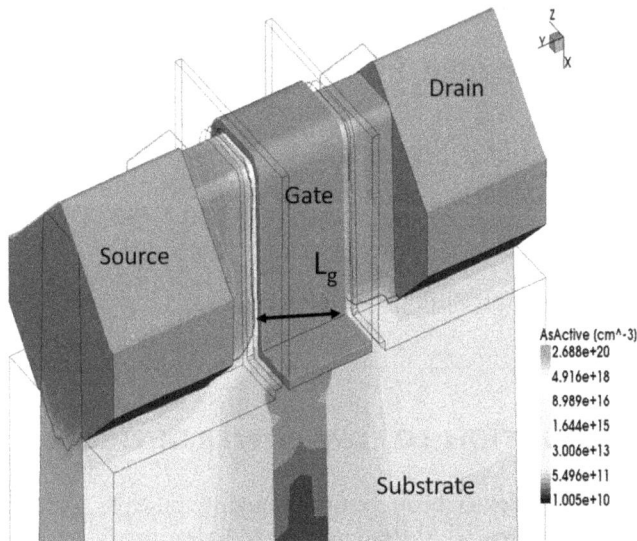

Figure 18.1 FinFET structure.

18.2.2 Device physics mechanism

It is observed that semi-classical theory is inadequate for short channel devices like FinFET. The reason is the confinement of carrier concentration inside the channel and threshold voltage shift. The carrier can tunnel through lower potential barrier as a result of thin gate oxide [2] called this phenomenon as quantum mechanical effect. This phenomenon occurs when the oxide layer thickness is equivalent to that of the inversion layer thickness in the fin. Additionally, a change in the threshold and the decline in the slope of the Id-Vgs can be seen. A thin layer of oxide places restrictions on the inversion carrier's proximity to the Si/SiO$_2$ interface layer. Both the Schrodinger equation and Poisson's equation can be used to calculate volume inversion.

18.2.3 Fabrication steps

The fabrication process for FinFETs depends on the type of the device whether it is bulk FinFET or a SOI FinFET. However, in the FinFET crucial part is fin geometry, and it is observed that fin thickness, height, and gate length play a major role to control the short channel device characteristics [3]. The process steps for SOI FinFETs [4] are as follows:

Step1: Silicon and SiN/SiO$_2$ hard mask layer
Step2: E-beam Lithography and etching for fin patterning
Step 3: Phosphorous doped polysilicon deposition for gate formation

Step 4: Chemical mechanical polishing (CMP) to uniform gate poly-Si layer

Step 5: Deposition and etching to form Si_3N_4 spacer

Step 6: Anisotropic etching of Si using Si_3N_4 as hard mask CMP

Step 7: Isolation oxidation using Si_3N_4 as the shielding layer or Gate Spacer

Step 8: Etching Si_3N_4 and tetraetoxysilane (TEOS) to form fins

Process Steps for Bulk FinFETs [5] are listed below:

Step 1: Photolithography, SiN etching, Poly-Si deposition and dry etching

Step 2: SiN removal using phosphoric acid

Step 3: SiO_2, SiN and SiO_2 etching dry process

Step 4: Fin etching dry process

Step 5: Thin oxide filling and densification

Step 6: Chemical mechanical polishing (CMP)

Step 7: Wet etch back

Step 8: Deposited doped poly Si as source/drain

Step 9: Pattern S/D fan out.

The difference in the process can be listed as shown in Table 18.1.

18.2.4 AC and DC characteristics

AC and DC characteristics and experimental results of the fabricated device are used to calibrate the TCAD-simulated FinFET design by activating appropriate models such as density gradient models, Lombardi model, etc. and tune the parameter to match the results [6]. The calibrated simulated model is then used to study the performance of device.

18.2.5 Issues and challenges

Like any new technology, FinFET with a sub 20 nm feature size has a lot of design problems. The majority of these issues are caused by short channel qualities that might be deteriorated by technological restrictions.

Undoped underlap zones must be employed to prevent raising channel resistance and lowering driving current. In recent years, strong field coupling between the gate and the undoped underlap area, which lowers the

Table 18.1 Comparison of Fabrication process steps for SOI and Bulk FinFET

SOI FinFET	Bulk FinFET
SOI substrate	Bulk substrate
FIN patterning and corner rounding	STI formation and field recess
No channel doping	Channel doping

high source/drain series resistance, has been made possible by high permittivity (k) spacer materials.

18.2.6 Applications

The Multigate MOSFET is a troublemaking device architecture for the circuit designers since the third dimension is expressly intended to mitigate SCEs and prevent off currents in VLSI technologies from extending above 45 nm technology node. To examine the advancements in the field of FinFET technology in relation to various circuit-related figures of merit (FOMs), particular layout and design considerations and the specifications for a future technology platform that will support system-level integration must consider the serious problem self-heating effect, trapped charges, flicker noise, in Hi-K, etc. in circuit applications such as Inverter [7], SRAM [8], Band reference circuits [9].

18.3 CARBON NANOTUBES

Carbon nanotubes (CNT) attract the researchers since the 1960s due to their remarkable mechanical, thermal, and electronic properties. The applications of CNTs in biosensors, circuits, optical devices, wireless technologies, etc. made it attractive to explore more about them. Researchers predicted that the CNTs-based transistors can offer better energy efficiency compared to conventional Si-based transistor. In this section, we will understand the fundamentals of CNTs and their types and will explore about the CNTs-based transistor and their deployment in various future applications.

18.3.1 Fundamentals of carbon nanotubes

CNTs are nanostructures of carbon that are hollow and cylindrical, with the length often being many orders of magnitude greater than the diameter. Single-walled CNTs (SWCNTs), which have a diameter of a few nanometers, are CNTs with only one shell of carbon atoms. Multi-walled CNTs are CNTs composed of several concentric carbon atom shells (MWCNT). The structure of single-walled CNTs is like rolled-up graphene sheet. CNTs contain one-dimensional bands due to the quantized wave vectors at the circumferential periodic boundaries.

18.3.2 Types of CNTs

Based on the number of rolled-up sheets stacked as concentric cylinders with various characteristics, CNTs can be divided into single-walled CNTs

(SWCNTs) and multi-walled CNTs (MWCNTs). Single-walled CNT structures are created as single-cylinder setups in the form of hexagons.

Three different types of SWCNTs, based on the wrapping mechanism are chiral, armchair, and zigzag patterns. This classification is done by pair of indices (n and m) that explain the vector mechanism of chirality and that affects the electrical properties. For armchair indices n and m are set to equal, whereas for zigzag m is set to zero. For chiral pattern n and m ranges to different values.

Multi-walled CNTs (MWCNTs): The sp2 property of carbon hybridization produces a layered pattern of organization with weaker plane bonding of Vander Waals forces at the outer plane limits and stronger forces at the inner plane bounds. Multi-walled CNTs are being fitted with interlayers with uniform spacing around the center hollow part as shown in Figure 18.2 The interlayer spacing, inner diameter, and outer diameters range from 0.35 to 0.40 nm, 0.40 nm to few nanometers and up to 25 nm, respectively, in real time spacing. The sides of the cylinder are closed by half-width fullerene molecules arranged as dome shape.

18.3.3 Properties of CNTs

The CNT is rich with various properties like physical, chemical, optical, and electromechanical properties. However, these properties are dependent on ambient conditions such as pH value and temperature.

The elasticity of CNT is a remarkable quality. It is able to bend, twist, kink, and eventually buckle under maximum force and high pressure by being subjected to larger compressive forces in the axial direction. As a result, the original geometric structure of nanocarbon tubes can be preserved. However, there are instances when elasticity has a limit, and as a result, under the impact of larger physical pressure forces, it may even temporarily bend into the shape of a nanotube. This property can be analyzed using transmission electron microscopy (TEM). In the experiment,

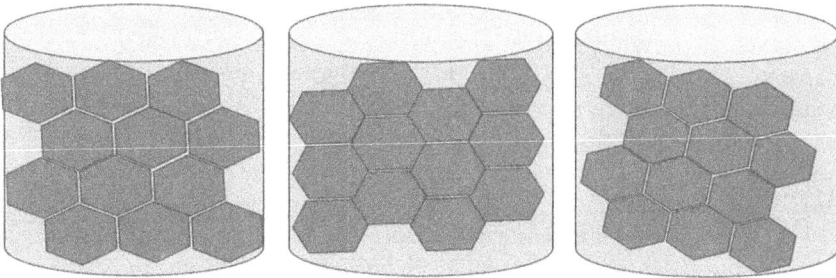

Figure 18.2 Different types of CNTs (a) Zigzag CNTs; (b) armchair CNT; and (c) chiral CNT.

molecular vibrations at both tube edges are examined for the measurement of elasticity. CNTs are good heat conductors in addition to withstanding high temperatures. They can resist temperatures of almost 800°C and normal pressure of 2,900°C.

18.3.4 CNT-based transistor

A coaxial CNT with source, drain, and gate formed carbon nanotube FET transistor (CNTFET) [10] applied strain to improve the performance. The CNTFET structures have two capacitances namely, Quantum Capacitance and electrostatic capacitances. If the quantum capacitance limit is reached in the structure, the semi-classical approach can yield extremely accurate results. The applied voltage at the drain end greater than the barrier height, resulting in the tunneling of the holes into the channel. The tunneling current through the barrier is altered as a result of the gate potential's reduction in barrier width. The CNTFET is attached with different materials, structure, and established sensors that can identify certain types of specific molecular and distinct testing responses of various sorts of samples in the near future.

18.3.5 Applications

In the past few decades, numerous research studies have suggested possible applications for CNTs and have remarkably shown promising outcomes when such recently developed materials are combined with conventional scientific outputs, for instance, the production of nanorods such as CNTs as reactive template materials. Nanotechnology, medical, environment, manufacturing, peripheral software and hardware, and electronics are only a few of the many disciplines and fields where CNTs are used as an actuator, composites with high strength, energy conversion, nano-sensors, nano-probes, and process catalysts. A significant amount of attention is being paid to the possible use of CNT as prospective building blocks for highly downscaled electronic devices with higher performance. Various researchers reported that the CNTs combined with the transistor offer application as low-cost, portable, and low-noise biosensors such as DNA, antibody bacteria, and glucose etch. Detection can be done by using different reactants [11–15], such as enzyme [16], biosensor, DNA biosensor [17], and immune sensor [17,18]. These biosensors required specific size, high purity, and helicity of CNT. Since it is difficult to control these factors which result in CNT, is expensive. An optical sensor has been developed by collecting [19–22] SWCNT emission and also detects DNA hybridization observation on absorption spectra. However, Modeling and simulation can help in the evaluation, optimization, and synthesis of functional CNTs at low cost. Biosensor challenges are to improve reliability. The detection ability and sensitivity of a sensor can be improved by a combination of nanocomposite material, CNT, and metals. The pharmacokinetics and biodistribution of aggregated nanoparticles, which in turn are governed by various physicochemical characteristics such as size, efficient fictionalization, shape, aggregation capability, surface solubility, and chemical are another major hurdle with CNT. Optical biosensors have great applications in gene vector therapy [19,20]. Multi-walled CNT-based immune sensors can detect the presence of atrazine pesticides [23]. Previous research has shown that CNT with water-soluble nature is considerably more biocompatible with the bodily fluids already present, and it also doesn't exhibit any hazardous side effects or aberrant mobility. The problem can be minimized by surface modification of the CNTs that help in stabilizing the functionalization of the surface with appropriate hydrophilic substitutes and reaction chemistries. The digital and analog circuits based on CNTFET such as adder [24], multiplier [25], SRAM [26], OpAmp [27], and amplifier [28] proved the bright future of CNTFETs.

18.4 NANOWIRE FETs

18.4.1 Introduction

In the ultra-deep submicron industry, it is necessary to have effective gate control to scale MOSFETs below 10 nm. This is why researchers are looking into silicon nanowires, which enable multigate or surrounding gate transistors. In nanowire structure, carriers are confined in two dimensions in such a way they are able to move in only 1-dimension (1D). Recent numerical simulations have shown the potential of silicon nanowire transistors (SNWTs), including electrostatic and manufacturable scaling below the limits of conventional MOSFET. It may be possible to scale silicon transistors below the scaling limit of planar MOSFETs thanks to superior electrostatic scaling for a given Si body thickness. SNWTs may provide superior gate control, easier manufacture, and more design flexibility as compared to planar double-gate MOSFETs. SNWT properties can be altered by high anisotropy, polarizability, and alignment method. Alignment is done by polarizability by electric field [29], microfluidic channels applied capillary force [30], and contact printing [31]. The nanowire geometries of linear and circularly fabricated [32] integrated circuit open opportunities in ultrasensitive sensor technology.

18.4.2 Device structure and its simulation

The nanowire MOSFET structure as shown in Figure 18.3 has a wire-like shape with the surrounding gate which provides excellent electrostatic control over the channel either horizontally or vertically oriented [33]. It

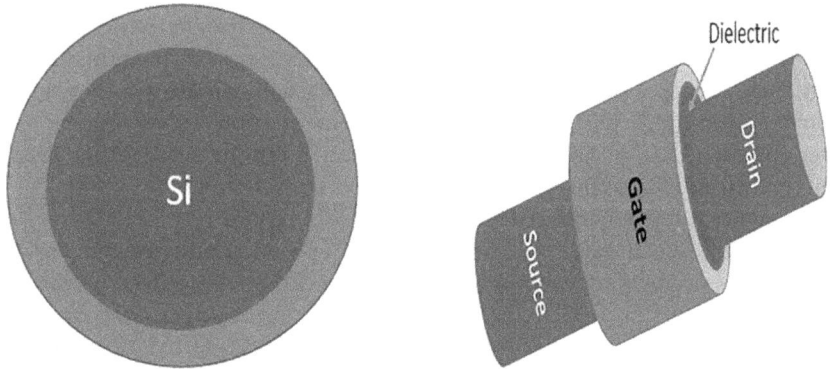

Figure 18.3 Surrounding gate or nanowire structure (a) Top view and (b) side view.

possesses quantum and dispersion effects and offers a large surface-to-volume ratio in addition to a smaller channel. The minimum gate length can be as follows [34]:

$$\lambda_{\text{gate}} = \frac{2\varepsilon_{\text{si}}\, t_{\text{si}}^2 \ln\ln\left(1+\dfrac{2t_{\text{ox}}}{t_{\text{si}}}\right) + \varepsilon_{\text{ox}}\, t_{\text{si}}^2}{16\varepsilon_{\text{ox}}}$$

where ε_{ox} and ε_{si} are permittivity of oxide and silicon, t_{ox} and t_{si} are oxide and silicon thickness respectively.

The coaxially gated nanowire capacitance is estimated as

$$C_{\text{gate}} = \frac{2\pi k\varepsilon_0}{\ln\left(\dfrac{2d_{\text{dielectric}} + d_{\text{wire}}}{d_{\text{wire}}}\right)} \text{ F/cm}$$

where d_{wire} and $d_{\text{dielectric}}$ are the diameter of nanowire and dielectric.

18.4.3 Device characteristics

Advanced effect models, such as velocity saturation, phonon dispersion, quantum effects, and SCE, are incorporated into the model in order to accurately simulate the electrical performances of nanowire. The uncoupled mode space NEGF approach was used to solve the transport on the assumption that it was ballistic [35]. The 3-D Poisson equation and the transport equation were solved using the self-consistent method [35]. The ballistic transport improves the ballistic efficiency. The band's region peak point is near the source points, and the electrons injected from the source region contribute to the drain current.

Drain current scalability is governed by wire diameter and parallelism. If the device diameter is very small, the nanowire shows potential for high linearity due to Id-Vgs. By using highly subtle metallic S/D Schottky contacts and injecting the majority of charge carriers into the inverted region, reconfigurable behavior can be introduced in the nanowire [36].

18.4.4 Performance analysis

The performance analysis of nanowire FET in the presence of interface trap charges (ITC) in [37] shows that higher positive ITC results in higher tunneling rates, but at the expense of lingering tunneling length. In consequence a transition from tunneling to diffusion current with a greater carrier recombining rate. The energy band slope can be controlled by ITC. The slope reduces with negative ITCs. With applied gate voltage, positive ITCs at the channel oxide aid in reversing the doping type of the channel.

18.4.5 Applications of nanowire FETs

Silicon nanowire-based piezoresistive airflow sensor reported by [38] offers better linearity, sensitivity, and hysteresis in addition to scalability features. This arrangement can be integrated with CMOS as well as other device for future applications.

With a ring gate, intrinsic f_t can reach 500 GHz at 30 nm, whereas simulation-only fmax can reach 800 GHz and intrinsic voltage gain gm/gds gate length nanowire at 30 nm with ring gate, up to 3 [39]. Silicon nanowire-based Memory and MRAM in [40] promising high-density memory and lower energy consumption.

REFERENCES

[1] Y. Taur, T. H. Ning, *Fundamentals of Modern VLSI Devices*, 2nd ed. New York: Cambridge University Press, 1998.

[2] H.-S. P. Wong, "Beyond the conventional transistor," *IBM Journal of Research and Development*, 46(2/3), 2002.

[3] N. Collaert et al., "Tall triple-gate devices with TiN/HfO2 gate stack," In *Digest of Technical Papers. 2005 Symposium on VLSI Technology, 2005*, Kyoto, Japan, pp. 108–109, 2005.

[4] Nuo Xu, Thin body process MOSFET's I, 2013 [Online]. Available: https://www-inst.eecs.berkeley.edu/~ee290d/fa13/LectureNotes/Lecture7.pdf.

[5] J. Kedzierski, D. M. Fried, E. J. Nowak, T. Kanarsky, J. H. Rankin, H. Hanafi, W. Natzle, D. Boyd, Y. Zhang, R. A. Roy, J. Newbury, C. Yu, Q. Yang, P. Saunders, C. P. Willets, A. Johnson, S. P. Cole, H. E. Young, N. Carpenter,

D. Rakowski, B. A. Rainey, P. E. Cottrell, M. Ieong, H.-S. P. Wong, "High-performance symmetric-gate and CMOS-compatible Vt asymmetric-gate FinFET devices," *IEDM Technical Digest*, 437–440, 2001.

[6] C. R. Manoj, V. R. Rao, "Impact of high-κ gate dielectrics on the device and circuit performance of nanoscale FinFETs," *IEEE Electron Device Letters*, 28(4), 295–297, 2007.

[7] A. Pandey, S. Raycha, S. Maheshwaram, S. Manhas, S. Dasgupta, A. Saxena, B. Anand, "Effect of load capacitance and input transition time on FinFET inverter capacitannce," *IEEE Transactions Electron Devices*, 61(1), 30–36, 2014.

[8] Ming-Hung Han et al., "Device and circuit performance estimation of junctionless bulk FinFETs", *IEEE Transaction on Electron Devices*, 60(6), 1807–1813, 2013.

[9] A. J. Annema, P. Veldhorst, G. Doornbos, B. Nauta, "A sub-1V bandgap voltage reference in 32nm FinFET technology," *IEEE International Solid-State Circuits Conference*, San Fransisco, CA, USA, pp. 332, 2009.

[10] Z. Kordrostami, M. H. Sheikhi, "Schottky barrier field effect transistors with a strained carbon nanotube channel," *Journal of Computational Theory and Nanoscience*, 6, 1571–1579, 2009.

[11] B. L. Allen, P. D. Kichambare, A. Star, Carbon nanotube field-effect-transistor- based biosensors," *Advanced Materials*, 19(11), 1439–1451, 2007.

[12] S. Liu, X. Guo, "Carbon nanomaterials field-effect-transistor-based biosensors," *NPG Asia Materials*, 4(8), e23, 2012.

[13] H. R. Byon, H. C. Choi, "Network single-walled carbon nanotube-field effect transistors (SWNT-FETs) with increased Schottky contact area for highly sen- sitive biosensor applications," *Journal of the American Chemical Society*, 128(7), 2188–2189, 2006.

[14] X. W. Tang, S. Bansaruntip, N. Nakayama, E. Yenilmez, Y. L. Chang, Q. Wang, "Carbon nanotube DNA sensor and sensing mechanism, *Nano Letters*, 6(8), 1632–1636, 2006.

[15] J. Oh, G. Yoo, Y. W. Chang, H. J. Kim, J. Jose, E. Kim, J. C. Pyun, K. H. Yoo, "A car- bon nanotube metal semiconductor field effect transistor-based biosensor for detection of amyloid- beta in human serum,". *Biosensors and Bioelectronics*, 50, 345–350, 2013.

[16] W. Feng, P. Ji, "Enzymes immobilized on carbon nanotubes," *Biotechnology Advances*, 29(6), 889–895, 2011.

[17] Y. C. Tyan, M. H. Yang, T. W. Chung, W. C. Chen, M. C. Wang, Y. L. Chen, S. L. Huang, Y. F. Huang, S. B. Jong, "Characterization of surface modification on self-assembled monolayer-based piezoelectric crystal immunosensor for the quantification of serum alpha-fetoprotein," *Journal of Materials Science: Materials in Medicine*, 22(6), 1383–1391, 2011.

[18] A. A. P. Ferreira, C. S. Fugivara, H. Yamanaka, A. V. Benedetti, *Preparation and Characterization of Imunosensors for Disease Diagnosis*, Biosensors for Health, Environment and Biosecurity, InTech, 2011.

[19] X. He, K. Wang, Z. Cheng, "In vivo near-infrared fluorescence imaging of cancer with nanoparticle-based probes," *Wiley Interdisciplinary Reviews: Nanomedicine and Nanobiotechnology*, 2(4), 349–366, 2010.

[20] F. Long, A. Zhu, C. Gu, H. Shi, "Recent Progress in Optical Biosensors for Environmental Applications," *Sensors*, 13(10), 13928–13948, 2013.

[21] S. Kruss, A. J. Hilmer, J. Zhang, N. F. Reuel, B. Mu, M. S. Strano, Carbon nano- tubes as optical biomedical sensors," *Advanced Drug Delivery Reviews*, 65(15), 1933–1950, 2013.

[22] C. Cao, J.H. Kim, D. Yoon, E.-S. Hwang, Y.-J. Kim, S. Baik, "Optical detection of DNA hybridization using absorption spectra of single-walled carbon nanotubes, tubes," *Materials Chemistry and Physics*, 112(3), 738–741, 2008.

[23] P. Norouzi, B. Larijani, M. Ganjali, F. Faridbod, "Admittometric electrochemical determination of atrazine by nano-composite immune-biosensor using FFT-square wave voltammetry," *International Journal of Electrochemical Science*, 7(11), 10414–10426, 2012.

[24] Rajendra Prasad Somineni, Y. Padma Sai, S. Naga Leela "Low leakage CNTFET full- adders," *Global Conference on Communication Technologies (GCCT)*, Thuckalay, India, pp. 174–179, 2015.

[25] B. Srinivasu, K. Sridharan, "Low-complexity multiternary digit multiplier design in CNTFET technology," *IEEE Transactions on Circuits and Systems II: Express Briefs*, 63(8), 753–757, 2016.

[26] José G. Delgado-Frias, Zhe Zhang, Michael A. Turi, "Near-threshold CNTFET SRAM cell design with removed metallic CNT tolerance." *2015 IEEE International Symposium on Circuits and Systems (ISCAS)*, pp. 2928–2931. Lisbon, Portugal: IEEE, 2015.

[27] Abhishek Puri, Ashwani Rana, "Performance analysis of CNTFET based low power operational amplifier in analog circuits for biomedical applications," *IEEE International Conference on Electronics, Computing and Communication Technologies (CONECCT)*, Bangalore, India, pp. 1–5, 2015.

[28] A. Taghavi, C. Carta, F. Ellinger, M. Haferlach, M. Claus, M. Schroter, "A CNTFET amplifier with 5.6dB gain operating at 460-590MHz," *2015 SBMO/IEEE MTT-S International Microwave and Optoelectronics Conference (IMOC)*, pp. 1–4. Porto de Galinhas, Brazil: IEEE, 2015.

[29] X. Duan, Y. Huang, Y. Cui, J. Wang, C. M. Lieber, "Indium phosphide nanowires as building blocks for nanoscale electronic and optoelectronic devices," *Nature*, 409, 66–69, 2001.

[30] A. Javey, S. Nam, R. S. Friedman, H. Yan, C. M. Lieber, "Layer-by-layer assembly of nanowires for three-dimensional, multifunctional electronics," *Nano Letters*, 7, 773–777.

[31] J. Martinez, R. V. Martinez, R. Garcia, "Silicon nanowire transistors with a channel width of 4nm fabricated by atomic force microscope lithography," *Nano Letters*, 8(11), 3636–3639, 2008.

[32] C. P. Auth, J. D. Plummer, "Scaling theory for cylindrical, fully-depleted, surrounding-gate MOSFETs," *IEEE Electron Device Letters*, 18, 72–74, 1997.

[33] C. Pan, Praveen Raghavan, Dmitry Yakimets, Peter Debacker, Francky Catthoor, Nadine Collaert, Zsolt Tokei, Diederik Verkest, Aaron Voon-Yew Thean, and Azad Naeemi, "Technology/system codesign and benchmarking for lateral and vertical GAA nanowire FETs at 5-nm technology node," *IEEE Transactions on Electron Devices*, 62(10), 3125–3132, 2015, doi: 10.1109/TED.2015.2461457.

[34] J. P. Colinge, "Multiple-gate soi mosfets," *Solid-State Electronics*, 48(6), 897–905, 2004.

[35] M. Shin, "Quantum simulation of device characteristics of silicon nanowire FETs," *IEEE Transactions on Nanotechnology*, 6(2), 230–237, 2007.

[36] Abhishek Bhattarjee, S. Dasgupt, "Impact of gate / spacer-channel underlap, gate oxide EOT and scaling on the device characteristics of a DG-RFET," *IEEE Transactions on Electron Devices*, 64(8), 3063–3070, 2017.

[37] Naveen Kumar, Ashish Raman, "Performance assessment of the charge-plasma-based cylindrical GAA vertical nanowire TFET with impact of interface trap charges," *IEEE Transactions on Electron Devices*, 66(10), 4453–4460, 2019.

[38] S. Zhang, L. Lou, C. Lee, "Piezoresistive silicon nanowire based nanoelectro-mechanical system cantilever air flow sensor," *Applied Physics Letters*, 100, 2012–2015, 2012.

[39] Seongjae Cho et al. "RF performance and small-signal parameter extraction of junctionless silicon nanowire MOSFETs", *IEEE Transactions on Electron Devices*, 58(5), 1388–1296, 2011.

[40] A. Veloso, T. Huynh-Bao, P. Matagne, D. Jang, G. Eneman, N. Horiguchi, J. Ryckaert, "Nanowire & nanosheet FETs for ultra-scaled, high-density logic and memory applications," *Solid-State Electronics*, 168, 107736, 2020.

Index

For Product Safety Concerns and Information please contact our EU
representative GPSR@taylorandfrancis.com
Taylor & Francis Verlag GmbH, Kaufingerstraße 24, 80331 München, Germany

www.ingramcontent.com/pod-product-compliance
Lightning Source LLC
Chambersburg PA
CBHW060812220326
41598CB00022B/2595

9 781032 604602